T0211513

Emerging Topics in Statistics and Biostatistics

More information about this series at http://www.springer.com/series/16213

Naitee Ting • Joseph C. Cappelleri • Shuyen Ho
(Din) Ding-Geng Chen
Editors

Design and Analysis of Subgroups with Biopharmaceutical Applications

 Springer

Editors
Naitee Ting
Biostatistics & Data Sciences
Boehringer Ingelheim Corporation
Ridgefield, CT, USA

Shuyen Ho
UCB Biosciences Inc.
Raleigh, NC, USA

Joseph C. Cappelleri
Pfizer Inc.
Groton, CT, USA

(Din) Ding-Geng Chen
School of Social Work
University of North Carolina
Chapel Hill, NC, USA

ISSN 2524-7735 ISSN 2524-7743 (electronic)
Emerging Topics in Statistics and Biostatistics
ISBN 978-3-030-40107-8 ISBN 978-3-030-40105-4 (eBook)
https://doi.org/10.1007/978-3-030-40105-4

This Springer imprint is published by the registered company Springer Nature Switzerland AG.
The registered company address is: Gewerbestrasse 11, 6330 Cham, Switzerland

Preface

Clinical development of new drugs started in the 1960s, and it has been over half a century up to the present. For the last three decades of the twentieth century (around 1970s to 1990s), most of the new drugs approved by the U.S. Food and Drug Administration (FDA) were developed to treat chronic conditions and the best drugs were those for which "one size fits all," with one drug for the entire patient population under the given diagnosis. In other words, for a given chronic condition such as high blood pressure, chronic pain, or depression, the ideal market leader would be one drug with a specific dose, once a day, and that applied to all patients with the indicated condition. Such a "one size fits all" drug would be easy to prescribe, easy to use, and tend to gain a major market share. However, with movement into the twenty-first century, many drugs have become available to treat most of the chronic conditions. The question regarding medical intervention has changed from "Is there a drug to treat this condition?" to "Which of the available drugs is best for this particular patient?"

Subgroup analysis of clinical trial data has been a common practice starting from the 1970s. Most of the clinical questions are about which subgroups of patients benefit most from the study drug or, conversely, which subgroups are most likely to experience adverse events. Another objective would be to see whether the treatment benefit could be consistent across all subgroups. Before the turn of the twenty-first century, the objectives of subgroup analysis were mostly about how the subgroups of patients respond to the study drug relative to control. Subgroup analyses were performed largely using pre-marketing clinical trials, and stakeholders were mainly sponsors (drug manufacturers) and regulatory agencies.

Nevertheless, in recent years, with the demand of personalized medicine or precision therapy, the objectives of subgroup identification and subgroup analysis turn into, for the given diagnosis and treatment indication, which drug is best for this pecific patient. In this new paradigm, subgroup analyses are performed on both pre-marketing studies and post-marketing studies (either based on a single clinical study, combined studies, or using meta-analysis). In addition to sponsors and regulatory agencies, other stakeholders such as payers (e.g., insurance companies,

Medicare), prescribing physicians, patients, and patient advocacy groups are all paying more attention to subgroups. One of the newer topics is about subgroup identification with the purpose of identifying the subgroups of interest.

Generally speaking, subgroup analysis is an exploratory practice—these analyses are performed post hoc and are data-driven. Results obtained from such analyses should not serve for confirmatory purposes, nor can they be the basis of drug approval or label change. If the benefit of a particular subgroup is considered before the clinical trial started, and the sponsor hopes to develop the study drug focus on this specific patient subpopulation, then, in order for the drug to receive approval within this subgroup, the pivotal study needs to be designed with this objective.

Given the history of drug development, the history of subgroup analysis, and the ensuing challenges in dealing with clinical trial design considerations, subgroup identification, as well as analysis, there is a need for a book devoted entirely to the design and analysis of subgroups with biopharmaceutical applications. We know of no other book dedicated solely to this important topic. In order to cover all these issues, this book is divided into three major parts—Part I: Subgroups in Clinical Trial Design and Analysis; Part II: Subgroup Identification and Personalized Medicine; and Part III: General Issues About Subgroup Analysis, Including Regulatory Considerations.

There are 18 chapters in this book—four chapters to cover the design and analytic considerations, nine chapters to discuss issues related to subgroup identification, and five chapters to deal with general topics about subgroup analysis. Contributors of this book are expert statisticians or methodological researchers with considerable experiences on the topic of subgroups; they are from academia, industry, and regulatory agencies. Examples are used, when possible, throughout the book; some are real-world applications, and others are simulated data sets. The computer code for these real-world or simulated examples is mostly provided in the book. If software code is not included in certain chapters, readers are encouraged to communicate with the chapter authors directly or go to their designated websites.

Part I of this book focuses on issues related to clinical trial designs and analysis. Part I includes four chapters. Chapter 1 entitled "Issues Related to Subgroup Analyses and Use of Intensive Stratification" discusses difficulties in interpretation of subgroup results, potential sources of confounding, and uncertainty. This chapter suggests the use of intensive stratification in a study design to help improve the quality and credibility of subgroup analyses. Chapter 2 entitled "Biomarker-Targeted Confirmatory Trials" reviews the alpha protection in treatment comparisons when either the entire population is used or only the marker-positive subpopulation is used. It covers various designs ranging from enrichment design to Enrichment versus All-Comers Designs.

Chapter 3 entitled "Data-Driven and Confirmatory Subgroup Analysis in Clinical Trials" covers the important topics in subgroups during drug development and approval. This chapter has a wide scope and touches on the key thinking from both regulatory and sponsor perspectives. It highlights not only design considerations but also data analysis and interpretation. Chapter 4 entitled "Considerations on Subgroup Analysis in Design and Analysis of Multi-Regional Clinical Trials" is about a recent trend in drug development: multi-regional clinical trial (MRCT). By the nature of MRCT, each region is a subgroup and sample sizes are not only considered from the overall study; they may often be dictated by local regulatory agencies. There are also challenges in data analysis, because regulatory decision of each country (or region) can be based on not only the entire data set but also the subgroup of patients recruited from a particular jurisdiction.

Part II deals with subgroup identification and personalized medicine, with nine chapters. Chapter 5 entitled "Practical Subgroup Identification Strategies in Late-Stage Clinical Trials" suggests some ideas to find subgroups from a failed late-stage trial when the reason of failure is lack of statistical significance. It presents a comprehensive overview of relevant considerations related to the selection of clinically candidate biomarkers, choice of statistical models, including the role of covariate adjustment in subgroup investigation, and selection of subgroup search parameters.

Chapter 6 entitled "The GUIDE Approach to Subgroup Identification" covers GUIDE, a multi-purpose algorithm for classification and regression tree construction with special capabilities for identifying subgroups with differential treatment effects. It is unique among subgroup methods in having all the following features: unbiased split variable selection, approximately unbiased estimation of subgroup treatment effects, treatments with two or more levels, allowance for linear effects of prognostic variables within subgroups, and automatic handling of missing predictor variable values without imputation in piecewise-constant models. One of the applications of GUIDE can be found in the next chapter (Chap. 7) entitled "A Novel Method of Subgroup Identification by Combining Virtual Twins with GUIDE (VG) for Development of Precision Medicines." In this chapter, the authors propose the VG method, which combines the idea of an individual treatment effect (ITE) from Virtual Twins with the unbiased variable selection and cutoff value determination algorithm from GUIDE. Simulation results show the VG method has less variable selection bias than Virtual Twins and higher statistical power than GUIDE Interaction in the presence of prognostic variables with strong treatment effects.

In Chap. 8 entitled "Subgroup Identification for Tailored Therapies: Methods and Consistent Evaluation," a number of methods (including SIDES, VT, and GUIDE) for identifying subgroups with enhanced treatment response are covered, and the authors expect many more to be developed in the coming years. In order for development programs for tailored therapeutics to be successful, it is imperative to determine the best method(s) for biomarker and subgroup identification to be applied in practice. Furthermore, it is expected that no single method can be optimal across all scenarios, so fully characterizing the properties of each methodology is of utmost importance.

Chapter 9 covers "A New Paradigm for Subset Analysis in Randomized Clinical Trials." The general subset identification methods are based on multiple hypothesis testing and using resubstitution estimates of treatment effect that are known to be highly optimistically biased. In this chapter, the authors describe a new paradigm for subset analysis. Rather than being based on multiple hypothesis testing, it is based on training a single predictive classifier and provides an almost unbiased estimate of treatment effect for the selected subset.

Chapter 10 is "Logical Inference on Treatment Efficacy When Subgroups Exist." The authors start with introducing the fundamental statistical considerations in the inference process when subgroups exist, followed by proposing suitable efficacy measures for different clinical outcomes and establishing a general logical estimation principle. Finally, as a step forward in patient targeting, the authors present a simultaneous inference procedure based on confidence intervals to demonstrate how treatment efficacy in subgroups and mixture of subgroups can be logically inferred. Chapter 11 entitled "Subgroup Analysis with Partial Linear Regression Model" recognizes that, in practice, some covariates are known to affect the response non-linearly, which makes a linear model not appropriate. To address this issue, the authors used a partial linear model in which the effect of some specific covariates is a non-linear monotone function, along with a linear part for the rest of the covariates. This approach makes the model not only more flexible than the parametric linear model but also more interpretable and efficient than the full nonparametric model. Illustrations are provided with both simulated data and real-life data.

Part II of this book continues with Chap. 12 entitled "Exploratory Subgroup Identification for Biopharmaceutical Development." This chapter provides a comprehensive review of general considerations in exploratory subgroup analysis, investigates popular statistical learning algorithms for biomarker signature development, and proposes statistical principles for the subgroup performance assessment. Chapter 13 covers "Statistical Learning Methods for Optimizing Dynamic Treatment Regimes in Subgroup Identification." It discusses the many statistical learning methods that have been developed to optimize multistage dynamic treatment regimens (DTRs) and identify subgroups that most beneficial from DTRs using data from sequential multiple assignment randomized trials (SMARTs) and for observational studies. These methods include regression-based Q-learning and classification-based outcome-weighted learning. For the latter, a variety of loss functions can be considered for classification, such as hinge loss, ramp loss, binomial deviance loss, and square loss. Furthermore, data augmentation can be used to enhance the learning performance.

Part III of the book covers some general issues about subgroup analysis, including some regulatory considerations. Chapter 14 entitled "Subgroups in Design and Analysis of Clinical Trials, General Considerations" is contributed by colleagues from FDA. This chapter addresses three topics in subgroup analysis in confirmative clinical trials: (i) general issues in subgroup analysis as part of the overall evaluation of a clinical trial, (ii) trial design considerations to establish treatment efficacy in specific subgroup of patients, and (iii) Bayesian subgroup analysis. Chapter 15 entitled "Subgroup Analysis: A View from Industry" discusses concerns in design and analysis of clinical trials with subgroup considerations. When planning and designing confirmatory trials of new medicines, discussion and agreement with regulatory and reimbursement authorities on the population is exceptionally valuable. Pre-identification on a small number of important biologically plausible subgroups which require exploration is helpful for interpretation.

Chapter 16 is "Subgroup Analysis from Bayesian Perspectives." This chapter examines subgroup identification and subgroup analysis from a Bayesian point of view. Identifying the subpopulation structures, along with the tailored treatments, for all groups plays a critical role for assigning the best available treatment to an individual patient. Subgroup analysis, a key component in personalized medicine, has become increasingly important over the past decade. Besides frequentist methods, there is a spectrum of methods developed from Bayesian perspectives to identify subgroups. In this chapter, the authors provide a comprehensive overview of Bayesian methods and discuss their properties. They further examine empirical performance of the two Bayesian methods via simulation studies and a real data analysis.

Subgroup analysis does not have to be based on a single clinical trial. Analysis can be performed on a combination of studies or can also be performed using meta-analysis. The last two chapters address this topic of meta-analysis of subgroups. Chapter 17 is "Power of Statistical Tests for Subgroup Analysis in Meta-Analysis." Typically, clinical trials of a new treatment also explore whether the treatment effect differs in subgroups of patients or subgroups of trials. This chapter discusses the power of statistical tests for subgroup analysis in order to help in both the planning and interpretation of subgroup tests in a meta-analysis. Chapter 18 is "Heterogeneity and Subgroup Analysis in Network Meta-Analysis." When treatment comparisons involve more than two interventions, the evidence base consists of multiple randomized clinical trials where each of the available studies involves a comparison on a subset of all the competing interventions of interest. If each of these trials has at least one intervention in common with another trial, such that the evidence base can be represented with one connected network, a network meta-analysis (NMA) can provide relative treatment effects between all competing interventions of interest. This chapter discusses subgroup analysis under NMA.

The entire content of this book is intended solely and strictly for educational and pedagogical purposes. The material herein expresses the views of the chapter authors and does not in any way reflect the views of the co-editors, their employers, or any other entity.

The co-editors are deeply grateful to those who have supported in the process of creating this book. We thank all the contributing authors to this book for their enthusiastic involvements and their kindness in sharing their professional knowledge and expertise. Our sincere gratitude goes to all the chapter reviewers for their expert reviews of the book chapters, which lead to a substantial improvement in the quality of this book. The co-editors thank all the reviewers for providing thoughtful and in-depth evaluations of the chapters contained in this book. We gratefully acknowledge the professional support of Ms. Laura Aileen Briskman from Springer who made the publication of this book a reality. The co-editors would also like to thank the support and encouragement from the editors of ICSA Book Series in Statistics, Professors Jiahua Chen and (Din) Ding-Geng Chen.

The co-editors welcome readers' comments, including notes on typos or other errors, and look forward to receiving suggestions for improvements to future editions. Please send comments and suggestions to any of the co-editors.

Ridgefield, CT, USA	Naitee Ting
Groton, CT, USA	Joseph C. Cappelleri
Raleigh, NC, USA	Shuyen Ho
Chapel Hill, NC, USA	(Din) Ding-Geng Chen

Contents

List of Contributors

Chakib Battioui Eli Lilly and Company, Indianapolis, IN, USA

Ilana Belitskaya-Lévy Cooperative Studies Program Palo Alto Coordinating Center, The US Department of Veterans Affairs, Palo Alto, CA, USA

Daniel J. Bratton GlaxoSmithKline Research and Development, Middlesex, UK

Pierre Bunouf Pierre Fabre, Toulouse, France

Ivan S. F. Chan Data and Statistical Sciences, AbbVie, North Chicago, IL, USA

Ming-Hui Chen Department of Statistics, University of Connecticut, Storrs, CT, USA

Yuan Chen Department of Biostatistics, Mailman School of Public Health, Columbia University, New York, NY, USA

Lu Cui UCB Biosciences, Raleigh, NC, USA

Aaron Dane Dane Statistics, Manchester, UK

Brian Denton Eli Lilly and Company, Indianapolis, IN, USA

Ying Ding Department of Biostatistics, University of Pittsburgh, 7133 Public Health, Pittsburgh, PA, USA

Alex Dmitrienko Mediana Inc., Overland Park, KS, USA

Li Ming Dong The Center for Devices and Radiological Health, U.S. Food and Drug Administration, Silver Spring, MD, USA

Lijiang Geng Department of Statistics, University of Connecticut, Storrs, CT, USA

Jean-Marie Grouin University of Rouen, Rouen, France

Yihua Gu Data and Statistical Sciences, AbbVie, North Chicago, IL, USA

Xin Huang Data and Statistical Sciences, AbbVie, North Chicago, IL, USA

Jeroen P. Jansen Precision Health Economics & Outcomes Research, Oakland, CA, USA
Department of Health Research and Policy (Epidemiology), Stanford University School of Medicine, Stanford, CA, USA

Jia Jia AbbVie Inc., North Chicago, IL, USA

Oliver N. Keene GlaxoSmithKline Research and Development, Middlesex, UK

Heng Li The Center for Devices and Radiological Health, U.S. Food and Drug Administration, Silver Spring, MD, USA

Ilya Lipkovich Eli Lilly and Company, Indianapolis, IN, USA

Yang Liu Department of Statistics, University of Connecticut, Storrs, CT, USA

Ying Liu Department of Psychiatry, Columbia University Irving Medical Center, New York, NY, USA

Wei-Yin Loh Department of Statistics, University of Wisconsin, Madison, WI, USA

Ying Lu Department of Biomedical Data Science, Stanford University School of Medicine, Stanford, CA, USA

Xuezhou Mao Biostatistics and Programming, Sanofi, Bridgewater, NJ, USA

Christoph Muysers Bayer AG Pharmaceuticals, Berlin, Germany

Terri D. Pigott Loyola University Chicago, Chicago, IL, USA
Georgia State University, School of Public Health and College of Education and Human Development, Atlanta, GA, USA

Hui Quan Biostatistics and Programming, Sanofi, Bridgewater, NJ, USA

Lei Shen Eli Lilly and Company, Indianapolis, IN, USA

Hollins Showalter Eli Lilly and Company, Indianapolis, IN, USA

Noah Simon Department of Biostatistics, University of Washington, Seattle, WA, USA

Richard Simon R Simon Consulting, Bethesda, MD, USA

Yan Sun Data and Statistical Sciences, AbbVie, North Chicago, IL, USA

Qi Tang Sanofi US, Bridgewater, NJ, USA

Ming T. Tan Department of Biostatistics, Bioinformatics and Biomathematics, Georgetown University, Washington, DC, USA

Ram Tiwari The Center for Devices and Radiological Health, U.S. Food and Drug Administration, Silver Spring, MD, USA

Hui Wang Cooperative Studies Program Palo Alto Coordinating Center, The US Department of Veterans Affairs, Palo Alto, CA, USA

Jun Wang Office of Biostatistics and Clinical Pharmacology, Center for Drug Evaluation, National Medical Products Administration, Beijing, China

Xiaojing Wang Department of Statistics, University of Connecticut, Storrs, CT, USA

Xinjun Wang Department of Biostatistics, University of Pittsburgh, 7121C Public Health, Pittsburgh, PA, USA

Yuanjia Wang Department of Biostatistics, Mailman School of Public Health, Columbia University, New York, NY, USA
Department of Psychiatry, Columbia University, New York, NY, USA

Yue Wei Department of Biostatistics, University of Pittsburgh, 7121C Public Health, Pittsburgh, PA, USA

Wangang Xie AbbVie Inc., North Chicago, IL, USA

Tu Xu Vertex Pharmaceuticals, Boston, MA, USA

Ao Yuan Department of Biostatistics, Bioinformatics and Biomathematics, Georgetown University, Washington, DC, USA

Lilly Q. Yue The Center for Devices and Radiological Health, U.S. Food and Drug Administration, Silver Spring, MD, USA

Donglin Zeng Department of Biostatistics, Gillings School of Global Public Health, University of North Carolina at Chapel Hill, Chapel Hill, NC, USA

Donghui Zhang Global Biostatistics and Programming, Sanofi US, Bridgewater, NJ, USA

Ji Zhang Biostatistics and Programming, Sanofi, Bridgewater, NJ, USA

Lanju Zhang AbbVie, Inc., North Chicago, IL, USA

Peigen Zhou Department of Statistics, University of Wisconsin, Madison, WI, USA

Yizhao Zhou Department of Biostatistics, Bioinformatics and Biomathematics, Georgetown University, Washington, DC, USA

Reviewers of the Subgroup Book

- Chung-Chou H. Chang (University of Pittsburgh)
- Thomas Debray (University Medical Center Utrecht)
- Qiqi Deng (Boehringer Ingelheim)
- Ying Ding (University of Pittsburgh)
- Haoda Fu (Eli Lilly and Company)
- Lei Gao (Vertex Pharmaceuticals)
- Xuemin Gu (Daiichi-Sankyo Inc.)
- Samad Hedayat (University of Illinois at Chicago)
- Jason Hsu (The Ohio State University)
- Peng Huang (Johns Hopkins University)
- Chia-Wen (Kiki) Ko (FDA)
- Shengchun Kong (Gilead Sciences)
- Sally Lettis (GSK)
- Gang Li (Johnson & Johnson)
- Xuefeng Li (FDA)
- Ilya Lipkovich (Eli Lilly)
- Shigeyuki Matsui (Nagoya University)
- Kaushik Patra (Alexion Pharmaceuticals)
- Joshua Polain (American Institutes for Research)
- Xin Qiu (Janssen Research & Development)
- Ian Saldanha (Brown University)
- Tao Sheng (Sanofi)
- Yu-Shan Shih (National Chung-Cheng University, Taiwan)
- Yan Sun (Abbvie)
- David Svensson (AstraZeneca)
- Ming Tan (Georgetown University Medical Center & Lombardi Comprehensive Center)
- Larry Tang (George Mason University)
- Shijie Tang (Karyopharm Therapeutics Inc.)
- Yuanyuan Tang (Saint-Luke's Health System)
- Peter Thall (University of Texas-MD Anderson Cancer Center)

- Lu Tian (Stanford University)
- Ryan Williams (American Institutes for Research)
- David Wilson (George Mason University)
- Wangang Xie (Abbvie)
- Lei Xu (Vertex Pharma)
- Xu (Sherry) Yan (FDA)
- Xuan Yao (Abbvie)
- Ron Yu (Gilead)
- Jason Yuan (Allergan Inc.)
- Sammy Yuan (Merck)
- Lanju Zhang (AbbVie Inc.)
- Ying Zhang (Eli Lilly and Company)
- Yinchuan Zhao (Georgia State University)

Part I
Subgroups in Clinical Trial Design and Analysis

Chapter 1
Issues Related to Subgroup Analyses and Use of Intensive Stratification

Lu Cui, Tu Xu, and Lanju Zhang

Abstract After the completion of a clinical study for the drug efficacy and safety, subgroup analyses are typically conducted. The analyses may yield supportive information for the main finding based on the overall population, or generate new hypotheses on the drug effect for further investigation. Although there are valid reasons to perform subgroup analyses, the warning has been given to caution the interpretation of subgroup results. There is a general doubt on the believability of subgroup analysis because of the potential confounding and uncertainty related to subgroup findings which could be anti-intuitive, inconsistent, unexpected, or unexplainable. The present work is to discuss potential sources of confounding in subgroup analyses which may bias interpretations and lead to erroneous claims. Solutions to the problem are discussed. A special attention is paid to the use of the intensive randomization stratification to improve the quality and believability of subgroup analyses.

1.1 Introduction

Subgroup analyses are often performed in clinical studies. The results of subgroup analyses can be answers to specific scientific questions or for hypothesis generations on the treatment effects in special patient populations. A recent survey of 97 clinical trials reveals that 59 studies or 61% of them employed subgroup analyses (Wang et al. 2007a, b). The new initiative on precision medicine announced by

L. Cui (✉)
UCB Biosciences, Raleigh, NC, USA
e-mail: lu.cui@ucb.com

T. Xu
Vertex Pharmaceuticals, Boston, MA, USA

L. Zhang
AbbVie, Inc., North Chicago, IL, USA

© Springer Nature Switzerland AG 2020
N. Ting et al. (eds.), *Design and Analysis of Subgroups with Biopharmaceutical Applications*, Emerging Topics in Statistics and Biostatistics,
https://doi.org/10.1007/978-3-030-40105-4_1

President Obama (Ashley 2015) further highlights the need and effort for more effective medical treatment via a better targeted patient population or a subgroup of patients. With the objective of the precision medicine, the quality of analysis of subgroup data becomes critically important. The guidance for conducting and interpreting subgroup analyses has been issued by European Medicine Agency and commented in other regulatory documents, for example, the ICH Guidance for Industry E9: Statistical Principles for Clinical Trials and the FDA Guidance for Industry: Enrichment Strategies for Clinical Trials to Support Approval of Human Drugs and Biological Products.

Despite the popularity of subgroup analyses, results from subgroup analyses have traditionally been viewed as suspicious and unreliable (Sleight 2000). This is because, if an analysis is handled improperly, an unexpected or inconsistent subgroup outcome can occur and the result can be hard to interpret and potentially misleading. Consequently, the believability of the subgroup analysis result suffers and an extra caution on interpreting the subgroup finding is warranted (Sun et al. 2010, 2012). There have been many discussions in the literature about issues related to subgroup analyses as given in the references.

Example 1.1 The most well-known example of issues of subgroup analysis is the result of retrospective analysis of the data of ISIS-2 trial (ISIS-2 1988) by patients' astrological star signs (Sleight 2000; Peto 2011). The trial (n = 17,187) was to investigate effectiveness of aspirin treatment as compared to placebo in reduction of mortality in patients with acute myocardial infarction. The result indicated a highly significant treatment difference in one month mortality rate in the overall population (p < 0.000001). The numbers of deaths were 804 (9.4%) in the aspirin group and 1016 (11.8%) in placebo, respectively. The observed treatment difference, however, diminished in the subgroup of patients whose astrological birth signs are Libra or Gemini. With 150 deaths in the aspirin group and 147 deaths in placebo, the treatment difference was not statistically significant. There was no scientific explanation to support this insignificant subgroup finding.

Example 1.2 European Carotid Surgery Trial (European Trialists 1998) was a large randomized controlled study (N = 3024) to assess the benefit of carotid endarterectomy vs. control without surgery in patients with recently symptomatic carotid stenosis. Eligible subjects had experienced, in the previous 6 months, one or more carotid-territory ischaemic events in the brain or eye, which were either transient (symptoms lasting minutes, hours, or days) or permanent but did not cause any serious disability. The trial allocated subjects to the surgical group and control at 3:2 randomization ratio. The results of the trial showed that the treatment difference between the surgical and control groups in the overall population was not statistically significant. There were a total of 202 events out of 1807 subjects (11.2%) in the surgical group and 136 events out of 1211 subjects (11.2%) in the control group (European Trialists 1998). While the subjects were arbitrarily grouped by their birth month into three subgroups: birth date between May and August, September and December, and January and April, no treatment difference, again, was detected (Rothwell 2005) in these subgroups. While further grouping the subjects according to their birth month plus whether they had $\geq$70% stenosis at the baseline, significant subgroup differences were found. In subjects born between May and August and with $\geq$70% stenosis, the surgical group performed much better than the control (13/168 vs. 32/105) with statistical significance at $\alpha = 0.05$, moderately better than the control (12/142 vs. 14/73) in subjects born between September and December, and almost equally (18/137 vs. 12/96) in subjects born between January and April. Such a trend was reversed in subjects with <70% stenosis. The surgical group performed much worse than the control (51/454 vs. 13/299) with statistical significance in subjects born between May and August, moderately worse (53/409 vs. 25/292) in subjects born between September and December, and almost equally (55/489 vs. 40/343) in subjects born between January and April.

Example 1.3 Bitopertin is a potent and selective GlyT1 inhibitor. With positive pre-clinical data, a placebo-controlled, double blinded proof-of-concept phase 2 study (Umbricht et al. 2014) was conducted to assess whether bitopertin could improve negative symptoms of schizophrenia. A total of N = 323 subjects were randomized at 1:1:1:1 ratio to receive placebo, 10 mg, 30 mg, or 60 mg of bitopertin once daily, respectively, for 8 weeks plus a 4-week follow-up. The primary efficacy endpoint was change in PANSS-NSFS score from baseline, and the secondary efficacy endpoints were the CGI-I-N score and percentage of responders. The reported trial results were based on per-protocol (PP) patient population. This outcome dependent sub-population consisted of N = 231 subjects or 72% all randomized subjects. The PP analyses indicated a significant difference in mean reduction of PANSS-NSFS from baseline at Week 8 between PBO and 10 mg (p = 0.049) and 30 mg (p = 0.03) dose groups of bitopertin. Statistically significant treatment difference in CGI-I-N score was also observed between PBO and 10 mg dose group. The seemingly positive PP analysis results led to three large phase III confirmatory studies: SunLyte, DayLyte, and FlashLyte. All three phase III studies later failed despite significantly increased sample sizes and treatment durations. Study SunLyte was declared futile before the patient enrollment completion. Study DayLyte (N = 621) and Study FlashLyte (N = 594) failed to demonstrate a treatment benefit on the primary efficacy endpoint, change from baseline in PANSS-NSFS score at Week 24. In the two studies, the p-values for comparison of an active dose with PBO ranged from p = 0.32 to p = 0.88. Viewing the PP analysis with 72% of the total population of the phase 2 study as a subgroup analysis, unlike the previous two examples with internal inconsistency of the subgroup findings, this example showed an across study inconsistency. The results of all three larger phase 3 studies were completely in the opposite direction of the significant PP analysis result of the phase 2 study.

The above examples from real clinical trials illustrate uncertainty of subgroup outcomes and difficulties in interpretation because of potential confounding introduced from arbitrarily subdividing an overall population and intentionally or unintentionally looking for eye-catching results after multiple subgroup analyses are performed. In the following, we will briefly touch upon the statistical issues related to the interpretation of subgroup analysis results, and then focus on the application of stratified randomization to improve the quality of subgroup analyses.

1.2 Issues in Interpreting Subgroup Data

Potential confounding factors in subgroup analyses have been well recognized. Cui et al. (2002) summarized some common confounding factors into several categories. The word "confounding" here is used in a relaxed sense to mean sources for potential confusions and misinterpretations in subgroup analyses, likely leading to incorrect claims including biased estimates of treatment effects and false positive or negative conclusions.

1.2.1 Confounding Due to Sampling Error

Just splitting an overall population into two subpopulations may lead to redistribution of the outcomes between the treatment groups. For example, while there is a treatment difference in the overall population, the treatment effect in the resulted subgroups may still be seen but the effect size can be smaller than the overall effect size in one subgroup and larger in the complement subgroup. The reversal can also be true. While there is no treatment effect presented and the treatment response is equal in the two treatment groups, splitting the overall population, which is easily uneven, may result in one subpopulation with the outcome in favor of the testing arm and the other subpopulation with the outcome in favor of the control arm with high likelihood. Such a forced change of the outcome distribution between the treatment groups can be viewed as a sampling error and the confounding associated, if any, as confounding due to sampling error.

Based on our experience, inconsistency in subgroup outcomes due to splitting of a parent population tends to be numerical but can be nominally statistically significant from time to time. In previous Example 1.1, arbitrarily grouping patients according to their astrology birth signs in a post-hoc fashion may introduce the confounding due to sampling error. Although it is hard to prove, such confounding is a likely source of the unexplainable and unexpected finding in patients with the birth signs Libra or Gemini. Similarly, such confounding is a likely source of the inconsistent and contradicting subgroup findings in Example 1.2 while the post-hoc splitting the patient populations continues. Statisticians should always be aware of the issue, and be cautious in interpreting numerical difference between treatment groups in subgroup analyses while there is no explanation of any other kinds. This is particularly true while there is no overall treatment effect but the effects in opposite directions are observed in a subgroup and its complement.

1.2.2 Confounding Due to Repeated Hypothesis Testing

While a hypothesis testing is performed in a subgroup, just by chance, a false positive outcome may arise. The probability of the false positive outcome or the type I error rate often is controlled at the level of $\alpha = 0.05$. While the same hypothesis is repeatedly tested in multiple subgroups, the chance of a false positive claim in at least one subgroup is increased. The more subgroups are tested, the higher chance for at least one false positive outcome in a subgroup is. The direction of the impact of the confounding due to repeated hypothesis testing across several subgroups is clear, i.e. increasing type I error rate.

In general, the inflation of type I error rate from multiple testing can be effectively controlled using multiple testing procedures. For example, if the number of subgroups to be tested is specified upfront in the protocol and the total number of subgroups analyzed is known, say, M, the usual Bonferroni test can be used by setting the statistical significance level as α/M for each subgroup analysis. The overall type I error rate for the claim of drug effectiveness in at least one subgroup is controlled under α. More sophisticated and efficient multiple testing methods can be used to analyze multiple subgroups depending on how the claims will be made. Good references on multiple testing methods can be found in Dmitrienko (2009).

In Example 1.2, multiple subgroups defined by the birth months and $\geq 70\%$ stenosis of patients are involved in the inference of the treatment effect. Confounding due to repeated testing may also be a contributing factor for a positive finding in patients born between May and August and with $\geq 70\%$ stenosis. It should be pointed out, for a similar reason of inflating type I error rate under the null, multiple testing, by chance, may also lead to a false negative outcome with a conclusion of no treatment effect or inflate type II error rate while in truth the treatment is efficacious. In Example 1.1, it is likely that the insignificant finding on the aspirin treatment in patients with birth signs Libra and Gemini actually is just an artifact from the multiple testing of the same hypothesis in many sub-patient populations defined by their birth signs.

1.2.3 Confounding Due to Lack of Statistical Power

A clinical trial typically is sized and powered based on the projected treatment effect in the primary efficacy endpoint in the overall patient population. The smaller sample size of a subgroup often provides insufficient statistical power for the subgroup analysis. Such confounding from insufficient sample size or lack of statistical power can inflate the type II error rate. The treatment may be efficacious for the subgroup but demonstrate no statistical significance due to lack of statistical power of the analysis.

The solution to address the low power issue in subgroup analysis essentially is to plan a sufficiently large sample size at the trial design stage for the subgroup analysis of interest. If it is difficult to make an upfront commitment, an adaptive sample size design (Cui et al. 1999, 2017; Lehmacher and Wassmer 1999) can be considered. In this case, a mid-course sample size increase is possible if the sample size and the power of the subgroup analysis are found insufficient in an interim analysis. Other techniques to mitigate the problem include enrichment design to make the subgroup more sensitive to the treatment through targeting at a special patient population, say with certain positive biomarkers. Such a biomarker positive subgroup can be specified in the protocol prior to the start of a trial and the analysis can be powered accordingly. A more responsive subgroup of patients can also be identified in an interim analysis instead of determined before the start of the trial. In this case, the sample size of the identified subgroup may be altered accordingly in order to power the subgroup analysis adequately. The mid-course subgroup enrichment in combination with the sample size adaptation requires the control of type I error rate. Depending on how the targeted subgroup is identified and the change to be made, it can be a complicated issue.

The typical sign of lack of statistical power of subgroup analysis is an overt large numerical treatment difference accompanied by an insignificant p-value. With that, in Example 1.1, the negative finding on the effect of aspirin in patients with birth signs Libra and Gemini does not appear to relate to the smaller sample size or lack of power of the subgroup analysis but indifferentiable responses between the two treatment groups. As pointed before, this negative outcome more likely is an artifact from the artificially splitting the overall population and/or a chance finding from the repeated testing.

1.2.4 Confounding Due to Baseline Incomparability

While baseline characteristics and prognostic factors, which impact on the treatment outcomes, are unevenly distributed between the treatment groups, confounding can be introduced to bias the outcome evaluation, leading to erroneous conclusions (Pocock and Simon 1975). A valid randomization helps to balance influential baseline factors across the treatment groups in the overall population. However, because of excluding subjects, a valid randomization within a subgroup may not be guaranteed and the imbalance of baseline factors is possible.

The confounding due to treatment group incomparability at baseline is a much more difficult issue to handle in subgroup analyses (Cui et al. 2002). First, unlike the confounding due to the repeated testing and the small subgroup sample size, the direction of the confounding from the baseline incomparability and the magnitude of the impact are unpredictable. Depending on the directions of the imbalance w.r.t. individual factors, the overall impact can be either in favor of or against the testing drug. Second, if there is the confounding due to baseline incomparability, it can be difficult to detect because the imbalance can be from influential factors which are not observed and/or from a joint effect of the imbalance of several influential factors. Third, even while the confounding is recognized, removing it via statistical analysis can be difficult. In theory, a model-based covariate adjustment can help to remove the confounding. In practice, it is difficult since for the analysis to be effective the complete knowledge about the prognostic factors and their functional relationship to the outcome is required. Such knowledge generally is incomplete or impossible.

Now let's go back to our previous examples. Excluding a large number of randomized subjects leads to significant doubts on the validity of randomization across the subgroups in Example 1.1 and Example 1.2, and subsequent concerns on the potential baseline incomparability. While confounding due to sampling error or repeated testing are suspected on the surface, the baseline incomparability could be the explanation underneath. Similarly there is a well-founded doubt about the validity of the randomization in the per-protocol analysis of the phase 2 study in Example 1.3 knowing that 28% randomized subjects are effectively excluded from the analysis. The potential baseline incomparability from the subject exclusion might confound the primary analysis of the phase 2 study leading to spurious outcomes that contradict the results of the later phase 3 studies.

Since there is no effective method to detect and correct the confounding due to baseline incomparability, the solution to the problem essentially relies on an upfront stratified randomization. Particularly, to address the issue, we advocate an intensive stratification approach as explained below.

1.3 Intensive Stratification in Subgroup Analysis

Although model-based statistical analyses adjusting for influential baseline factors may help to minimize the impact of the confounding due to baseline incomparability, the approach may not be fully effective in practice. This is because often it is hard to identify such influential factors for the adjustment and their functional relationship to the outcome measure is unknown.

More proactively, a stratified randomization by subgroups can be planned in the protocol and implemented in the trial. With the stratification, the randomization nested in individual subgroup strata can balance the observable and unobservable influential baseline factors across the treatment groups in each stratum. Consequently, the subgroup analyses in the individual strata and their combinations are less likely to subject to the confounding due to baseline incomparability.

The design and implementation of randomization stratification involve the determination of stratification factors and the minimum stratum size allowed or the maximum number of strata allowed given the fixed total trial sample size. For the former, the general agreement is that the stratification factors should be those which are measured at the baseline and influential on the efficacy outcomes of the trial. For the latter discussions on the proper number of stratification factors and strata continue (Hallstrom and Davis 1988; Therneau 1993; Kernan et al. 1999).

While people tend to prefer a few stratification factors in order to have a large number of patients in each stratum, we advocate intensive randomization stratification or having as many stratification factors as possible if the factors affect the treatment outcome and the number of patients in the trial permits. The more stratification factors and strata are used, the more subgroups will be immune from the confounding due to baseline incomparability as shown in the following example.

Often a phase III cardiovascular clinical trial to study a new drug in treating patients with myocardial infarction requires tens of thousands of subjects. Consider such a two-arm clinical trial with a total of 18,000 subjects. If the strategy of intensive stratification is used and at least 8 subjects per stratum are required, a maximum 11 binary baseline stratification factors can be accommodated. This is because 11 binary factors will generate $2^{11} = 2048$ strata involving minimally a total of 16,384 subjects. The scheme will accommodate many analyses of subgroups, supported by a valid randomization, from the arbitrary combinations of some of the 2048 strata. While the common risk factors in the disease population (Antman et al. 2006) involve age (<=65, >65), smoke (Y, N), diabetes (Y, N), BMI (<=30, >30), hypertension (Y, N), Pre-MI (Y, N), disease risk score (low, high), pre-treatment (Y, N), the intensive stratification may cover all mentioned risk factors. As so, there should be a less concern on the confounding due to baseline incomparability in the subgroup analyses based on the aforementioned disease risk factors.

The major concern for including many stratification factors is potential treatment group imbalance due to incomplete strata. Blocked randomization can help to minimize the imbalance within a stratum. Kernan et al. (1999) have suggested that the minimum size of a stratum be at least 4B, where B is the randomization block size. In the above example, a block size B $= 2$ is assumed and thus the minimum size of a stratum is set to 8.

For a two-arm clinical trial with K binary stratification factors there are 2^K strata generated from the combinations of the different levels of the K factors. If the total number of subjects of the trial is N and the minimal size of a stratum is 4B, the maximum number of strata allowed is $\lfloor \frac{N}{4B} \rfloor$. The maximal number of stratification factors allowed K^* then is the maximum value of K such that $2^K \leq \lfloor \frac{N}{4B} \rfloor$. Let $L = L(N) = 2^{K^*}$ be the maximum number of the strata, given the total sample size N. There are $\sum_{j=1}^{L-1} \binom{L}{j} = 2^L - 2$ subgroups, from a single stratum or a combination of multiple strata, supported by the valid stratified randomization. The intensive stratification ensures a valid randomization for a large number of subgroup analyses.

1.4 Statistical Simulations

1.4.1 Setting

In this section, we examine the performance of the proposed intensive stratification strategy via a simulated example. The numerical performance of the intensive stratification strategy (IS) will be compared to the simple randomization without stratification (SR). For the illustration purpose, we also include the intensive stratification with complete blocks (ISC) as a benchmark for the comparison.

The generation of simulated data is based on a hypothetical two-arm clinical trial with N subjects per treatment group and the randomization assignment at 1:1 ratio. We assume that the continuous efficacy measure Y for the subject i is $Y = Y_i$ which is generated through the following regression model

$$Y_i = a * trt + 0.1 * (1000 - g(age)) + \epsilon,$$

where $\epsilon \sim N(0, 10)$, $trt = 1$ for the testing drug arm and $trt = 0$ the control arm. Under the null and alternative hypothesis, the treatment effect size a is set as

$$H_0 : a = 0 \text{ and } H_1 : a = 5.6.$$

The age is considered as an influential prognostic factor with the range $(20, 60)$, and $g(age)$ is defined as

$$g(age) = \begin{cases} 0.25 * age, & if \ age \leq 45; \\ \\ (age - 45)^3 + 11.25, & if \ age > 45. \end{cases}$$

Further, assume that the density of the *age* factor is linearly increasing as there are more elderly patients, that is, $f(x) = \frac{x}{1600}$, $20 \leq x \leq 60$. Thus age_i for the subject i is generated by the inverse transform sampling method. For the subject i, a *gender_i* is also generated using the Bernoulli distribution (0.5). In each simulation sample, the treatment outcomes $Y_i, i = 1, \ldots, N$, are generated based on the normal distribution following $N (a * trt_i + 0.1 * (1000 - g(age_i)), 10)$. A total of 50,000 simulated samples are generated for each randomization scheme to evaluate the performance of the intensive stratification (IS and ISC) as compared to the simple randomization (SR).

Under the proposed intensive stratification strategy, 10 age levels (20 to <24, 24 to <28, ..., 56 to <60) and the gender factor (male, female) are used in the stratified randomization, which leads to 20 strata in total. As suggested in Kernan et al. (1999), in order to maintain the benefit of the stratified randomization, the number of strata should be no more than $N/(B \times 4)$, where N is the total sample size and B is the block size. We set the total sample size $N = 160$ and $B = 2$.

In simulated samples with simple randomization (SR), to reflect the reality, the subjects in the overall population are sequentially randomly assigned to ether the active treatment or placebo groups with an equal probability 0.5. The assignments are nested within each stratum under the intensive stratification (IS) scheme. In both types of randomization assignments, incomplete blocks are allowed. As a benchmark, simulation samples based on intensive stratification without incomplete blocks (ISC) are also generated. In this case, the equal number of samples (8 subjects) is assigned to each stratum such that all the randomization blocks are completely filled.

For illustration purpose, the simulated outcomes are analyzed based on four regression models with different levels of covariate adjustments in the female subgroup. The four analysis models are

$$\text{Model 1}: Y = b + c * trt,$$

$$\text{Model 2}: Y = b + c * trt + e * age,$$

$$\text{Model 3}: Y = b + c * trt + e * age + f * age^2,$$

$$\text{Model 4}: Y = b + c * trt + e * g(age).$$

The partial covariate adjustments based on Models 1–3 reflect the practical scenarios in which either the important covariate is likely missed (Model 1) or there is only incomplete knowledge of the functional relationship between the covariate and the outcome (Models 2 and 3). In contrast, Model 4 represents the full adjustment of *age* based on the true underlying data model.

1.4.2 Results

From the underlying data generation model, it is clear that age is a highly influential prognostic factor, especially for those older than 45. The estimated treatment differences under H_0 in female subgroup from 50,000 simulation samples for each combination of the 3 randomization schemes and 4 analysis methods are plotted in Fig. 1.1. In the figure, the rows from the top to the bottom are for IS, SR, and ISC, respectively. The columns from the left to the right are for analysis Model 1, Model 2, Model 3, and Model 4, respectively. The treatment differences for a combination of the randomization scheme and analysis method in the plot is organized according to the percent of age imbalance which is the percentage of the difference in the number of patients who is older than 50 between the treatment groups. More precisely, it is defined as

age imbalance

$$= \frac{\#of\ female\ with\ age{>}50\ in\ the\ treatment - \#of\ female\ with\ age{>}50\ in\ placebo}{\#of\ female\ in\ the\ treatment}.$$

In addition to Fig. 1.1, the variation of the estimated treatment difference in female subgroup in terms of SD, Q_1, Q_3, Minimum and Maximum are summarized in Table 1.1.

Based on the simulation outcomes, the following observations can be made.

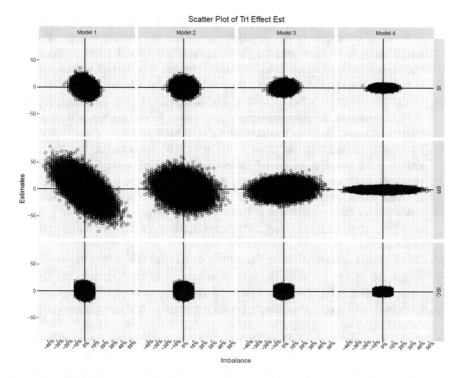

Fig. 1.1 Estimated treatment difference in females with or without covariate adjustments

	Method	Model	Mean	SD	Q1	Q3	(Min, Max)
Table 1.1 Summary of estimated treatment difference in females with or without covariate adjustments	SR	1	0.0	13.2	−8.8	8.9	(−53.6, 54.8)
		2	0.0	9.6	−6.6	6.5	(−39.9, 38.4)
		3	0.0	5.9	−4.0	4.0	(−21.0, 24.7)
		4	0.0	1.6	−1.1	1.1	(−6.3, 6.5)
	IS	1	0.0	5.0	−3.4	3.4	(−21.1, 18.4)
		2	0.0	4.3	−2.9	2.9	(−18.3, 16.5)
		3	0.0	3.3	−2.1	2.2	(−14.0, 12.2)
		4	0.0	1.6	−1.1	1.1	(−6.4, 6.4)
	ISC	1	−0.1	3.7	−2.6	2.5	(−15.3, 14.6)
		2	0.0	3.4	−2.4	2.3	(−14.5, 14.3)
		3	0.0	2.8	−1.9	1.9	(−12.0, 11.0)
		4	0.0	1.6	−1.1	1.1	(−6.5, 6.4)

(a) The intensive stratifications (IS and ISC) reduce the variation of the estimated treatment difference as compared to the simple randomization (SR) while there is no or only a partial covariate adjustment. This can be seen by comparing the vertical ranges of the plots in Fig. 1.1 and SD column in Table 1.1 with three randomization schemes under each of Mode 1, Model 2, and Model 3.

(b) The variation of the estimated treatment difference becomes smaller with the increase of the degree of the completeness of the covariate adjustment in each randomization scheme, and the full covariate adjustment leads to the same variation regardless of the randomization schemes. The former can be seen by comparing the vertical ranges of the plots from Model 1 to Model 4 within each of the three randomization schemes. The latter can be seen by comparing the vertical ranges of the plots in three randomization schemes under Model 4. It is illustrated numerically by the SD column in Table 1.1.

(c) The intensive stratification plays a similar role of the covariate adjustment in reducing the variation of the estimated treatment difference. The variation of the estimated treatment difference under the intensive stratification (IS and ISC) is much smaller than that of the simple randomization (SR). With the intensive stratification, the contribution of the covariate adjustment diminishes. The SD is reduced from 13.2 to 1.6 with full covariate adjustment for SR while from 3.7 to 1.6 for ISC.

(d) The intensive stratifications (IS and ISC) reduce the chance of a large percent difference in age between the treatment groups as compared to the simple randomization (SR). This can be seen by compare the horizontal ranges of the plots in different randomization schemes under individual covariate adjustment models.

(e) For all 3 randomization schemes, the mean of the estimated treatment difference in females appears around zero or the estimate of the treatment difference is unbiased in the female subgroup. This is expected since the valid randomization in the female group is finely supported by the age stratified randomization within the females.

(f) However, without intensive stratification or under SR, the estimated treatment difference can be biased or significantly biased in a subgroup of the female population in which the age factor is not evenly distributed between the two treatment groups. For example, while there is about 10% age imbalance, without the covariate adjustment (Model 1), the estimated treatment difference with the simple randomization (SR) is about 20 which is significantly biased.

(g) The intensive stratification generally can significantly minimize the impact of or the chance of bias due to baseline incomparability. Under both ISC and IS, the estimated treatment difference appears unbiased with the mean value at or close to zero in subgroups with various levels of percent difference. For example, while there is about 10% age imbalance, without the covariate adjustment (Model 1), the estimated treatment differences under the intensive stratifications (IS, ISC) are around zero or unbiased.

(h) Without intensive stratification or under SR, the analysis with the covariate adjustment of age reduces both variation and bias of the estimated treatment difference. The analysis with the full covariate adjustment removes the bias completely though knowing the underlying data model is generally impossible in practice.

(i) The Intensive stratification works like a model-based adjustment but requiring no knowledge of the functional relationship between the covariate and the outcome. By finely matching the age categories between the treatment groups and by preventing the baseline imbalance in age upfront, the contribution of the model-based covariate adjustment under IS and ISC is fairly limited.

As a note, our simulations give similar results under the alternative assumption of the treatment effect, and therefore the results are not presented.

1.5   Conclusion

This article touches upon the likely sources for confounding in subgroup analyses. The potential confounding due to repeated testing without multiplicity adjustment or due to lack of statistical power in subgroup analyses with sizable estimated treatment difference but achieving no statistical significance can be easily identified. The directions of the impacts are also clear. The former inflates Type I error rate while the latter inflates Type II error rate. The remedy measures can be implemented through a multiplicity adjustment, or increasing subgroup sample size and/or focusing on a more responsive patient population.

Confounding may arise if the treatment groups are not comparable with respect to certain influential baseline factors due to lack of a valid randomization from excluding subjects in subgroup analyses. The impacts on the outcomes of such confounding can be severe and often are undetectable and unpredictable.

To address the baseline incomparability issue in subgroup analyses, the proactive implementation of the intensive randomization stratification is proposed. The immediate benefit of the approach is the guaranteed valid randomization for a large number of analyses based on the subgroups from a single stratum or an arbitrary combination of the strata of the stratification factors. In the hypothetic example of the cardiovascular trial, with the intensive stratification, a total of 2048 strata from 11 influential baseline factors are allowed. Consequently, a vast number of subgroups from the combinations of the strata are supported by a valid randomization, and the results of the corresponding subgroup analyses tend to be trustful. This is particularly true while the outcome of a subgroup analysis is statistically significant and the claim is limited to this specific subgroup.

Our simulations show that the intensive stratification helps to remove the potential bias and reduce the variation in estimation. Under the ideal scenario where all stratification blocks are complete, the intensive stratification performs well in producing the unbiased analysis result. In this case, the subgroup analysis without the covariate adjustment performs equally well as the analysis with the full covariate adjustment based on the true underlying data model. The subgroup analyses without the covariate adjustment under the intensive stratification still perform better than the covariate adjusted analyses while the model is misspecified due to lack of knowledge of the dependency of the outcome on the covariates. Without loss of generality, our simulations are limited to one subgroup of female patients and one influential baseline factor age. The extension of the results to situations with multiple subgroups and influential baseline factors is apparent.

After the discussion of the idea of intensive stratification and its potential benefits, it is worthy to comment on its implementation. In general, the candidate stratification factors can be identified based on the clinical knowledge on the disease and the effects of the drugs of the same class. Such information might be found in literature or through analyses of historical data. In general, for a confirmatory clinical trial, the strategy of a stratified randomization needs to be pre-specified in the protocol, and detailed in the statistical analysis plan before the database unblinding. Although a general guidance on the maximum number of strata allowed is given in the previous section, the statistician needs to decide the extent of the intensive stratification taking all things into considerations. For example, if there are too many strata, incomplete blocks become more likely. In such a case, one may consider dropping some less critical stratification factors in the randomization or ignored them in the later analyses. Either ways, however, the analyses based on the stratified subgroups remain valid. In general, the influential or likely influential factors can be identified either based on clinical knowledge or based on historical data. Occasionally, there can be cases in which influential factors are not easy to be identified. In such cases, the intensive stratification can be either implemented based on well accepted factors (e.g. age, gender, race, and etc.) or factors such as those generally related to patients wellbeing (e.g. BMI, smoking status, diabetes, hypertension and etc.) and those about treatments previously received (drugs received, response status, drug classes and etc.).

In terms of implementation, one should also consider potential challenges associated with the operation, for example, the added complexity with respect to randomization, drug supply, patient recruitment, and others. Whether or not such issues will become limiting factors for the implementation may depend on individual clinical trial setting. For a large cardiovascular trial, the concern could be much less because the stratification information in general is well collected at the baseline and the patient recruitment is not an issue due to the availability of large number of patients. The situation, however, can be quite different for an oncology study because of the limited patient availability. In such a case, the best one may do is to maximize the number of the critical stratification factors within the practical limit. As for every design of a clinical trial, it is the statistician's responsibility to evaluate pros and cons of the intensive stratification to make the best implementation decision to meet the design objectives.

References

Antman EM et al (2006) Enoxaparin versus unfractionated heparin with fibrinolysis for st-elevation myocardial infarction. NEJM 354:1477–1488

Ashley EA (2015) The precision medicine initiative - a new national effort. JAMA 313:2119–2101

Cui L, Hung HMJ, Wang SJ (1999) Modification of sample size in group sequential clinical trials. Biometrics 55(3):853–857

Cui L, Hung HMJ, Wang SJ, Tsong Y (2002) Issues related to subgroup analysis in clinical trials. J Biopharm Stat 12:347–358

Cui L, Zhang L, Yang B (2017) Optimal adaptive group sequential trial design with flexible timing of sample size determination. Contemp Clin Trials 63:8–12

Dmitrienko A, Tamhane AC, Bretz F (2009) Multiple testing problems in pharmaceutical statistics, 1st edn. Chapman and Hall/CRC

European Carotid Surgery Trialists' Collaborative Group (1998) Randomised trial of endarterectomy for recently symptomatic carotid stenosis: final results of the MRC European Carotid Surgery Trial (ECST). Lancet 351:1379–1387

Hallstrom A, Davis K (1988) Imbalance in treatment assignments in stratified blocked randomization. Control Clin Trials 9:375–382

ISIS-2 Collaborative Group (1988) Randomized trial of intravenous streptokinase, oral aspirin, both or neither among 17817 cases of suspected acute myocardial infarction: ISIS-2. Lancet 332:349–360

Kernan W, Viscoli C, Makuch R, Brass L, Horwitz R (1999) Stratified randomization for clinical trials. J Clin Epidemiol 52:19–26

Lehmacher W, Wassmer G (1999) Adaptive sample size calculations in group sequential trials. Biometrics 55(4):1286–1290

Peto R (2011) Current misconception 3: that subgroup-specific trial mortality results often provide a good basis for individualising patient care. Br J Cancer 104:1057–1058

Pocock S, Simon R (1975) Sequential treatment assignment with balancing for prognostic factors in the controlled clinical trial. Biometrics 31:103–115

Rothwell PM (2005) Treating individuals 2. Subgroup analysis in randomised controlled trials: importance, indications, and interpretation. Lancet 365:176–186

Sleight P (2000) Subgroup analysis in clinical trial: fun to look at – but don't believe them! Curr Control Trials Cardiovasc Med 1:25–27

Sun X, Briel M, Walter SD, Guyatt GH (2010) Is a subgroup effect believable? Updating criteria to evaluate the credibility of subgroup analyses. BMJ 340:c117

Sun X et al (2012) Credibility of claims of subgroup effects in randomised controlled trials: systematic review. BMJ 344:e1553

Therneau TM (1993) How many stratification factors are "too many" to use in a randomization plan? Control Clin Trials 14:98–108

Umbricht D et al (2014) Effect of bitopertin, a glycine reuptake inhibitor, on negative symptoms of schizophrenia a randomized, double-blind, proof-of-concept study. JAMA Psychiatry 71:637–646

Wang R, Lagakos S, Ware J, Hunter D, Drazen J (2007a) Statistics in medicine — reporting of subgroup analyses in clinical trials. N Engl J Med 357:2189–2194

Wang SJ, O'Neill R, Hung HMJ (2007b) Approaches to evaluation of treatment effect in randomized clinical trials with genomic subset. Pharm Stat 6:227–244

Chapter 2
Biomarker-Targeted Confirmatory Trials

Hui Wang, Ilana Belitskaya-Lévy, and Ying Lu

2.1 Introduction

The design of confirmatory phase III trials involving predictive biomarkers has a
wide range from enrichment design to all-comers design. The enrichment design
only enrolls patients tested positive for the biomarker and is often the most
efficient and ethical approach when there is established evidence that treatment
benefit is restricted to the biomarker positive group. On the other hand, when
there is inadequate evidence that treatment benefit is restrictive and no concern
for the safety of patients in the biomarker-negative group, an all-comer design that
enrolls all patients regardless of their biomarker status will have the advantage of
evaluating treatment effect concurrently in the overall population and the biomarker
subgroups. An all-comer design also offers the opportunity for maximizing the size
of patient population, often an important consideration for drug developers. The
recent pivotal trials of PD-L1 inhibitor drugs including nivolumab, pembrolizumab,
and atezolizumab for non-small cell lung cancer all adopted the all-comer design
and enrolled both PD-L1 positive and negative patients (Brahmer et al. 2015;
Borghaei et al. 2015; Fehrenbacher et al. 2016; Rittmeyer et al. 2017; Langer et al.
2016; Bylicki et al. 2018). The thresholds and utility of the PD-L1 biomarker were
studied in these trials and companion diagnostic tools were developed. Patients with
higher PD-L1 expression benefited more from the treatment, while treatments for

H. Wang · I. Belitskaya-Lévy
Cooperative Studies Program Palo Alto Coordinating Center, The US Department of Veterans
Affairs, Palo Alto, CA, USA

Y. Lu (✉)
Department of Biomedical Data Science, Stanford University School of Medicine, Stanford, CA,
USA
e-mail: ylu1@stanford.edu

© Springer Nature Switzerland AG 2020   19
N. Ting et al. (eds.), *Design and Analysis of Subgroups with Biopharmaceutical
Applications*, Emerging Topics in Statistics and Biostatistics,
https://doi.org/10.1007/978-3-030-40105-4_2

patients negative on PD-L1 expression showed similar benefit as the standard of care (docetaxel). The drug was approved for the overall population regardless of the PD-L1 status, considering the better safety profiles among all patients (Borghaei et al. 2015). This chapter is devoted to statistical issues related to all-comer designs. A comprehensive review of related methods for other designs such as enrichment and biomarker-strategy designs can be found in Freidlin and Korn (2014) and Ondra et al. (2016).

2.2 Inference Errors in an All-Comer Design

In the all-comer design, the treatment effect is evaluated in both the overall and the prespecified biomarker-positive group. A long-standing statistical problem is how to optimize the power for testing more than one population while controlling the family-wise error rate (FWER). As an example, Eichhorn et al. (2001) and Liggett et al. (2006) presented a genetic sub-study of the Beta-Blocker Evaluation of Survival Trial (BEST) trial. The BEST trial enrolled 2708 patients between May 1995 and December 1998 to evaluate the effectiveness of bucindolol in improving survival of patients with Class III/IV heart failures. The genetic sub-study consisted of 1040 BEST study participants and considered polymorphisms in the β_1-adrenergic receptor (β_1-AR), a β-blocker target, as candidate pharmacogenomic loci. The sub-study found that the Arg-389 homozygotes carriers of β_1-AR treated with bucindolol had an age-, sex-, and race-adjusted 38% reduction in mortality vs. placebo ($p = 0.03$). In contrast, the Gly-389 carriers of β_1-AR had no clinical response to bucindolol compared with placebo. The statistical challenge is how to design a pivotal trial to confirm this finding with simultaneous tests in the overall population and the subpopulation of the Arg-389 homozygotes carriers while controlling the FWER.

The complexity of hypothesis testing increases with the number of populations to be tested due to the increase in the number of possible decisions and in the corresponding risk of false positive conclusions. For instance, under the assumption that the efficacy in the marker positive subgroup is not worse than that in the marker negative subgroup based on biology, when efficacy is assessed in both the overall population and the marker-positive subgroup, there are three possible scenarios, (1) no efficacy in any population; (2) efficacy in the marker-positive subgroup only; and (3) efficacy in the overall population. Efficacy in marker-negative subgroup is not an *a priori* hypothesis to be confirmed. Accordingly, a false positive conclusion is possible for decisions (2) and (3) and a false negative conclusion is possible for decisions (1) and (2). The correctness of decision depends on the associated composite null and alternative hypotheses.

For illustration, let us assume that there are two pre-defined subpopulations, referred to as marker-positive ($M+$) and marker-negative ($M-$) groups, and that randomization leads to an equal distribution of biomarkers among treatment (T) and control (C) groups. Let $\mu = \mu_T - \mu_C$, $\mu_+ = \mu_{T_+} - \mu_{C_+}$ and $\mu_- = \mu_{T_-} - \mu_{C_-}$

denote the treatment effects in the overall population, the $M+$ subgroup, and the $M-$ subgroup. Without loss of generality, we assume that a positive mean difference between the treatment arms T and C indicates a desired treatment effect and that the hypothesis tests of interest are one-sided, and the treatment is more effective in the $M+$ subgroup based on prior clinical evidence. Then, there is a monotone ordering of the treatment effects:

$$\mu_+ \geq \mu \geq \mu_-. \tag{2.1}$$

The conventional null and alternative hypotheses in the overall population are

$$H_{0O} : \mu \leq 0 \text{ versus } H_{AO} : \mu > 0;$$

the corresponding hypotheses in the marker-positive subgroup are

$$H_{0+} : \mu_+ \leq 0 \text{ versus } H_{A+} : \mu_+ > 0.$$

A composite null hypothesis for simultaneous testing in both the overall population and the $M+$ subgroup can be written as the intersection of the two null hypotheses H_{0O} and H_{0+}.:

$$H_{0C} : \mu \leq 0 \ AND \ \mu_+ \leq 0,$$

and the corresponding composite alternative hypothesis is that the treatment efficacy is present in at least one of these populations:

$$H_{AC} : \mu > 0 \text{ or } \mu_+ > 0.$$

The composite alternative hypothesis H_{AC} is the union of the following two mutually exclusive alternative hypotheses:

$$H_{A1} : \mu \leq 0 \text{ and } \mu_+ > 0 \text{ OR } H_{A2} : \mu > 0,$$

where H_{A1} indicates that desired treatment effect is present in the marker-positive subgroup only and H_{A2} indicates that the desired treatment effect is present in the overall population including the marker-positive subgroup. H_{A1} implies an undesired treatment effect in the marker-negative subgroup, i.e. $\mu_- < 0$. Therefore, when H_{A1} is true, the desired decision is to reject the null hypothesis H_{0+} and accept the null hypothesis H_{0O} in the overall population. When H_{A2} is true and as long as the treatment is not harmful in the marker-negative group (i.e. $\mu_- \geq 0$), the desired decision is to reject both H_{0O} and H_{0+}.

A conventional hypothesis testing framework involves a binary inference decision based on the test statistics. When the null hypothesis is incorrectly rejected, a type I error, i.e., a false positive error, is made. When the null hypothesis is incorrectly accepted, a type II error, i.e., a false negative error, is made. An optimal

Table 2.1 The error structure of the joint hypothesis testing problem in a decision framework

		Truth		
			H_{0C} is False (H_{AC} is True)	
		H_{0C} is True $\mu \leq 0$	H_{A1} is True	H_{A2} is True
Decision		and $\mu_+ \leq 0$	$\mu \leq 0 < \mu_+$	$\mu > 0$
Accept H_{0C}		Correct decision	II* (β_{12})	II (β_{13})
Reject H_{0C}	Reject H_{0+} only	I* (α_{21})	CD (π_1)	II* (β_{23})
	Reject H_{0O} (and H_{0+})	I (α_{31})	I* (α_{32})	CD (π_2)

I = Type I Error; II = Type II Error; I* = Type I-like Error; II* = Type II-like Error; CD = Correct Decision

π_1, π_2, are the probabilities of Correct Decision

$\pi_1 = $ P(Accept $H_{A1}|H_{A1}$)

$\pi_2 = $ P(Accept $H_{A2}|H_{A2}$)

α_{ij} denotes the probabilities of type I and type I-like errors

β_{ij} denotes the probabilities of type II and type II-like errors

statistical test maximizes the statistical power (i.e., minimizes the false negative error) under the constraint that the FWER is under a pre-specified significance level.

To perform simultaneous tests of multiple hypotheses, incorrect acceptance of any one of the multiple alternative hypotheses can result in either a false positive (type I-like) error or a false negative (type II-like) error. As illustrated in Table 2.1, when $\mu_+ \leq 0$, we can only make type I-like (false positive) errors, denoted by α_{21} and α_{31}. When $\mu \leq 0 < \mu_+$, however, both type I-like and type II-like errors are possible: when the composite null H_{0C} is accepted, a type II-like error is committed (β_{12}); on the other hand, if H_{0O} (and H_{0+}) is rejected with the conclusion that treatment works for the overall population, a type I-like (false positive) error is committed (α_{32}). When H_{A2} is true, only type II-like (false negative) errors are possible (β_{13} or β_{23}).

2.3 Multiple-Step Inferences for the Overall Population and Marker Subgroups

Several methods have been proposed for testing in both the overall population and the pre-defined subgroups, with a focus on controlling the FWER α. One approach is to test specific subgroups with Bonferroni correction on α (parallel subgroup-specific design, Freidlin et al. 2010, 2013). This approach when strong belief exists that the positive group will benefit from the treatment more than the negative group. One can also carry out the test in a sequential manner that provides more design efficiency (Douillard et al. 2010) where the positive group is tested at the level of α and, if the test is significant, the negative group is tested also at the level of α. Other strategies include testing the treatment effect in the overall patient population followed by testing in the marker positive group. This strategy offers the opportunity for sponsor to claim efficacy in a patient population as large as possible and is effective when treatment is also beneficial for the marker-negative group, although the treatment effect can be much smaller. A well-known example is the fallback procedure (Simon 2008): the treatment effect is tested in the overall population at a reduced level of α_0; if significant, the procedure stops and an overall effect is claimed; otherwise, the treatment effect is tested in the marker-positive group at α_+ for claiming efficacy in a sub-population, where $\alpha_0 + \alpha_+ = \alpha$. Song and Chi (2007) proposed an approach that takes into account the correlations between the test statistics derived from the overall population and the subgroup. They define a conservative threshold α_0 and relaxed threshold α_0^* for the overall p-value, where $\alpha_0 < \alpha < \alpha_0^*$. If the overall p-value is less than α_0, the treatment effect is considered significant in the overall population and tested in the marker-positive group at level α; if the overall p-value falls between α_0 and α_0^*, the treatment effect in the overall population is borderline and to be tested in the marker-positive group at a more stringent level of α_+ ($\alpha_+ < \alpha$). If this test is significant, one goes back to test the overall treatment effect at level α, where α_+ is selected conditional on the value of the overall test statistics to control for the FWER. Freidlin et al. (2014) proposed a marker sequential test (MaST) that involves all three groups (overall, marker positive and negative). It begins with testing the treatment effect in the marker-positive group at a reduced level of α_+. If treatment effect is significant in the marker-positive group, the marker-negative group is tested at α; otherwise, treatment effect is tested in the overall population at $\alpha - \alpha_+$. The MaST prioritizes the testing in a subgroup, while the fallback procedure and Song and Chi's approach prioritize on the overall treatment effect.

A general class of multiple testing procedures called the chain procedure can be applied to subgroup analysis with great flexibility (Dmitrienko and D'Agostino 2013; Millen and Dmitrienko 2011). The procedure is governed by an α allocation rule described by the proportions of α allocated to each hypothesis up-front and an α propagation rule described by the proportions of transferrable α among hypotheses upon rejection. For example, when treatment efficacy in the overall population and the marker-positive group is of equal interest, the allocation and propagation rule can be specified as $\omega_O = \omega_+ = 0.5$ and $g_{O+} = g_{+O} = 1$. The ω_O and ω_+ are the initial weights allocated to the FWER (α) for testing the null hypothesis in the overall population and the marker-positive population. The g_{O+} (or g_{+O}) is the fraction of the α that can be carried over for testing the null hypothesis in the marker-positive population (or overall population) when the null hypothesis is rejected in the overall population (or marker-positive population). The treatment effect is first tested in the overall population with $\omega_O \alpha$. If the overall test is significant, the allocated α to the overall test (i.e. $\omega_O \alpha$) is transferred to the subgroup test, and the treatment effect is tested in the marker-positive group at level $(\omega_O g_{O+} + \omega_+)\alpha = \alpha$; if not significant, the treatment effect is tested in the marker-positive group at $\omega_+ \alpha$ and if significant one tests again in the overall population at level $(\omega_O + \omega_+ g_{+O})\alpha = \alpha$. Testing the subgroup in the first place reaches the same conclusion because the allocation of α is equal between the overall and subgroup test. The fallback procedure and the sequential subgroup-specific test can be viewed as a simplified version of the chain procedure.

In Johnston et al. (2009), the treatment effect is first tested in the marker-positive group at level α, and the overall treatment effect is tested at α, too, but only if statistical significance is achieved for the subgroup test. This testing strategy can lead to unnecessary or false treatment for marker-negative patients when the overall population is dominated by the marker-positive population.

Figure 2.1 provides graphic representations and illustrations of rejection regions of these multi-step sequential tests.

The rejection regions of the sequential tests are illustrated based on $(Z-, Z+)$, the normalized test statistics. In all tests, the FWER is $\alpha = 0.025$. The Direct Test

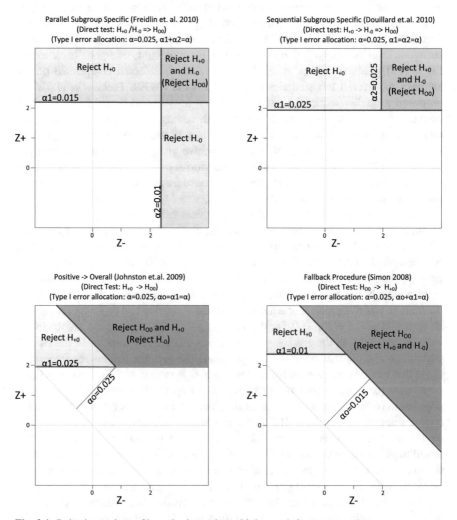

Fig. 2.1 Rejection regions of hypothesis test in multiple-step inference procedures

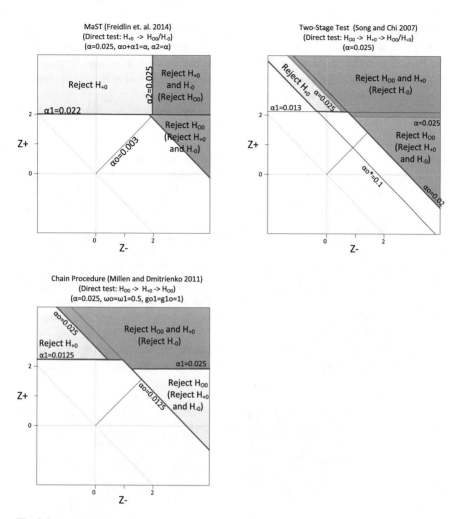

Fig. 2.1 (continued)

arrows show the order of the hypothesis tests performed in each procedure. The α_1 denotes the allocated α to the first test, and α_O denotes the α allocated to the overall test.

2.4   Simultaneous Inferences for the Overall Population and Marker-Positive Subgroup

In Belitskaya-Lévy et al. (2016), a simultaneous test in the overall population and the marker-positive subgroup was proposed. The proposed Confirmatory Overall-Subgroup Simultaneous Test (COSST) is based on partitioning the sample space of the test statistics Z+ and Z− in the marker-positive and marker-negative subgroups. Let Z, Z_+ and Z_- be the test statistics for the treatment effects defined previously in the overall study population, the $M+$ subgroup and the $M-$ subgroup, respectively. Note that Z_+ and Z_- are independent. The following relationship holds between the overall and the subgroup test statistics:

$$Z = \sqrt{K_+}Z_+ + \sqrt{K_-}Z_-,$$

where K_+ and $K_- = 1 - K_+$ depend on biomarker prevalence and standard errors in the overall sample and the subgroups. For instance, when the variances are equal among the groups, $K_+ = p_+$ and $K_- = p_-$, where p_+ is the biomarker prevalence (the proportion of marker-positive patients in the general patient population) and $p_- = 1 - p_+$.

The two-dimensional sample space of $(Z-, Z+)$ can be divided into two rejection regions and one acceptance region (Fig. 2.2). A safety boundary (SB) is also incorporated. If the test statistic drops below SB, i.e. $Z_- \leq SB$, the treatment may have an undesired effect in the $M-$ subgroup, and the treatment efficacy should not be claimed in the overall population. There is no need to perform any inference on SB. All it does is to redistribute the weight in the probability space to assure the rejection region is safe for $M-$ subgroup.

Rosenblum et al. (2014) proposed a simultaneous test within a Bayesian frame-work with three null hypotheses: H_{0+}: treatment effect in the marker-positive group

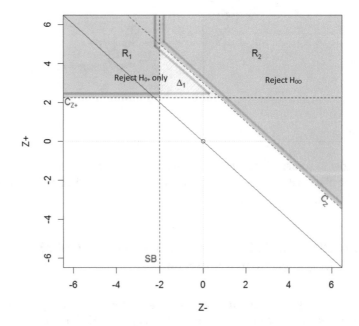

Fig. 2.2 Rejection regions of hypothesis tests in COSST

is less than 0 ($\mu_+ \leq 0$); H_{0-}: treatment effect in the marker-negative group is less than 0 ($\mu_- \leq 0$); and H_{00}: treatment effect in the overall population is less than 0 ($\mu \leq 0$). The test minimizes an objective function with respect to a multiple testing procedure $M(Z_+, Z_-)$ based on the test statistics in the marker-positive and negative subgroups:

$$\int E_{\mu_+, \mu_-} L\left(M\left(Z_+, Z_-\right); u_+, u_-\right) d\Lambda\left(u_+, u_-\right),$$

where L is a pre-specified loss function and $\Lambda(u_+, u_-)$ is the prior distribution of the treatment effects u_+ and u_-, under the type I error constraint

$$Prob_{\mu_+, \mu_-} \text{ (reject any null hypothesis)} \leq \alpha$$

and the power constraint

$$Prob_{\mu_+, \mu_-} \text{ (reject } H_{00} : \mu \leq 0) \geq 1 - \beta$$

We illustrate the COSST with plots of rejection regions (Fig. 2.2). The red dashed diagonal line is $Z = \sqrt{K_+}Z_+ + \sqrt{K_-}Z_- = C_Z$, and any region above this line is the marginal rejection region for the overall null hypothesis. The red dashed horizontal line is $Z_+ = C_{Z+}$, and any region lying above this line is the marginal rejection region for the subgroup null hypothesis. The vertical black dashed line is the safety boundary for the biomarker negative subgroup. If Z_- is less than the safety boundary, the overall null hypothesis will not be rejected. The thick blue and magenta lines outline the two rejection regions based on (Z_-, Z_+):

(1) Reject the subgroup null and accept the overall null ($R_1 + \Delta_1$, thick blue line);
(2) Reject the overall null which implies rejecting the subgroup null (R_2, thick magenta line).

The regions R_1, Δ_1 and R_2 are defined as following:

$$R_1 = \{(Z_-, Z_+) : Z_+ > C_{Z+} \text{ and } Z_- \leq SB_-\},$$

$$\Delta_1 = \left\{ (Z_-, Z_+) : C_{Z+} < Z_+ \leq \frac{C_Z}{\sqrt{K_+}} - SB_-\sqrt{\frac{K_-}{K_+}} \text{ and} \right.$$

$$\left. SB_- < Z_- \leq \frac{C_Z}{\sqrt{K_-}} - Z_+\sqrt{\frac{K_+}{K_-}} \right\},$$

$$R_2 = \left\{ (Z_-, Z_+) : Z = \sqrt{K_+}Z_+ + \sqrt{K_-}Z_- \geq C_Z \text{ and } Z_- \geq SB_- \right\},$$

2.5 Extension to Sequential Analysis

Another interesting question for marker defined subgroup clinical trial is to use a two-stage clinical trial design. Based on the interim subgroup analysis of the first stage, a decision can be made for stopping the trial either for efficacy or futility or to continue the trial using overall population or enriched subgroup in the second stage. In Matsui and Crowley (2018), the authors proposed adaptive subgroup selection strategies for sequential assessment across marker-defined subgroups. In the proposed design, superiority and futility boundaries are defined for interim analysis for marker-positive and marker-negative groups. Four possible decisions can be made for the trial based on interim look: (1) If superiority boundary is crossed in the marker-positive group and futility boundary is crossed in the marker-negative group at interim analysis, the trial will be closed for treatment efficacy in the M+ group. (2) If superiority boundary is crossed but the futility boundary is not, treatment efficacy will be claimed in M+ group and the trial will continue in M− group. (3) If superiority boundary is not crossed but futility boundary is crossed, then the trial will be continued in M+ group but stopped for M− patients; (4) if neither the superiority nor the futility boundary is crossed, the trial will continue in the overall group. Figure 2.3 illustrates the sequential testing procedures.

In Lai et al. (2014), instead of pre-defined subgroups, several patient subgroups are chosen adaptively by partitioning the parameter space and defining corresponding type I and type II errors. The subgroups are not known at the design stage but can be learned statistically from the data collected during the trial. These patient subgroups can be defined by biomarkers, brain imaging, or other risk factors measured at baseline. The authors propose a novel 3-stage group sequential design that incorporates adaptive choice of the patient subgroup among several possibilities which include the entire patient population as a choice. The goal is to reject the null hypothesis for the largest possible subgroup for which the null hypothesis of no treatment difference is false. At the first interim analysis, the efficacy and futility are tested in the overall population. If early stopping for efficacy occurs, the trial is terminated and efficacy of the new treatment is claimed over the entire population. If stopping occurs for futility, then the overall hypothesis is accepted and the trial is continued with the most promising patient subgroup, that is, the subgroup that maximizes the generalized likelihood statistic, but with the sample sizes re-estimated. The future enrollment of the trial will include patients of this subgroup only, while the maximum total sample size N remains the same. The same procedure is repeated at the second interim analysis and at the final stage. If the overall test is not stopped for efficacy or futility, then the trial is continued to the next stage of the 3-stage design and the procedure is repeated. Lansberg et al. (2016) used simulations, based on real-world patient data, to demonstrate that adaptive subgroup selection has merit in endovascular stroke trials at it substantially increases power when treatment effect differs among subgroups in a predicted pattern.

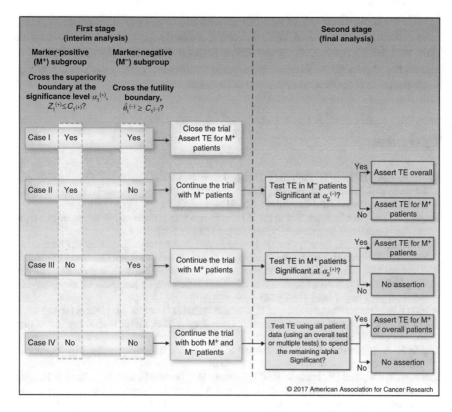

Fig. 2.3 Matsui and Crowley (2018). Biomarker-Stratified Phase III Clinical Trials: Enhancement with a Subgroup-Focused Sequential Design

2.6   Discussion

In this chapter, we reviewed several statistical tests for confirmative biomarker trials. We focused on the trial with intention to make inference for either the entire population or a biomarker indicated subgroup. Multiple-step inference methods and simultaneous inference methods were presented. We also presented sequential clinical trial designs that based on interim analysis to decide the subsequent use of entire population or enrich biomarker-based subpopulations.

In the above multi-step inference methods, sequential tests are multi-step and the decision in each step is binary, regardless of the number of hypotheses. As a result, they are typically laid out in a framework of rejecting the null hypothesis of no treatment effect either in the overall population or the subgroup regardless of the specific components of composite alternatives, which may lead to a loss of power and less accurate decisions. But sequential tests are widely used due to its ease of implementation. The simultaneous tests are typically more powerful as well as more complex and often require specialized software.

With the rapid discovery of new biomarkers and development of drugs targeting these biomarkers, the use of these inference methods for biomarker confirmative trials will be increasing. It is an active research area for efficient designs to address the challenges of biomarker-targeted confirmatory trials such as inferences in subgroups and multiple comparison adjustment.

References

Belitskaya-Lévy I, Wang H, Shih MC, Tian L, Doros G, Lew R, Lu Y (2016) A new overall-subgroup simultaneous test for optimal inference in biomarker-targeted confirmatory trials. Stat Biosci. https://doi.org/10.1007/s12561-016-9174-8

Borghaei H, Paz-Ares L, Horn L et al (2015) Nivolumab versus docetaxel in advanced nonsquamous non-small-cell lung cancer. N Engl J Med 373(17):1627–1639

Brahmer J, Reckamp KL, Baas P, Crinò L, Eberhardt WE, Poddubskaya E, Antonia S, Pluzanski A, Vokes EE, Holgado E, Waterhouse D, Ready N, Gainor J, Arén Frontera O, Havel L, Steins M, Garassino MC, Aerts JG, Domine M, Paz-Ares L, Reck M, Baudelet C, Harbison CT, Lestini B, Spigel DR (2015) Nivolumab versus docetaxel in advanced squamous-cell non-small-cell lung cancer. N Engl J Med 373(2):123–135

Bylicki O, Paleiron N, Rousseau-Bussac G, Chouaïd C (2018) New PDL1 inhibitors for non-small cell lung cancer: focus on pembrolizumab. Onco Targets Ther 11:4051–4064. https://doi.org/10.2147/OTT.S154606

Dmitrienko A, D'Agostino RB (2013) Tutorial in biostatistics: traditional multiplicity adjustment methods in clinical trials. Stat Med 32:5172–5218

Douillard JY, Shepherd FA, Hirsh V, Mok T, Socinski MA, Gervais R, Liao ML, Bischoff H, Reck M, Sellers MV, Watkins CL, Speake G, Armour AA, Kim ES (2010) Molecular predictors of outcome with gefitinib and docetaxel in previously treated non-small-cell lung cancer: data from the randomized phase III INTEREST trial. J Clin Oncol 28(5):744–752

Eichhorn et al (2001) A trial of beta-blocker bucindolol in patients with advanced chronic heart failure. N Engl J Med 344(22):1659–1667

Fehrenbacher L, Spira A, Ballinger M, Kowanetz M, Vansteenkiste J, Mazieres J, Park K, Smith D, Artal-Cortes A, Lewanski C, Braiteh F, Waterkamp D, He P, Zou W, Chen DS, Yi J, Sandler A, Rittmeyer A, POPLAR Study Group (2016) Atezolizumab versus docetaxel for patients with previously treated non-small-cell lung cancer (POPLAR): a multicentre, open-label, phase 2 randomised controlled trial. Lancet 387(10030):1837–1846

Freidlin B, Korn EL (2014) Biomarker enrichment strategies: matching trial design to biomarker credentials. Nat Rev Clin Oncol 11(2):81–90. PMID: 24281059

Freidlin B, Korn EL, Gray R (2014) Marker sequential test (MaST) design. Clin Trials 11(1):19–27. PMID: 24085774

Freidlin B, McShane LM, Korn EL (2010) Randomized clinical trials with iomarkers: design issues. J Natl Cancer Inst 102(3):152–160

Freidlin B, Sun Z, Gray R, Korn EL (2013) Phase III clinical trials that integrate treatment and biomarker evaluation. J Clin Oncol 31(25):3158–3161

Johnston S, Pippen J Jr, Pivot X, Lichinitser M, Sadeghi S, Dieras V, Gomez HL, Romieu G, Manikhas A, Kennedy MJ, Press MF, Maltzman J, Florance A, O'Rourke L, Oliva C, Stein S, Pegram M (2009) Lapatinib combined with letrozole versus letrozole and placebo as first-line therapy for postmenopausal hormone receptor-positive metastatic breast cancer. J Clin Oncol 27(33):5538–5546

Lai TL, Lavori PW, Liao OYW (2014) Adaptive choice of patient subgroup for comparing two treatments. Contemp Clin Trials 39(2):191–200

Langer CJ, Gadgeel SM, Borghaei H et al (2016) Carboplatin and pemetrexed with or without pembrolizumab for advanced, non-squamous non-small-cell lung cancer: a randomised, phase 2 cohort of the open-label KEYNOTE-021 study. Lancet Oncol 17(11):1497–1508

Lansberg MG, Bhat NS, Yeatts SD, Palesch YY, Broderick JP, Albers GW, Lai TL, Lavori PW (2016) Power of an adaptive trial design for endovascular stroke studies: simulations using IMS (Interventional Management of Stroke) III Data. Stroke 47:2931–2937

Liggett SB, Mialet-Perez J, Thaneemit-Chen S, Weber SA, Greene SM, Hodne D, Nelson B, Morrison J, Domanski MJ, Wagoner LE, Abraham WT, Anderson JL, Carlquist JF, Krause-Steinrauf HJ, Lazzeroni LC, Port JD, Lavori PW, Bristow MR (2006) A polymorphism within a conserved β_1-adrenergic receptor motif alters cardiac function and beta-blocker response in human heart failure. Proc Natl Acad Sci 103(30):11288–11293

Matsui S, Crowley J (2018) Biomarker-Stratified Phase III Clinical Trials: enhancement with a subgroup-focused sequential design. Clin Cancer Res 24(5):994–1001

Millen BA, Dmitrienko A (2011) Chain procedures: a class of flexible closed testing procedures with clinical trial applications. Stat Biopharm Res 3:14–30

Ondra T, Dmitrienko A, Friede T, Graf A, Miller F, Stallard N, Posch M (2016) Methods for identification and confirmation of targeted subgroups in clinical trials: a systematic review. J Biopharm Stat 26(1):99–119

Rittmeyer A, Barlesi F, Waterkamp D, Park K, Ciardiello F, von Pawel J, Gadgeel SM, Hida T, Kowalski DM, Dols MC, Cortinovis DL, Leach J, Polikoff J, Barrios C, Kabbinavar F, Frontera OA, De Marinis F, Turna H, Lee JS, Ballinger M, Kowanetz M, He P, Chen DS, Sandler A, Gandara DR, OAK Study Group (2017) Atezolizumab versus docetaxel in patients with previously treated non-small-cell lung cancer (OAK): a phase 3, open-label, multicentre randomised controlled trial. Lancet 389(10066):255–265

Rosenblum M, Liu H, Yen E-H (2014) Optimal tests of treatment effects for the overall population and two subpopulations in randomized trials, using sparse linear programming. J Am Stat Assoc (Theory Meth) 109(507):1216–1228. PMID: 25568502

Simon R (2008) The use of genomics in clinical trial design. Clin Cancer Res 14:5984–5993

Song Y, Chi GYH (2007) A method for testing a pre-specified subgroup in clinical trials. Stat Med 26:3535–3549

Chapter 3
Data-Driven and Confirmatory Subgroup Analysis in Clinical Trials

Alex Dmitrienko, Ilya Lipkovich, Aaron Dane, and Christoph Muysers

Abstract In this chapter we provide an overview of the principles and practice of subgroup analysis in late-stage clinical trials. For convenience, we classify different subgroup analyses into two broad categories: data-driven and confirmatory. The two settings are different from each other primarily by the scope and extent of pre-specification of patient subgroups. First, we review key considerations in confirmatory subgroup analysis based on one or more pre-specified patient populations. This includes a survey of multiplicity adjustment methods recommended in multi-population Phase III clinical trials and decision-making considerations that ensure clinically meaningful inferences across the pre-defined populations. Secondly, we consider key principles for data-driven subgroup analysis and contrast it with that for a guideline-driven approach. Methods that emerged in the area of principled data-driven subgroup analysis in the last 10 years as a result of cross-pollination of machine learning, causal inference and multiple testing are reviewed. We provide examples of recommended approaches to data-driven and confirmatory subgroup analysis illustrated with data from Phase III clinical trials. We also illustrate common errors, pitfalls and misuse of subgroup analysis approaches in clinical trials often resulting from employing overly simplistic or naive methods. Overview of available statistical software and extensive bibliographical references are provided.

A. Dmitrienko (✉)
Mediana Inc., Overland Park, KS, USA
e-mail: admitrienko@medianainc.com

I. Lipkovich
Eli Lilly and Company, Indianapolis, IN, USA
e-mail: ilya.lipkovich@lilly.com

A. Dane
Dane Statistics, Manchester, UK
e-mail: aarondane@danestat.com

C. Muysers
Bayer AG Pharmaceuticals, Berlin, Germany
e-mail: christoph.muysers@bayer.com

© Springer Nature Switzerland AG 2020
N. Ting et al. (eds.), *Design and Analysis of Subgroups with Biopharmaceutical Applications*, Emerging Topics in Statistics and Biostatistics,
https://doi.org/10.1007/978-3-030-40105-4_3

3.1 Introduction

Subgroup analyses are commonly performed throughout all phases of clinical development. As any data analysis in a broad sense, they naturally fall into exploratory (or hypothesis generation) and confirmatory.

In the early stages this will be to identify potential biomarkers which may help to explain an enhanced treatment effect for future development, and in proof of concept studies the aims of subgroup analysis may be more targeted in order to define the patient population(s) to study in pivotal Phase III clinical trials.

For late-stage trials, confirmatory clinical trials subgroup evaluation is important to be confident that the drug under study is effective for the entire patient population studied. If this is not the case, and the drug effect is only apparent in a subset of the overall population, that may need to be noted. Equally, if there is an enhanced effect for one group of patients this may also need to be stated in the regulatory label. Therefore, given the importance of any conclusions regarding subgroup analysis at this stage of clinical development, it is important that the risk of making the wrong conclusions regarding subgroup effects are well understood, and indeed this is a topic specifically addressed by the EMA guidance on subgroup analysis (EMA 2014).

It is important to understand these risks, particularly because the rate of incorrect identification of a differential treatment effect within a subgroup can be inflated when many subgroup analyses are undertaken. In addition, given that these trials are often not designed to detect subgroup differences, any evaluation of differential treatment effects may miss detecting true differences due to low power. However, as it is still necessary to understand possible subgroup differences, statistical methods are required which provide a robust evaluation while addressing the dual issues of the inflated risk of incorrect subgroup identification and of the low power to detect true differences in treatment effect across subgroups.

During these later phase studies, the subgroups evaluated can be either intrinsic factors (such as age, race, gender, severity and type of disease) or extrinsic factors (such as environmental factors or standard of care differences such as the different use of concomitant medications or differences in clinical practice regionally).

There are a range of reasons for performing subgroup investigation in late stage trials, all of which require careful design and interpretation of the results in order to be sure such interpretation appropriately accounts for the subgroup search strategy when making any conclusions. The key types of subgroup investigation often undertaken in confirmatory clinical trials are defined as follows:

- For patient subgroups with anticipated differential treatment effects, it is possible to design a trial to investigate effects within specific subgroups. This can be accomplished either by recruiting patients from a target subpopulation (e.g., biomarker-positive patients) or by defining several analysis populations (e.g., the overall and biomarker-positive populations with an appropriate multiplicity adjustment strategy to control the Type I error rate).

- The assessment of a range of subgroups important for regulatory or clinical purposes, where the aim is to show broad consistency with the overall trial result. Such types of subgroup investigations are considered predominantly in meta-analysis, e.g., in the context of an ISS (integrated summary of safety) or ISE (integrated summary of efficacy) of a submission package for marketing authorization. In this case it is also possible to determine the reproducibility of subgroup effects across studies to help assess whether such effects are real. However, it is often the case that a new therapy will make a regulatory submission on the basis of a single pivotal trial. Thanks to recent developments of statistical methodologies and the regulatory guidelines these investigations are now increasingly considered within individual late-stage clinical trials. Formally, strong control of the Type I error rate is not required in this setting but it is still important to understand the Type I and Type II error rates to be sure that appropriate conclusions are made regarding the subgroups of interest. For this type of strategy, it is important to pre-specify the patient subgroups to be explored, which will help design a subgroup selection strategy that appropriately accounts for the number of subgroups investigated.
- A subgroup approach to address questions related to unanticipated (or unexpected) subgroup effects, or requests from regulatory agencies after the trial has been completed. In this case it will be more challenging to account for all subgroups explored but is still important to have strategies to address questions such as these and maintain some discipline with respect to the subgroup selection strategy.
- Last but not least, the presence of a rich set of data collected in the course of clinical trials, which is often underutilized, calls for application of modern methods of machine learning, which have been rarely used in the analysis of clinical data. Such methods assume a rather broad context of candidate subgroups defined by multiple biomarkers and their combinations. Interest in such analyses is motivated, on one hand, by the desire of sponsors to salvage "failed" studies, and, on the other, by recent growing interest in personalized/precision medicine.

In the following sections of this chapter we will clarify these broadly defined types and refer to them as *confirmatory, exploratory, post-hoc*, and *biomarker and subgroup discovery*, respectively.

Given the strong regulatory context, the gap between post-hoc (or subgroup discovery) and confirmatory stages of subgroup analyses in clinical trials were historically somewhat exaggerated and represented wildly different approaches and practices. For exploratory analyses, little concern was given to careful planning and pre-specification of the strategy, arguing that no control for multiplicity is needed when evaluating subgroups for "internal decision making." Confirmatory subgroup analyses were performed with strong control of the Type I error rate and complete pre-specification, often employing overly conservative multiple comparison procedures.

Here we advocate for a more balanced approach based on the understanding that whether exploratory or confirmatory, subgroup analysis should be always performed with a goal of decision making in mind. Therefore the risks and benefits should be properly evaluated within a well-planned analytic strategy. As an example, an "unconstrained" brute force subgroup search to assess subgroup effects in a failed Phase III trial is not recommended. From a regulatory perspective, such effects would be interpreted with extreme caution and it would be unlikely that they were accepted for regulatory approval. Even for the purposes of exploratory analysis aimed at hypothesis generation, caution is needed to ensure that future resources are not invested based upon spurious findings driven by chance.

One may argue that the best way of avoiding misguided "chasing" of subgroups would be through always employing prospectively planned confirmatory analysis methods. However, there are many hurdles to using such approaches consistently in the framework of subgroup analysis. The lack of prior knowledge about predictive biomarkers for an investigational treatment also requires pre-planned exploratory assessments. Even post-hoc analyses to examine the homogeneity of the results or to identify subgroups with remarkably different efficacy compared to the complementary group require thorough planning. As a result, the different objectives in combination with multiple hurdles have promoted the development of a significant number of statistical methods for subgroup analysis, but also a significant number of considerations in regulatory guidance documents. The inclusion of subgroup analysis in regulatory guidance supports a more consistent assessment across different applications, but cannot solve all problems. To avoid data dredging with subgroup analyses, all guidelines indicate that the described approaches are either exploratory in nature and interpret the results very cautiously, or there is a request for pre-specified confirmatory approaches; e.g., controlling the familywise Type I error rate.

When it comes to decision making in the regulatory context, simplified as approval or non-approval, the typical scenarios with the right decision are an approval when a therapy is truly effective, and a non-approval when a therapy is truly ineffective. In addition to the classic false-positive and false-negative decision, there is a risk especially in the situation of analysis of patient subgroups. A subgroup that does not benefit from treatment might be overlooked. This would lead, after marketing of the product, to unnecessary exposure of the drug, and thus, to a potential risk of side effects without benefit for that patient subpopulation. The reason for such a situation may be a heterogeneity that has not been detected, e.g., an overwhelming effect in another, much more favorable subgroup. In combination with the non-benefitting subgroup, there might still have been an acceptable effect for the overall patient population. The most desired result, i.e., the promotion of the most advantageous subgroup and the contraindication for the non-benefitting subgroup, would be missed. The corresponding counterpart to this risk is based on a spurious subgroup and the erroneous exclusion of this subpopulation from the product label due to the apparent lack of efficacy. Such a wrong decision is certainly made more frequently in clinical trials with a large number of prospective and/or post-hoc subgroup analyses without appropriate multiplicity adjustment. In this case, the subpopulation will be deprived of an available beneficial treatment.

In practice, a clear-cut decision can hardly be done based only on the observed effect in subgroup analysis for one or the other patient population. However, an appropriate benefit-risk assessment would limit the risks, either to withdraw efficatious drugs from patient subpopulations or to protect patient subpopulations from unnecessary exposure to an ineffective drug. Despite the fact that uncertainties arise from small subgroups and the usually descriptive nature of the analyses in the appropriate benefit-risk assessment, it is important to note that ultimately a decision must be made that either includes or excludes the appropriate subgroup of patients. Unfortunately, the level of correct vs. wrong decisions typically remains unknown, as it cannot be quantified that more subjects are prevented from ineffective treatment in this group compared to the number of subjects wrongly excluded from benefits. In some cases, it might happen that additional post-hoc analysis or post-marketing studies provide new evidence for a previous wrong or unclear situation.

The objective of this chapter is to introduce a broad range of topics related to subgroup analysis in clinical trials that are divided into three relatively independent parts that can be read separately: regulatory guidance, data-driven subgroup analysis, and confirmatory subgroup analyses. Owing to the diversity of the topics covered, we are targeting a diverse audience. While some readers may naturally find it more useful to focus on the section(s) of their special interest and skip the rest, we believe that many readers will benefit from learning about approaches that until now may have not been on their "radar screen". Indeed, from our literature review we found that, as was noted by Leo Breiman in his famous article on two cultures in statistical modeling (Breiman 2001), there are still boundaries separating statisticians trained in different cultures, e.g., statistical/machine learning versus hypothesis testing. An important goal of this review chapter is to help overcome these barriers by sharing the wealth of knowledge and methodologies developed within the disparate communities of multiple testing, causal inference and statistical learning.

The rest of chapter is organized as follows. Section 3.2 contains an overview of regulatory guidance for subgroup analysis. Section 3.3 provides a discussion of key principles of data-driven subgroup analysis following the taxonomy of methods from Lipkovich et al. (2017a) that are illustrated with a case study from Phase III trial with time to event outcome. Section 3.4 gives an overview of designs and multiple comparison procedures that are encountered in confirmatory subgroup analyses. We conclude the chapter with a brief discussion in Sect. 3.5.

3.2 Overview of Regulatory Guidance for Subgroup Analysis

This section is organized as follows. First, we briefly describe the global ICH guidelines and illustrate some aspects of subgroup analysis for a Multi-Regional Clinical Trials (MRCT) using a case study, PLATO. Then we divide the rest of the overview into subsections covering guidelines in three geographical regions: United States and Europe with the Food and Drug Administration (FDA) and European

Medicines Agency (EMA), respectively, as well as the Chinese and Japanese regions represented by the China National Medical Product Administration (NMPA) (formerly known as the China Food and Drug Administration, CFDA, and the State Food and Drug Administration, SFDA) and Pharmaceuticals and Medical Devices Agency (PMDA), respectively.

Apart from formal guidance documents from health authorities, various published articles were developed out of the surroundings from health authorities or statisticians explicitly dealing with the regulatory context of subgroup analyses. The spirit of certain guidance text dealing with subgroup analyses was influenced or afterwards explained in more detail. For instance, in Rothwell (2005), Sun et al. (2010), Carroll and Le Maulf (2011), Alosh and Huque (2013), Hemmings (2014), Koch and Framke (2014), Koch and Schwartz (2014), Wang and Hung (2014) and Alosh et al. (2015), Alosh et al. (2016).

3.2.1 International Framework: ICH Guidelines

Taking the regulatory framework across all regions into account, it is appropriate to have a closer look into documents of The International Council for Harmonisation of Technical Requirements for Pharmaceuticals for Human Use (ICH, https://www.ich. org/home.html). Only a few aspects of subgroup analysis are mentioned in the ICH E9 Statistical Principles for Clinical Trials guidance (1999). As a general statement, "the subjects in confirmatory trials should more closely mirror the target population," which certainly sounds reasonable. Nevertheless, when more sophisticated trial designs with adaptive approaches and enrichment strategies are considered, it raises the question of how closely the initial trial population should match the final prescribed population because of adaptations that take place after trial planning or even after trial start. In the later Sect. 3.3.4, "Adaptive designs in multi-population trials," especially adaptive trials with data-driven subpopulations, offer, per definition, the option to deal with a flexible description of the trial population. Furthermore, the ICH E9 guideline is requesting the balance between most flexible inclusion and exclusion criteria for a broad target population and "maintaining sufficient homogeneity to permit precise estimation." Here it is reasonable to clarify which level of homogeneity translates into "sufficient homogeneity." In practice, decisions regarding homogeneity should be made on a case-by-case basis, with an interdisciplinary discussion of relevant factors that define the population. It is recommended to have a distinct reflection of this discussion in the trial protocol. However, the statistical analysis after the conduct of the trial might indicate heterogeneity for the investigated target population when subgroup analyses are made.

For the concrete analysis of subgroups, clear warning about careless usage are provided. Section 5.7 of the ICH E9 guidance (1999) stated that "when exploratory, these analyses should be interpreted cautiously; any conclusion of treatment efficacy (or lack thereof) or safety based solely on exploratory subgroup analyses are unlikely to be accepted." Furthermore, Alosh et al. (2016) mentioned, in the same context, "the danger of carrying out too many post-hoc analyses and over-interpreting their findings. Subgroup analyses that are not supported by a scientific rationale can lead to spurious findings that are prone to bias and consequently may lead to misleading interpretations." These statements are based on scientific rationale and on many examples where obvious chance findings were picked out of numerous unadjusted post-hoc analyses. However, Alosh et al. (2016) state, "while concerns about post-hoc subgroup findings are justified, there are also many success stories where findings from a subgroup analysis were critical for discovering new treatments or for revising the population for treatment use after learning that the treatment is beneficial for only a certain subgroup." This can be considered as another specific situation described in Section 6.5 of the EMA subgroup analysis guideline (EMA 2014). As also stated in this guidance document, it can be doubtless categorized as a rare situation. Nevertheless, it seems that with the availability of large computing capacity, combined with thorough consideration of a clinical rationale, the exploratory investigation offers additional insights. This also follows the trend for more sensitivity analyses.

An extension to the ICH E9 guidance is under preparation ("Choosing Appropriate Estimands and Defining Sensitivity Analyses in Clinical Trials" ICH 2014). Even though this initiative is primarily triggered by the recent developments in the handling and prevention of missing data, it supports some ideas of the EMA approach of gaining more information from sensitivity analyses. The final concept paper states that "it has become standard in all regions to pre-specify a primary statistical analysis for efficacy, but it has also been common practice to investigate the extent to which the outcomes of other approaches to the analysis lead to consistent findings." This leads directly to a more extensive investigation of subgroup effects considering factors for regional and/or ethnical factors. The different regions take care for this aspect also in their own guidance documents, as described in the following subsections. Two other ICH guidelines, namely E5 and E17, play a relevant role in this context.

The ICH E5 Ethnic Factors in the Acceptability of Foreign Clinical Data (1998) provides a general framework for a faster and more cost-effective approach to worldwide drug development. The idea is to avoid repetition of the entire development program in a new region if no intrinsic and extrinsic ethnic differences argue against bridging the data to show consistency across regions. Nevertheless, the guidance text does not explicitly describe how to perform this consistency assessment. Apart from various publications in that area, which go beyond the usual framework of subgroup analyses, the Japanese Health authorities PMDA provide further information on their Regulatory Science internet homepage (https://www.pmda.go.jp/english/index.html).

The assessment of consistency across regions is meanwhile regulated in the finalized ICH E17 guidance (2017) on general principles for planning and design of multi-regional clinical trials (MRCT). The approaches are dealing with the achievement of internal consistency in one global trial, in contrast to a separate local trial for external consistency. Although this guidance did not make explicit recommendations regarding statistical methods, it requests that "subgroup findings should take into consideration biological plausibility, internal consistency (e.g., similar patterns of regional variability observed for other secondary endpoints) and/or external consistency (e.g., similar patterns observed in another clinical trial of the same investigational treatment), the strength of evidence, the clinical relevance, and the statistical uncertainty." Koch and Framke (2014) highlighted aspects for the analysis of treatment effect across regions in multi-regional clinical trials. In this paper, different constellations of heterogeneity and homogeneity in (non-)significant trials are discussed against the backdrop of the regulatory decision-making process. Some of these thoughts go beyond the multi-regional context and can be certainly generalized for other single-region trials.

3.2.2 Multi-Regional Clinical Trials: Case Study

The PLATO trial was a randomized, double-blind, multicenter, multinational, Phase III trial in more than 18,000 patients, described in Mahaffey et al. (2011). The primary analysis intended to show superiority of ticagrelor vs the active comparator clopidogrel in patients with an acute coronary syndrome. The endpoint for the primary analysis was a composite endpoint consisting of death from vascular causes, myocardial infarction, and stroke.

While superiority of ticagrelor over clopidogrel could be established through the primary analysis (with the hazard ratio of 0.84), an important pre-specified subgroup analysis triggered discussions due to heterogeneous results. The investigational centers were clustered into four regions: Europe, Middle East, and Africa with almost 75% of all patients; Asia and Australia with roughly 9%; Central and South America close to 7%; and finally North America with almost 10%. All regional subgroup analyses (with the hazard ratios ranging between 0.80 and 0.86) except the North American subgroup were roughly consistent with the primary analysis. The North American subgroup demonstrated a more favorable effect of the active comparator (clopidogrel) versus the investigational treatment (ticagrelor) with the hazard ratio of 1.25.

It is worth mentioning that the pre-planned regional subgroup analysis was descriptively fixed in the statistical analysis plan among more than 30 other factors that defined the other subgroup analyses. No adjustment had taken place since these subgroup analyses were only descriptively planned. An interaction test for treatment and the four regions yielded a p-value which was only slightly below 5% ($p = 0.045$). Nevertheless, these findings initially led the FDA not to approve the investigational ticagrelor. Later on, and based on additional data analysis by independent statistical groups, it was concluded that the trial was appropriately conducted and that the observed heterogeneity might have been a chance finding because of a long list of pre-specified unadjusted subgroup analyses. However, while chance could not be ruled out entirely, the outcome appeared substantially relevant, since a reasonable rationale could be established. The appearance of the regional heterogeneity was potentially a manifestation of an underlying interaction with concurrent aspirin medication. When use of lower dosages of aspirin were analyzed, ticagrelor actually demonstrated a benefit in the United States population (the vast majority of the North American patients in the trial), similar to that seen in the other regions, e.g., in Mahaffey et al. (2011). Finally, the FDA approved the investigational ticagrelor, whereas high-dose aspirin was contraindicated. More details about the statistical aspects of the PLATO trial can be found in Carroll and Fleming (2013).

In summary, MRCTs can provide major advantages compared to small regional trials, e.g., higher recruitment opportunities with greater statistical power, investigation of homogeneity within one larger trial in contrast to homogeneity investigation across different trials, faster submission, and availability of innovational treatments on the market. On the other hand, heterogeneity might be more easily hidden in larger trials where intrinsic and extrinsic factors were not appropriately taken into account. When not analyzed, this heterogeneity is just simply overlooked. However, long lists of factors to be used for reckless and unadjusted subgroup analyses increase the likelihood for false positive signals.

While the PLATO trial can be considered as a relevant multi-regional trial based on the ICH definition, it is also affected by different aspects of the regional regulations. Apart from the advantages of an MRCT to solve or at least describe problems, the same problems can occur in a large regional study. In order to counter unexpected pre-planned or post-hoc results due to a very long list of analyzed factors, it is advisable to discuss all relevant factors with a cross-functional team in advance to identify the most relevant factors be evaluated at the analysis stage. This approach helps remove the need for a long list of factors to explore and the probability of a false signal can be reduced by focusing on a smaller set of factors.

Section 3.2.4 describes different scenarios that are fundamental for the dedicated subgroup guidance document issued by the EMA (EMA 2014). The PLATO trial can be considered as an interesting case study for these scenarios even though the guidance is not intended for different scenarios in parallel for one study. While consistency could be not established for the different regions in this trial and, in fact, the North American population even demonstrated a directionally different result, the trial's credibility could have been questioned if regulators followed the decision tree presented in Annex 1 of the EMA subgroup guidance. Most probably, due to the clear consistency across all other regions, the issue could be restricted to the North American region. A clear separation of subgroups listed in the study protocol into important and rather exploratory subgroups would have been supportive in the discussion of the unfavorable results in the PLATO trial when the rules defined in the decision tree are applied.

When focusing on the North American population alone, it is worthwhile to consider the second scenario presented the EMA guidance. This is accompanied again by a decision tree described in Annex 2. Since the Aspirin group was not a-priori considered, it would have required a strong compelling plausibility explanation according to the EMA guidance and indeed it required a long pathway for the FDA approval in the relevant subgroup. For example, Carroll and Fleming (2013) described extensive additional analyses to provide a convincing reliable post-hoc explanation of the subgroup findings in the trial.

3.2.3 FDA Regulations

Despite the fact there is still no dedicated guideline on subgroup analysis available from the FDA, this health authority already considers subgroup analysis very thoroughly. More recently, the FDA issued the guidance on "Enrichment Strategies for Clinical Trials," FDA (2019) with a focus on confirmatory analyses controlling the familywise error rate. The Type I error rate for the test can be shared between the corresponding enriched subgroup and the total trial population. In general, this guidance deals mostly with the composition of the trial population with respect to prognostic and predictive factors to support personalized medicine approaches in clinical trials. Remarkably, consistency assessments are considered predominantly in the context of post-hoc analysis rather than in prospective planning. This appears to be a major difference compared to the EMA approach in the EMA draft guidance (2014), and as discussed in Dmitrienko et al. (2016) and Hemmings (2015).

Since specific statistical methodologies and approaches are rarely recommended in guidance documents, it makes sense to review additional recent publications by the authors of the white paper. This includes Alosh and Huque (2013) and Wang and Hung (2014). For the approach of prospective confirmatory subgroup analysis, Alosh and Huque (2013) defined a criterion for concluding that the treatment effect in the least-benefitted (complementary) subgroup exceeds a certain minimum threshold. The criterion is based on testing the effect in the complementary subgroup and showing that the estimated treatment effect in the complementary subgroup is in the right direction. To compensate a lower statistical power, an alpha level higher than the usual two-sided 0.05 is acceptable and might even be up to 0.5 for a safe treatment. This technology formalizes a more objective criterion to assess consistency. On the other hand, the exact choice of the so-called consistency alpha might still be a debatable question. In addition, the approach would be difficult to apply outside the specific situation of the complementary subgroup in a confirmatory subgroup analysis. Wang and Hung (2014) introduced an approach based on the interaction-to-overall-effects ratio, which might lead, in certain cases, to a recommendation for a label restriction. Apart from the specific statistical approach to describe the likelihood of a baseline covariate that may be predictive of a treatment effect in a subgroup, they provide general regulatory review recommendations. The recommendations include a decision tree concept, which, among others, result in prospective design planning or post-hoc analysis with corresponding statistical measures.

In general, all trials should report descriptive statistics for outcomes of interesting subgroups, including at least a point estimate and an estimate of variance or standard deviation. This holds true independent from the corresponding subgroup size and the potentially limited statistical power or conclusion. More specifically, the FDA Guidance on the Evaluation of Sex-Specific Data in Medical Device (2014) always request gender-specific subgroup analyses, where the data should be analyzed for clinically meaningful gender differences in the primary and secondary effectiveness and also safety endpoints.

Most recently, the FDA has released a guidance on the "Evaluation and Reporting of Age-, Race-, and Ethnicity-Specific Data in Medical Device Clinical Studies," (2017a) which also applies to post-approval trial submissions and postmarket surveillance trials. The primary goals are structured into: (1) diverse participation which requires collection and consideration during the trial design stage of relevant covariates, (2) consistent analysis based on analyses of subgroup data, considering especially demographic data when interpreting overall trial outcomes, and (3) transparency with specified expectations for reporting demographic specific information in summaries and labeling.

Albeit the above-mentioned guidance documents trace back to different FDA divisions, e.g., drugs, biologics or medical devices, it is a good idea to consider the logic behind the guidance and reflect it for other areas, unless it is, content-wise, clearly related to a specific division's area.

The FDA has established a working group within the Office of Biostatistics, which intended to prepare a white paper to provide guidelines for subgroup analysis. The concept was presented in 2014 at an EMA workshop on the investigation of subgroups in confirmatory clinical trials, Russek-Cohen (2014) and published by Alosh et al. (2015). As an outlook within the white paper, the following specific topics were mentioned: Bayesian subgroup analyses, including shrinkage concepts, noninferiority margin aspects, and personalized medicine, subgroup misclassification, and safety considerations, such as detection of signals in subgroups or contribution to benefit risk assessments.

3.2.4   EMA Regulations

Following discussions across various health authorities, and with academic and industry representatives, the EMA compiled guidance for subgroup analyses parallel to the FDA's working group (Alosh et al. 2015), FDA enrichment guideline (FDA 2019), and other initiatives. In 2014, the EMA released a draft guideline on the investigation of subgroups in confirmatory clinical trials (EMA 2014). While the guideline has a focus on planning and prospective analysis aspects in the setting of a confirmatory (pivotal) trial, it also requests the investigation of (in)consistency and homogeneity (heterogeneity) in a descriptive manner. An important aspect of the guideline is the discrimination of confirmatory versus exploratory analysis and their relevance under certain circumstances. A structured approach for such circumstances is reflected in the following scenarios, which are described in Sections 6.3–6.5 of the guidance document. This structure is not only a recommendation for trial sponsors in how to plan and conduct clinical trials, but also to provide assessors in European regulatory agencies with guidance on assessment of subgroup analyses.

Scenario 1 intends to establish credibility in a situation where the available clinical data are generally favorable and consistency should be considered. The focus is here to collect circumstantial evidence of consistent efficacy and safety across the relevant subgroups.

Scenario 2 demonstrates the situation with less convincing clinical data, but borderline results in certain subgroups. This might be the case when the statistical test is formally positive, but clinically not sufficiently stable across the whole trial population. Here, the intention is to establish credibility and find a subgroup with clinically relevant efficacy or improved risk-benefit. In this scenario, the demand for concrete evidence is, compared to scenario 1, even greater. This comprises issues of multiplicity and selection bias.

Scenario 3 describes the rather exceptional case in which the clinical data failed to establish statistically evidence, but there is, e.g., the medical need in identifying a subgroup, which has a relevant positive treatment effect.

Even though the identification of the relevant subgroup in Scenario 3 is channeled towards one specific subgroup, the question remains which subgroups need to be considered and analyzed for Scenario 1 and 2, i.e., screening across all "relevant" factors for consistency in scenario 1 and identifying within all "relevant" factors a subgroup with improved benefit risk in Scenario 2. Collecting all information from the guidance text which subgroup types can be considered "relevant," a remarkably long list evolved. This comprises subgrouping factors used for stratified randomization, demographic factors, including genomic factors, factors that might be predictive for different response to treatment, such as stage, severity or phenotype of disease, use of concomitant medications, and possibly region, country, or center, and clinical characteristics, and complement subsets of investigated factors. This implies, obviously, an expectation for the sponsor to consider a huge list of parameters to perform a thorough characterization of subgroup effects. The characterization of the treatment effects consistency across all defined subgroups can consequently not be done in a confirmatory approach with corresponding multiplicity adjustments, but is at least, in major parts, based on an exploratory approach. In the case of a more homogenous trial population and consistent explorative results, it should not be a particular challenge. Otherwise, the less homogenous the trial population, the greater the number of investigations required to explore the degree of heterogeneity. In the best case, a prospective plan for suspicious subgrouping factors was set up.

The above-mentioned decision trees in Scenarios 1 and 2 may be applied to several covariates from the huge list described in the previous paragraph. Thus, it is important to recognize that the probability of an incorrect conclusion increases remarkably. This is addressed in the EMA guideline recommending to predefine key covariates for which biological plausibility of an interaction can be expected, whereas the remaining candidate covariates would be treated as "truly exploratory."

The EMA guideline (2014) refers to several previously issued guidance documents, since it describes principles and does not regulate a specific statistical methodology for the analysis of the treatment effect in subgroups of the trial population. Apart from straightforward tools for consistency assessments, e.g., forest plots with subsequent visual inspection, strict statistical principles in conjunction with subgroup analysis were considered in earlier EMA guidelines. Specifically the following guidance documents reflect largely the current state-of-the-art statistical concepts: Draft points to consider on multiplicity issues in clinical trials (EMA 2017), Guidance documentation for the consideration of adjustment for baseline covariates (EMA 2015) and Reflection paper on methodological issues in confirmatory clinical trials planned with an adaptive design (EMA 2007). In contrast to this, the draft guideline on the investigation of subgroups in confirmatory clinical trials (EMA 2014) deals predominantly with the conceptual aspects of subgroup analysis.

While the FDA also mentions exploratory subgroup analysis in their guidance documents as outlined in the previous section, the extent of such analysis is much more prominent in the European guideline. The differences in both regional approaches are condensed in Hemmings (2015). It seems that the suspicion behind the EMA guideline (2014) is a potential treatment interaction on the investigated factors, which can never be assumed to be non-existent and therefore triggers a large list of factors to be considered in an all-embracing explorative set of consistency checks. As pointed out by Hemmings (2015), the topic goes beyond a purely statistical problem, and is set into the context of licensing decisions and putting a structure around the design, analysis, and interpretation of subgroup analyses that informs and facilitates risk-benefit decision making. His statement, "If we do not, or if we do not know, it may be misleading to assume that we do!" is apparent when the guideline emphasizes, in its executive summary, that ignoring subgroup analysis is not an option, and not sufficient to dismiss all subgroup findings that indicate heterogeneity of response as being spurious.

Following the attitude of the EMA guideline (2014) going beyond the statistical framework, and considering the assessment also of the biological plausibility, it consequently requires thorough interdisciplinary discussions. As described above, it might necessitate not only the trial data, but also other circumstances, such as historical data. These discussion should be reflected in the trial protocol and analysis plan as part of the planning stage, but also in the end of the clinical trial report, where the trial results are discussed.

Further reflections on regulatory considerations in late-stage trials with pre-planned and post-hoc subgroup assessments are highlighted in Hemmings (2014) and Koch and Framke (2014). Both publications are available in the special issue of Journal of Biopharmaceutical Statistics on subgroup analysis in clinical trials guest-edited by Dmitrienko and Wang (2014). The ideas presented in these papers are consistent with the general framework presented in EMA (2014). In addition to Hemmings (2015), he provided, in 2014, an overview of statistical and regulatory issues with respect to three stages; namely, the trial planning, analysis planning, and reporting stages. This article covers a broad spectrum of topics and provides, beside a debate, also practical recommendations. The background information presented in this article sheds additional light on the key principles of the EMA guidance on subgroup analysis, and helps explain why consistency assessments play a predominant role in that guidance document.

3.2.5   China National Medical Product Administration (NMPA) and Japanese Pharmaceuticals and Medical Devices Agency (PMDA) Regulations

Regulatory guidance documents, especially from Chinese and Japanese health authorities, are closely connected to the ICH E5 guideline (Ethnic Factors in the Acceptability of Foreign Clinical Data) (ICH 1998). The formalization in these guidance texts provides a rough concept to avoid simple use of foreign trials, and transfers the results to the corresponding region. It rather regulates the bridging process, and thereby addresses intrinsic and extrinsic ethnic differences. This is in line with the fact that the China National Medical Product Administration (NMPA) and the Japanese Pharmaceuticals and Medical Devices Agency (PMDA) have certainly a much stronger focus on the evaluation of ethnic differences. However, the ICH E5 guidance does not describe concrete statistical methodologies. Also, it was suffering from a time-consuming successive conduct of clinical trials, since the bridging concept is based on separate trials, rather than subgroup analysis in one single trial of corresponding Asian and non-Asian populations. This procedure results in a remarkable time delay of market access to the target patients.

With the bridging of safety and efficacy from other regions, such as the US or Europe based on ICH E5 (1998), the experiences have been steadily accumulated with respect to how Japanese and foreign data can be appropriately evaluated. On the other hand, it resulted in a relevant time lag in approving experimental treatments for Japanese patients, as outlined in PMDA (2007). To enable simultaneous investigation of Japanese patients, and subsequently an earlier submission and potential approval, large multi-regional clinical trials are proposed. A reasonable size of the subgroup of Japanese subjects is expected in such multi-regional clinical trials to support subgroup evaluations to demonstrate consistency with the overall trial population. Again, and according to the ICH E9 (1999), the question of sufficient homogeneity is a precaution. A global multiregional clinical trial including Japanese subjects should be only conducted if homogeneity can be assumed.

To clarify what a reasonable size of Japanese patients in such a multi-regional trial means, the PMDA (2007) recommends calculating a sample size considering the number of regions to be included, the scale of trial, target disease, and the relevant ratio between the total and Japanese subject numbers. To obtain consistent results between the entire trial population and the Japanese population when designing such a trial, taking, as an example, a placebo-controlled trial using quantitative endpoints, the following two methods are defined. The first method requires a sufficient number of Japanese subjects to show that at least half of the overall effect is retained in the Japanese subgroup. The second method requires a sufficient number of Japanese subjects to show the same positive trend as in the overall population. Since no method has been currently established as generally recommended, it is worthwhile to reflect some publications on this topic. Important statistical considerations such as sample-size requirements, as well alternative criteria, are discussed in Carroll and Le Maulf (2011). Apart from the standard ethnic factors, there is a trend toward considering other background factors or influential factors that need to be evaluated and discussed as part of subgroup investigation. Ikeda and Bretz (2010) proposed an alternative method with better operating characteristics than the current approaches for the first method, with thorough formalized aspects accompanied with simulation studies. For the second method, Liu et al. (2016) investigated the specific situation of discrete endpoints approaches with mixed models. It should be mentioned that the proposed methods are exemplary for a quite often but also concrete situation, and can be considered as a first guidance; eventually, a consultation of the PMDA health authority is recommended.

As supplemental explanation to the "Basic Principles on Global Clinical Trials," the PMDA (2012) released the reference cases to encourage Japanese subject participation in global trials. Hirakawa and Kinoshita (2017) provides a broad overview of such global trials with practical experience. They investigated the proportion of Japanese patients in MRCTs and further compared the efficacy results from the overall population to that of the Japanese population. This provides an impression of encouragement after the PMDA (2007) initiative.

Depending on the specific circumstances, different trial types in local Chinese population are requested from NMPA (formerly CFDA, until 2013). The circumstances are basically the treatment approval status in other regions, which trigger a thorough investigation of the treatment effect in Chinese population. The desired replication of the overall treatment effect in the Chinese subgroups is described in "Provisions for Drug Registration" (CFDA 2007). In 2016, the NMPA published a draft with a remarkable revision of this guidance predominantly focusing on fundamental changes in the administration of the treatment approval process. Nevertheless, it comes without any additions to the statistical aspects of subgroup analysis or, more specifically, consistency assessments. The NMPA advice often in other documents includes references to "consistency assessments" comparable to the EMA guidance (2014) on subgroup analysis. In 2015, the NMPA released a guideline on International Multicenter Clinical Trials (IMCTs) (NMPA 2015). The intention was to define the requirements for IMCTs involving Chinese sites and subsequent analysis of subgroups of Chinese subjects, compared to the overall trial population, to investigate the consistency of treatment effects.

3.3   Data-Driven Subgroup Analysis

3.3.1   Key Principles of Data-Driven Subgroup Analysis

The term "exploratory subgroups analysis" is often used to cover a variety of situations when subgroups are evaluated without strictly controlling the Type I error rate. In this section, we will refer to "exploratory subgroup investigation" in a broad sense thus covering any data-driven subgroup analyses where the set of hypotheses tested is random rather than fixed/pre-specified (see a typology of different data-driven analyses in the next section). The problem of subgroup selection is a special case of a more general problem of model selection. In this situation, an additional challenge is that the modeling targets the causal treatment effect which is unobservable at the individual patient level except with a cross-over design.

The key principles of data-driven subgroup analysis can be derived from the fields of machine learning, causal inference and multiple testing were the novel method of subgroup/biomarker evaluation originated (a detailed discussion of these topics can be found in Lipkovich et al. 2017a, 2018). The principles include

- Applying complexity control to prevent data overfitting and selection bias, e.g., bias due to selecting the best patient subgroup from a large set of candidate biomarkers (patient characteristics) and associated cutoffs. Tuning parameters controlling the subgroup search process often need to be determined in a data-driven fashion, e.g., via cross-validation.
- Evaluating the Type I error rate for the entire subgroup search strategy, e.g., by using resampling under the null hypothesis of no subgroup effects. Subgroup analyses are often performed in clinical trials using a multi-stage strategy as described in the Introduction where a multiplicity correction is applied to the last stage but is not applied at earlier stages.
- Obtaining "honest" estimates of the treatment effect within identified subgroups, expected if evaluated in an independent (future) data set. In the absence of independent data this can be approximated by using resampling methods or Bayesian model averaging/Empirical Bayes. Again, uncertainty associated with the subgroup identification should be taken into account.

3.3.2 Types of Data-Driven Subgroup Analyses

Following the framework in a survey of current industry practices (Mayer et al. 2015) we consider the following three types of data-driven or exploratory (in the broad sense) subgroup analyses in clinical trials:

- Exploratory (in the narrow sense).
- Post-hoc.
- Biomarker and subgroup discovery.

This taxonomy is not prescriptive but descriptive reflecting current clinical practice (and language) rather than what may be desired in a "perfect world."

Exploratory Subgroup Analysis (in the Narrow Sense) This includes strategies specified in the exploratory analysis section of statistical analysis plans. The number of evaluated biomarkers is relatively small, typically limited to known prognostic variables (some of them are included as stratification covariates in the primary analysis). Often it is conducted using multi-stage strategies. For example, a subset of biomarkers is selected by fitting separate regression models with one biomarker at a time and comparing the biomarker-by-treatment interaction p-value with an arbitrary pre-specified cutoff. Then each selected continuous biomarker is examined further by choosing an optimal cutoff to define patient subgroups. Therefore, although the candidate biomarkers are pre-specified, the selection of subgroups is only partially pre-specified owing to the data-driven selection of the cutoffs. The Type I error rate is typically controlled only for some elements of the multi-stage strategy (e.g., selection of the biomarker-specific cutoffs and ignoring the multiplicity due to multiple candidate biomarkers). While this description reflects the existing practices, full control of multiplicity could have been achieved by using resampling methods (e.g. replicating entire subgroup search strategy on each sample drawn from the null distribution).

Post-hoc Subgroup Analyses This covers subgroup investigations that are unanticipated prior to data unblinding and therefore are not pre-specified in a statistical analysis plan. Typically, these are subgroups with unanticipated post-hoc findings after data unblinding that may have trigged regulatory inquiries or subgroups that raise regulatory or sponsor's concerns after approval of an investigational treatment. The Type I error rate is typically not controlled or only partially controlled within the selected subgroups (ignoring the fact that they were chosen post-hoc, in a data-driven manner). Although the evaluation of subgroups may appear to be limited to pre-specified subgroups, the set of subgroups is conditional on post-hoc findings and is data-driven in this sense. For example, if a certain country showed unusually small or large treatment effect, an analysis shrinking the country effect towards a common mean can be entertained, ignoring a multitude of other covariates where heterogeneity in treatment effect was not found. Although it is possible to account for the multiplicity within the subgroup identified (for example, by taking account of the number of countries evaluated), this does not account for the post-hoc nature of the initial subgroup identification. As a result, methods which can somehow reflect the post-hoc nature of the initial identification would be valuable. A somewhat different situation may occur when a post-hoc finding is in regards to a treatment-related safety effect where, contrary to efficacy findings, there is a tradition to avoid any multiplicity adjustment as "anti-conservative." However, as with any signal detection, it is important to attempt to understand how likely we are to see a given safety signal by chance given the number of safety variables we have looked at, as this could help us interpret the result.

Biomarker and Subgroup Discovery This includes data mining of large sets of available candidate biomarkers and is not limited to those anticipated as potential predictors prior to data unblinding. Traditionally, false positive rates are not controlled and external validity is established using cross-validation or replication based on independent data sets. However, some form of the overall Type I error rate control or the false discovery rate control can be incorporated (this approach is emphasized in this chapter).

Some features distinguishing different subgroup analyses discussed above are reflected in Rows 2–4 of Table 3.1. The first row presents confirmatory subgroup analyses (as a "limiting" case when subgroup selection is not data-driven) which will be discussed in Sect. 3.4.

Table 3.1 Types of subgroups analyses in clinical trials

Subgroup analysis type	Strategy (Preplanned/ Post-hoc)	Multiplicity control (Yes/Partially/No)	Data scope (Broad/Narrow)	Subgroups (Pre-specified/ Data-driven)
Confirmatory	Preplanned	Yes (often ensuring the strong Type I error control)	Narrow (limited to subgroups previously identified)	Pre-specified
Exploratory	Preplanned	Partially (only some elements are controlled)	Narrow (limited to pre-specified biomarkers/ subgroups)	Pre-specified (often up to unknown cutoffs)
Post-hoc	Post-hoc	Partially (controls for multiple testing among subgroups selected post-hoc)	Narrow (limited to biomarkers became known after data lock)	Pre-specified (after "unanticipated" subgroups identified)
Biomarker/ subgroup discovery	Preplanned	Partially (often controls Type I error in weak sense or FDR)	Broad (learning from a large set of biomarkers)	Data-driven (learning from the data)

3.3.3 *"Guideline Driven" Versus Data-Driven Subgroup Analysis*

Subgroup analysis in the context of clinical trials has been a controversial topic as a potential tool that may be used by sponsor for making unsubstantiated efficacy claims. Many authors made a point about the inflation of Type I error rates as a result of "undisciplined" or haphazard subgroup analyses typically followed a failed primary analysis. To promote principled subgroup analyses some authors came up with checklists of good practices or guidelines that if followed would assure integrity in subgroup analyses and prevent discovering false subgroups. For example, Brookes et al. (2001) provides a list of 25 recommendations, Rothwell (2005) proposed a guideline with 21 rules, Sun et al. (2010) listed the existing 7 plus 4 additional criteria for assessing credibility of subgroup analysis. The general theme in these rules is summarized below along with some critical points.

- "Subgroups should be pre-specified." While allowing application of well-established multiple comparison procedures (MCP), this does not fit into discovery spirit of statistical science.
- "Subgroups should be biologically plausible." This is obviously one of the most important pre-requisites; however, it is not often the case that the biological mechanism driving the heterogeneity of treatment effects is known in advance.

- "All significance tests should be multiplicity adjusted." While appealing in general, one should bear in mind that many classical MCP are overly conservative and have low power. The key is how to combine efficient subgroup selection with appropriate MCP. A popular option to improve power is to provide some form of multiplicity adjustment, but use more liberal Type I error rates. For example, it is often recommended to use a significance level of 0.1 when testing for biomarker-by-treatment interactions, as the interaction test has notoriously low power. While this strategy may be particularly appealing in a more confirmatory setting, any improvement in power comes at the cost of greatly increased Type I error. A more powerful procedure could be entertained by incorporating multiplicity adjustment not after selection but rather as part of selection/estimation. As an example, using shrinkage estimators such as *lasso* may result in selecting a subgroup different from the apparent winner, which will pay-off by requiring a lesser degree of post-selection adjustment. Note that significance testing for subgroup effects should still account for model selection, e.g. via *lasso*, which became possible thanks to recent advances in post-selection inference (see, for example, Lee et al. 2016; Taylor and Tibshirani 2017).
- "No testing in a subgroup unless the interaction test is significant." The idea that the interaction test be used as a gatekeeper is just one example of possible search strategy which has low power; also there are many ways to operationalize interaction testing when the biomarkers are continuous (e.g. by considering interactions of treatment with many subgroup indicators that can be created by dichotomizing a continuous biomarker or using other options for non-parametric modeling of treatment by covariate interactions).
- Sometimes: "No testing in a subgroup unless the overall effect is significant." This does not agree with the spirit of personalized medicine, as the experimental treatment may be not for everyone. Rather than using a test for the overall treatment effect as a "gatekeeper," some forms of shrinkage towards the overall treatment effect may prove more reasonable.
- Finally, a common guideline is to "interpret the results with caution" which is also hard to operationalize. Clearly, lack of pre-specification often causes concerns, and rightfully so. For example, subgroup findings based on a large set of candidate covariates would intuitively be considered prone to model selection error. However, as stated in the first bullet, complete pre-specification of hypotheses removes the discovery element from statistical science.

The EMA guideline on subgroup analysis (EMA 2014) provides useful points to consider when planning subgroup investigation activities. Briefly, the guideline attempts to discourage reviewers and trial sponsors from making wrong decisions at two extremes. The guideline warns against dismissing subgroup analysis, which is seen in the context of current practices often "creating disincentive to properly plan the investigation of subgroups," and also warns against "reckless" subgroup analysis (that does not exercise caution). In particular, the EMA guideline encourages discussion about potential subgroups at the trial design stage arguing that done properly,"this should minimize the need for data-driven investigations, relying instead on a well-reasoned pre-specified strategy."

The use of "data-driven" as a negative, perhaps, stresses that the present guideline relates specifically to the situation when a pivotal trial has been conducted and a regulatory decision regarding labeling is required and thus discourages from excessive hypothesis generation. However, there may still be the need to understand any unexpected findings. As such, the subgroup investigation strategy would need to be principled (and even pre-specified). We argue that such principled approaches can be developed which are a special case of model selection in the presence of a large number of biomarkers.

This allows us to connect this task with a wealth of relevant methods that have been developed in the areas such as statistical learning, causal inference and multiple testing. Indeed, we have witnessed a surge of publications on data-driven subgroup analysis coming from a cross-pollination of these areas. However, as any conclusions affect regulatory labeling or the need for significant future investment in expensive clinical trials, the type I error rates need to be clearly understood.

In the next section we will examine several classes of methods that were recently proposed under the heading of "principled data-driven subgroup analysis." The key principles are those listed in Sect. 3.3.1.

3.3.4 Case Study

A case study that will be used throughout the rest of Sect. 3.3 is based on a Phase III trial conducted to investigate the effect of an experimental therapy for the treatment of colorectal cancer. The subjects in the treatment arm ($N = 353$) received the experimental treatment in addition to the best supportive care (BSC) whereas the subjects in the control arm ($N = 177$) received BSC only.

This example illustrates a retrospective approach to biomarker discovery and subgroup identification. The overall outcome of the trial, defined as progression-free survival (PFS), was negative with a one-sided log-rank p-value of 0.324. The sponsor was hoping that a positive treatment effect might be found within a reasonably sized subpopulation defined by nine biomarkers identified as potentially having prognostic effects prior to randomization (see Table 3.2). These included important clinical biomarkers, such as tumor grade and diagnostic site, and genetic markers, e.g., KRAS gene mutation. Most of the biomarkers in the candidate set were continuous, one biomarker was ordinal (X_1) and two biomarkers were measured on a nominal scale (X_2 and X_5). Because a substantial number of patients had their tumor grade assessment unknown, we treated X_1 as a nominal variable with five levels labelled as G1, G2, G3, G4 and G5.

Table 3.2 Candidate biomarkers in the case study

Biomarker	Description	List of values or range
X_1	Tumor grade	1, 2, 3, 4, unknown
X_2	Primary diagnostic site	Colon, rectum, colon and rectum
X_3	Time from the initial diagnosis to the metastatic disease (months)	0–126
X_4	Time from the initial diagnosis to the start of treatment (months)	12–189
X_5	KRAS mutation status	Wild type, mutated
X_6	Protein expression marker	1–32
X_7	Protein expression marker	40–236
X_8	Protein expression marker	1–38
X_9	Protein expression marker	0.6–7.4

3.3.5 Typology of Data-Driven Subgroup Analysis Methods

Following Lipkovich et al. (2017a), we briefly outline four classes of methods that emerged in the recent literature on data-driven subgroup analysis.

3.3.5.1 Global Outcome Modeling

This approach aims at constructing regression models relating outcome with patient-level biomarkers for each treatment arm. This can be implemented with a single regression incorporating main (prognostic) effects and treatment by covariate interactions (predictive effects) or through separate models within each treatment arm. Identifying subgroups from such models typically requires multistage procedures. For example, in Virtual twins method of Foster et al. (2011) at the first stage a global outcome model is estimated using a Random forest with all candidate biomarkers $(X_1, \ldots, X_p)$ and treatment indicator, T, included as covariates. Then the fitted forest is used to compute hypothetical individual treatment differences. That is, for each patient two potential outcomes (for the case of two alternative treatments) are predicted from the fitted model: one assuming an experimental treatment ($T = 1$) and the other assuming a control treatment ($T = 0$) (hence the name, Virtual twins). Thus, a treatment contrast can be computed for each patient (e.g. as the difference between the fitted values $Z_i = \widehat{Y}(\mathbf{x} = \mathbf{x}_i, t = 1) - \widehat{Y}(\mathbf{x} = \mathbf{x}_i, t = 0)$ in case of continuous outcome). These values $Z_i, i = 1, \ldots, n$ are then used as an outcome variable of the seconds stage and modeled using the classification and regression trees (CART) to construct subgroups capturing heterogeneity in individual treatment effects (as the tree leaves). The approach allows for variations in implementation. In general, any "black box" modeling can be used at the first stage (see a comparative study in Lu et al. 2018) and any method of predictive modeling—at the second stage.

To illustrate this approach with a survival outcome from our case study, we will apply the Gradient boosting algorithm of Friedman (2002) using the R package **gbm** (Ridgeway 1999). The gradient boosting model is an extension of the proportional hazards Cox model that can be represented as $h_0(y) \exp(f(\mathbf{x}, t))$, where $h_0(y)$ is the baseline hazard as a function of time (Y), and $f(\mathbf{x}, t)$ is a function of baseline covariates and treatment. We fitted the gbm model using up to third order interactions among included variables (i.e., candidate biomarkers from Table 3.2 and a treatment indicator); the shrinkage parameter was set as 0.001 and the number of trees in the model was estimated by a tenfold cross validation. Each subsequent tree was estimated using a random sample of 50% of the data. Then we computed predicted $\widehat{f}(\mathbf{x}, t), t = 0, 1$ for each subject, assuming s/he is treated with experimental and control treatments, respectively. The difference between the two represents an individual hazard ratio on log scale (as the baseline hazard cancels out). Figure 3.1 displays the density plot (the upper panel) of the estimated treatment differences on a hazard scale. Clearly, the distribution is bimodal suggesting heterogeneity of treatment effect; particularly, a subgroup of subjects (with hazard ratios < 1) may benefit from the experimental treatment. The scatter plot in the lower

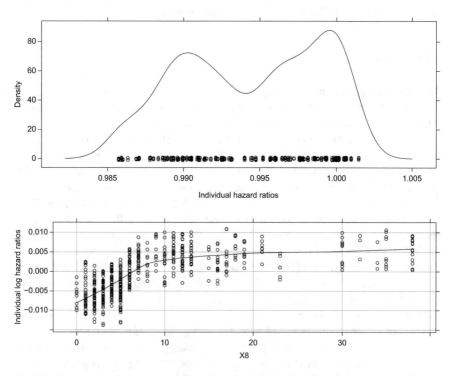

Fig. 3.1 The graph in the upper part shows the distribution of individual hazard ratios estimated by gradient boosting. The lower graph shows a scatter plot (with a smooth spline) for individual treatment differences on the log hazard scale against protein expression marker X_8 (random jitter was added to break the ties in the y-axis)

Fig. 3.2 Pruned regression
tree fitted to the individual
log hazard ratios estimated by
gradient boosting. The values
displayed for each terminal
node are: sample sizes and
hazard ratios estimated with
Cox model from observed
data

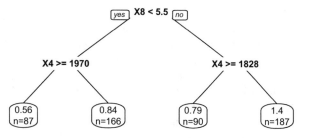

part of the figure suggests that patients with lower levels of protein expression X_8
may benefit from the treatment.

At the second stage, we fitted a regression tree (using **rpart** R package) to the
estimated individual treatment differences. The resulting tree (after pruning via
cross-validation) is shown in Fig. 3.2. The tree suggests that patients with lower
levels of X_8 ($X_8 < 5.5$) and higher levels of X_4 ($X_4 \geq 1970$) may have treatment
benefits. The values shown in the terminal nodes of the tree are sample sizes and
hazard ratios estimated from the original data. These are given for illustration. One
should bear in mind that resubstitution of the same data used for model search
would result in estimated treatment effect in subgroups prone to optimism bias.
As we saw from the distribution of predicted treatment differences by gbm, they are
rather modest, perhaps reflecting both lack of fit and the fact that Gradient boosting
(like other methods of ensemble modeling such as Random forest) does not overfit
the data provided its parameters are tuned appropriately. More accurate estimates
of treatment effect in the identified subgroups can be obtained using resampling
approaches (see Foster et al. 2011).

It is worth noting that in recent years methods for estimating and making
inference about individual treatment effects—the outcomes of the first stage of
the Virtual Twins method—received a lot of attention (irrespective of using them
for subsequent subgroup identification). Lamont et al. (2018) reviewed two broad
classes of methods for the identification of "predicted individual treatment effects"
(PITE) from randomized clinical trials, namely, parametric multiple imputation
and recursive partitioning methods. They emphasized that the individual treatment
effects should be predicted for out-of-sample individuals using models estimated
on training data. Although one of the potential outcomes needed for computing the
treatment difference is observed in the test sample, the predicted values rather than
the observed ones should be used to minimize overfitting. Ballarini et al. (2018)
applied penalized regression methods (lasso) to computing PITE and used recently
developed theory of post-selection inference for the lasso to construct confidence
intervals for the individual effects. Su et al. (2018) utilized Random forests of
Interaction trees (RFIT) to estimate individual treatment effects and obtained
standard errors using the infinitesimal jackknife method of Wager et al. (2014).
Wager and Athey (2018) developed a new procedure Causal random forest (CRF)
to estimate individual treatment effects as a function of covariates, building on the
Causal trees of Athey and Imbens (2016). To construct each tree, this method first

extracts a random sample without replacement from the data which is further divided into two subsets of equal size; then one subset is used for selecting splits whereas the other for estimating treatment effects. The fact that for each tree no outcome value is used for both splitting and estimation satisfies the "honesty" condition. They show that valid standard errors for the individual treatment differences computed from CRF can be obtained using the infinitesimal jackknife method. Lu et al. (2018) compared several methods for obtaining individual treatment effects from observational data, including various types of random forests and BART (Bayesian Adaptive Regression Trees, Chipman et al. 2010; Hill 2011).

3.3.5.2   Global Treatment Effect Modeling

Approaches of this class obviate the need to include prognostic effects that effectively "cancel out" so that only predictive effects need to be modeled. This often results in more robust estimates of predictive effects as the overall model is not prone to a misspecification of prognostic effects. Examples of this class are Interaction trees (Su et al. 2008, 2009), Gi method (Loh et al. 2015, implemented within the GUIDE package), Model-based recursive partitioning (Seibold et al. 2016, 2018), Causal trees and forests (Athey and Imbens 2016; Wager and Athey 2018), and the "Modified covariate method" (Tian et al. 2014).

We illustrate this class of approaches using the method of Interaction trees (IT) extended to survival outcomes in Su et al. (2008). Briefly, IT constructs a decision tree similarly to a classification and regression tree (CART, Breiman et al. 1984, for the case of binary or continuous outcomes), or to a survival tree (Leblanc and Crowley 1993) with one crucial difference. Specifically, it replaces the splitting criterion based on the reduction of heterogeneity in the outcome (after splitting a parent node into two child nodes) with the one based on the reduction in heterogeneity in treatment effect. This is achieved by choosing the split that maximizes the treatment-by split interaction in the model fitted for a parent node. Specifically, when evaluating each candidate split of a parent node, two models are entertained: one including the binary indicators for the candidate split and treatment, respectively; the other including additionally treatment by split indicator. In the context of survival outcomes, this is the proportional hazards Cox model. The split maximizing the difference in $-2\times$ log (partial) likelihood for the two models is selected. As a result, a decision tree dividing the covariate space into segments with piecewise constant treatment effects is obtained.

Note that, unlike the IT approach that directly incorporates the treatment variable in the splitting criterion, the Virtual Twins method uses the CART method at the second stage. CART employs the usual splitting criterion based on the reduction in the heterogeneity of the outcome variable due to the split. That is why computing individual treatment differences (here, the log hazard ratios) were needed at the first stage of VT, so that they could be used as the input for the usual tree regression.

To analyze the data from the case study, we used a suite of R functions developed and kindly provided by Xiaogang Su. A slightly pruned interaction tree (to the

Fig. 3.3 Pruned interaction
tree (IT). The values
displayed for each terminal
node are: sample sizes and
hazard ratios estimated with
Cox model from observed
data

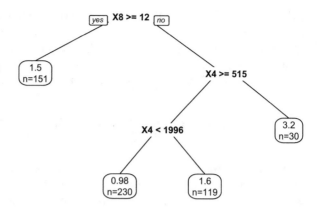

four terminal nodes) is presented in Fig. 3.3. It is similar to the tree based on the
Virtual twins method, in that it also selected X_8 and X_4 as splitters for the first
and second level, however the cut-offs are different. Of note, pruning based on a
bootstrap approach (Su et al. 2008) suggested an empty tree (with no splits) for this
data set, which underscores the uncertainty inherent in the model selection process.

3.3.5.3 Modeling Individual Treatment Regimes

Broadly, this includes any approach that determines which patients should receive
which of a set of candidate treatments, based on patient-level data. For example,
Qian and Murphy (2011) formulated finding optimal individual treatment regime
(ITR) by using the following two-stage procedure. At the first stage it estimates (via
penalized regression) the conditional mean response given the treatment choice and
a large number of candidate biomarkers as predictors; and at the second stage derives
the estimated treatment rule from the model for conditional mean obtained at stage
1. In our taxonomy this approach would fall into the first class (global modeling).
Note, however, that all that is needed for assigning patients to optimal treatment
regime is to evaluate the sign of the predictive score, if positive they should receive
the experimental treatment and if negative—the control (in case of two treatment
options). Of course, some patients may fall within an "indifference zone" with no
clear benefits for choosing one treatment over the other, which may be accounted
for by assessing uncertainty of treatment assignment and this "indifference" could
form part of any future treatment algorithm.

An important subclass of modeling ITR reduces the problem of identifying an
optimal treatment rule to directly modeling the sign of the hypothetical treatment
difference. Such methods were pioneered by Zhang et al. (2012) and Zhao et al.
(2012) who showed that estimating optimal treatment regimes can be framed as
a classification problem where the optimal classifier (minimizing an outcome-
weighted classification loss) corresponds to the optimal treatment regime. Other
examples include the ROWSi method (Xu et al. 2015; Huang and Fong 2014).

Laber and Zhao (2015) and Fu et al. (2016) developed frameworks for estimating an optimal ITR within a class of interpretable rues (e.g. tree-structured or "rectangles").

The idea of outcome weighted learning (OWL) is that an optimal individual treatment assignment rule $d^*(\mathbf{X})$ can be estimated as a function of biomarkers that returns values 1 or 0, by minimizing a weighted classification loss,

$$d^*(\mathbf{X}) = \operatorname*{argmin}_{d(\mathbf{X})} \sum_{i=1}^{n} \left[I(t_i \neq d(\mathbf{x}_i)) \frac{y_i}{Pr(T = t_i | \mathbf{x}_i)} \right]$$

The subject weights $y_i / Pr(T = t_i)$ are proportional to the outcome (here, time to event) and inversely related to the probability of having been assigned treatment that was actually observed (a known constant in a randomized clinical trial). Intuitively, patients with longer survival would have higher weights and the optimal rule is likely to reproduce treatment they actually received in the clinical trial. Conversely, patient with shorter survival would get smaller weights, hence the price of misclassifying is low, so they are likely to be assigned to the treatment different from actually received ("misclassified").

As minimizing 0–1 loss is unwieldy, typically it is replaced by an appropriate smooth loss function (e.g. "exponential loss" as in logistic regression, or a "hinge loss" as in support vector machines). For our illustrative example, we chose to fit a classification tree so that the ITR will be approximated as a decision tree. Specifically, we used **rpart** to fit a classification tree with the treatment indicator as an outcome variable and the biomarkers $X_1, \ldots X_9$ as candidate covariates. We excluded all records with censored outcomes (by assigning zero weights). To correct for possible selection bias due to including only uncensored observations, we incorporated additional weights computed as the inverse probability that the patient remains uncensored by the time the event had occurred, i.e., the censoring time is greater than the event time. The probabilities were estimated in a preliminary step using a separate gradient boosting model. This proposal was recently made by Fu (2018), along with a more complex doubly robust method. Therefore, our weights were computed as follows.

$$w_i = \frac{y_i \delta_i}{Pr(T = t_i) S_c(y_i | \mathbf{x}_i)},$$

where y_i is time to event or censoring δ_i is event indicator and $S_c(y | \mathbf{x})$ is the (estimated) survival function for censoring process, that is, $S_c(y_i | \mathbf{x}_i) = Pr(C > y_i | \mathbf{x}_i)$ and C is (partially observable) time to censoring.

The fitted classification tree, after pruning by cross-validation contained a single split by biomarker X_8, so that the estimated ITR is: assign to the experimental treatment, if $\{X_8 < 11.5\}$, otherwise assign to control.

It is always important to estimate the expected gain if using the optimal rule versus treating everyone with one of the available treatments. Like with evaluating treatment benefits within specific subgroup identified by a data-driven method, the gain associated with ITR is prone to optimism bias and resampling methods should be used to obtain a bias-corrected estimates (see, for example, Xu et al. 2015).

3.3.5.4 Local Modeling (Direct Subgroup Search)

The last class of subgroup search methods focuses on a direct search for treatment by covariate interactions and selecting subgroups with desirable characteristics. For example, the search for subgroups with improved treatment effect. This approach obviates the need to estimate the response function over the entire covariate space and instead focuses on identifying specific regions with a large differential treatment effect. Some of the approaches under this heading were inspired by Bump hunting (also known as PRIM) by Friedman and Fisher (1999) which is a method of predictive modeling that aims at estimating only regions where a target function is large. They argued that it may be better to search directly for such "interesting" regions in the covariate space rather than estimating first in the entire space and then discarding the regions that are "uninteresting." Examples include extensions of Bump hunting to subgroup analysis by Kehl and Ulm (2006), Chen et al. (2015). Other approaches of this class include SIDES (by Lipkovich et al. 2011), and its further developments, SIDEScreen (Lipkovich and Dmitrienko 2014a) and Stochastic SIDEScreen (Lipkovich et al. 2017b).

Various methods that seek forming appropriate inference on findings (often post-hoc) from a collection of patient subgroups can be attributed to the same category. While a usual practice was to present such findings using graphical tools like forest plots, little attention was payed to properly accounting for multiplicity and selection bias inherent in such displays. A resampling-based graphical method to present the SEAMOS (Standardised Effects Adjusted for Multiple Overlapping Subgroups) method was recently proposed by Dane et al. (2019).

To illustrate this class of methods, we will present and briefly discuss the results of the SIDEScreen method using so-called Adaptive screening. Briefly, the regular (base) SIDES method is a recursive partitioning method for generating a collection of subgroups derived from the candidate biomarkers. At each level of recursion it essentially utilizes exhaustive search through all possible splits to find optimal spits of the current parent population into two child groups and then applies the search to the best of the two child groups (the one with largest treatment effect) while abandoning the child with the smaller treatment effect. To generate a large collection of subgroups not only the best splitting variable is retained but a specified number M of best splitters (the *width* parameter). Therefore, the search is repeated recursively within M best child groups (resulting from the M best splits). The number of levels in recursion is called the *depth*, denoted by L.

The Adaptive SIDEScreen method can be thought of as a three stage procedure that uses the base SIDES as a building block through the following steps (see Lipkovich and Dmitrienko 2014a,b for details):

1. The base SIDES is applied to generate a collection of subgroups. The variable importance scores, denoted VI(X), are computed for each candidate biomarker X to measure its predictive properties. Like in other tree-based methods, the VI score for a biomarker X is essentially an average contribution of that biomarker across all subgroups (a biomarker receives a zero score if it is not involved in a given subgroup, otherwise its VI-score is a function of the size of the splitting criterion).
2. The candidate biomarkers are screened by applying a threshold v to their variable importance scores, so that only biomarkers with VI $> v$ are passed. The v is data-driven, and computed from the null distribution of the maximal VI score (across all candidate biomarkers), typically as $v = E_0 + kS_0$, where E_0 and S_0 are the mean and the standard deviation of the null distribution, respectively, and k is a user-defined multiplier.
3. If any biomarkers are selected, base SIDES is applied again to the selected biomarkers that passed the screening.

The multiplicity adjusted p-values are computed for the subgroups selected at the third step by using resampling methods. Specifically, a large number K of reference data sets are produced by permuting the treatment variable, then the above 3-step procedure is applied to each, using the same threshold v as for the observed data. Let the p-value for the best subgroup identified in the jth null set be p_{j0} (if no biomarkers are selected in the null set, set $p_{j0} = 1$). The multiplicity adjusted value for an observed p_{obs} for a given subgroup identified by the Adaptive SIDEScreen is computed as $p_{adj} = \sum_{j=1}^{K} I(p_{j0} < p_{obs})/K$, where $I(\cdot)$ is the indicator function.

The following parameters were used in the subgroup search algorithm:

- The *differential splitting criterion* was utilized to find optimal cutoffs during subgroup search. In the context of survival data it is based on a scaled difference between the Z statistics from the log-rank test evaluated in the two child subgroups resulting from the split.
- The maximum number of promising subgroups for each parent group (*width*) was set to 5 ($M = 5$). That is, starting from the full data set, five promising child groups were pursued, resulting from splitting on five top biomarkers with the optimal value of splitting criterion.
- The maximum number of biomarkers in the definition of a subgroup (*depth*) was set to 3 ($L = 3$). That is each final subgroup could be defined by up to three biomarkers.
- The minimal number of subjects within a child subgroup was set to 30 ($n_{min} = 30$).
- The multiplier for the Adaptive SIDEScreen procedure was set to 1 (corresponding to about 16% chance probability of selecting a noise biomarker at the screening stage, assuming no biomarker has predictive effect)

To speed up subgroup search, each continuous biomarker was discretized by converting it into a categorical covariate with 15 levels based on the 15 percentile groups.

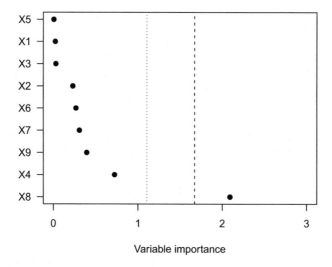

Fig. 3.4 Ordered variable importance scores in the SIDES subgroup search algorithm. The dotted and dashed lines indicate benchmarks from the null distribution drawn at E_0 and $E_0 + 1S_0$, respectively

Figure 3.4 shows variable importance scores ordered from the smallest (on the top) to the largest (at the bottom). These are computed based on a total of 50 subgroups generated at the first stage of the SIDEScreen procedure. As we can see, only biomarker X_8 passed the threshold based on 1 standard deviation above the mean of the null distribution for the maximal VI score (shown with the vertical dashed line on the graph) at the second stage of the procedure. Of note, none of the remaining candidate biomarkers even reached the null's mean (shown with the dotted line). The subgroup signature based on the single selected biomarker X_8 obtained at the third stage was $\{X_8 \leq 9\}$ with associated multiplicity adjusted p-value $p_{adj} = 0.081$, $N = 340$.

3.3.5.5 Summary of Key Features of Data-Driven Methods

Table 3.3 summarizes some of the features of novel methods classifying them within the four types considered in previous sections (in the rows) while each method is further characterized with key features (listed below) in the columns of the table. The last column contains information on available software for subgroup identification. Many of these can be accessed from the Biopharmaceutical Network web site at

```
http://biopharmnet.com/subgroup-analysis/
```

1. Modeling type: F (Frequentist), B (Bayesian), P (parametric), SP (semiparametric), NP (nonparametric).

Table 3.3 Key features of commonly used subgroup identification methods

Method	Modeling type (1)	Dimensionality (2)	Biomarker selection (3)	Control of false positive rate (4)	Complexity control (5)	Selection control (6)	Honest estimate of treatment effect (7)	Software implementation (8)
Global outcome modeling								
STIMA (Dusseldorp et al. 2010)	Freq/NP	Medium	S	No	Yes	No	No	R [stima]
Virtual Twins (Foster et al. 2011)	Freq/NP	High	P,S	No	Yes	No	Yes	R [aVirtualTwins]
FindIt (Imai and Ratkovic 2013)	Freq/P	High		No	Yes	No	No	R [FindIt]
Bootstrap-corrected estimation after model selection (Rosenkranz 2016)	Freq/P	Low	S	Yes		Yes	No	R [subtee]
Bayesian linear models (Dixon and Simon 1991; Hodges et al. 2007)	Bayes/P	Low	P	No	Yes	No	Yes	R [DSBayes]
Bayesian trees (Henderson et al. 2020; Zhao et al. 2018)	Bayes/SP	High	P	No	Yes	No	Yes	R [AFTrees]
Global treatment effect modeling								
STEPP (Bonetti and Gelber 2000)	Freq/NP	Low	P	Yes		No	No	R [stepp]
Multivariable fractional polynomials (Royston and Sauerbrei 2004, 2013)	Freq/NP	Low	P	Yes		No	No	R [mfp]
Interaction trees (Su et al. 2009)	Freq/NP	Medium	S	No	Yes	No	No	B
Modified covariate method (Tian et al. 2014)	Freq/P	High	P	No	Yes	No	No	B
QUINT (Dusseldorp and Mechelen 2014)	Freq/NP	Medium	S	No	Yes	No	No	R [quint]
Gi as part of GUIDE (Loh et al. 2015, 2016)	Freq/NP	Medium	B,S	No	Yes	Yes	Yes	B
Model-based trees and forests (Seibold et al. 2016, 2018)	Freq/NP	High	B,P,S	Yes	Yes	Yes	Yes	R [model4you]
Causal random forests (Wager and Athey 2018)	Freq/NP	High	B,P	No	Yes	No	Yes	R [grf]

Optimal treatment regimes

Biomarker selector (Gunter et al. 2011)	Freq/P	High	B	Yes	Yes	No	No	
Qian and Murphy (2011)	Freq/P	High	P;T	No	Yes	No	No	
AIPWE estimator by Zhang et al. (2012)	Freq/SP	Medium	T	No	Yes	No	No	R [DynTxRegime]
Zhang et al. (2012), Xu et al. (2015)	Freq/P	High	P;T	No	Yes	No	No	R [personalized]
Tree- and list-based ITR (Laber and Zhao 2015; Zhang et al. 2016; Fu et al. 2016)	Freq/NP	Medium	P;T	No	Yes	No	No	B,R [listdtr]

Local modeling

SIDES (Lipkovich et al. 2011);	Freq/NP	Medium	B,S	Yes	Yes	Yes	Yes	B,R [SIDES]
SIDEScreen (Lipkovich and Dmitrienko 2014a)								
Adaptation of PRIM (Chen et al. 2015)	Freq/NP	Medium	S	No	Yes	No	No	R [SubgrpID]
Sequential-Batting (Huang et al. 2017)	Freq/NP	Medium	S	Yes	Yes	Yes	Yes	R [SubgrpID]
TSDT (Battioui et al. 2018)	Freq/NP	Medium	B,S	Yes	Yes	No	Yes	R [TSDT]
Bayesian model averaging (Berger et al. 2014; Bornkamp et al. 2017)	Bayes/NP	Low	S	Yes	Yes	No	Yes	P,R [subtee]

2. Dimensionality of the covariate space: L (low; $p \leq 10$), M (medium; $10 < p \leq 100$), H (high; $p > 100$).
3. Results produced by the method: B (selected biomarkers or biomarker ranking based on variable importance scores that can be used for tailoring), P (predictive scores for individual treatment effects), T (optimal treatment assignment), S (identified subgroups).
4. Assessment of Type I error rate/False discovery rate for the entire subgroup search strategy: Yes/No.
5. Application of complexity control to prevent data overfitting: Yes/No.
6. Control (reduction) of selection bias when evaluating candidate cut-offs: Yes/No.
7. Availability of "honest" estimates of treatment effects (i.e., the estimates are corrected for the optimism bias) in identified subgroups: Yes/No.
8. Availability of software implementation: R ["name"] (R package available on the CRAN website), B (R code or reference to implementation by the developers available on the Biopharmaceutical Network website), P (proprietary).

3.4 Confirmatory Subgroup Analysis

The main goal of this section is to present an overview of the general topic of confirmatory subgroup analysis. The setting that will be assumed in this section includes pivotal Phase III trials with a very small number of prospectively defined populations of patients. The efficacy and safety profiles of a novel treatment will be evaluated in the overall population of patients who meet the trial's inclusion and exclusion criteria, also known as the population of all comers, as well as one or two pre-set subsets of the overall population that will be referred to as subpopulations. These subsets are defined using binary classifiers derived from relevant baseline patient characteristics, e.g., demographic, genetic and clinical characteristics, that are typically identified in earlier Phase II trials or in historical trials (as in Sect. 3.3, these characteristics will be referred to as biomarkers). For example, if a patient characteristic is continuous, a binary classifier is constructed using a cut point and the overall trial population is partitioned into a biomarker-low subgroup (patients with biomarker values below the cut point) and biomarker-high subgroup (patients with biomarker values above the cut point).

A comprehensive characterization of the treatment effect within the pre-defined subpopulations in pivotal trials provides the foundation for tailored therapies and targeted agents (FDA 2019). There are numerous examples of biomarker-driven trials that have employed confirmatory subgroup analysis strategies. These examples include the trastuzumab trials in the subpopulation of patients with breast cancer whose tumors over-expressed HER2 (human epidermal growth factor receptor 2) (Piccart-Gebhart et al. 2005) and the erlotinib trial for the treatment of advanced non-small-cell lung cancer that investigated the treatment's efficacy in the overall population as well as a pre-specified subpopulation of EGFR (epidermal growth factor receptor) immunohistochemistry-positive patients (Cappuzzo et al. 2010).

A broad set of subgroup analysis topics arising in a confirmatory setting will be presented in this section. The section will begin with a discussion of designs commonly used in clinical trials for targeted agents or tailored therapies (Sect. 3.4.1). An overview of multiplicity issues arising in trials with several pre-defined patient populations will be provided in Sect. 3.4.2. This section will introduce basic settings with a single source of multiplicity and more complex settings with two or more sources of multiplicity. Important considerations in biomarker-driven trials such as the development of a decision-making framework that facilitates the interpretation of overall and subgroup effects will be defined in Sect. 3.4.3. Lastly, adaptive designs in multi-population trials with an option to select the most promising patient populations at an interim look will be discussed in Sect. 3.4.4.

3.4.1   Multi-Population Trials

Confirmatory subgroup analysis strategies are employed in pivotal Phase III trials aimed at the development of tailored therapies (Wang et al. 2007; Freidlin et al. 2010). A variety of trial designs have been developed to incorporate biomarker information and facilitate the process of evaluating subgroup effects. We will restrict our attention to the two most common approaches to designing biomarker-driven trials known as *subpopulation-only* designs and *multi-population* designs.

Subpopulation-only designs, also known as enrichment designs, utilize relevant historical data to restrict patient enrollment to a certain subset of the general population. A well-known example of subpopulation-only designs is the development of trastuzumab as a treatment for breast cancer. Only women with HER2-positive tumors were enrolled in the pivotal trials and, within this subpopulation, trastuzumab was shown to be highly beneficial (Piccart-Gebhart et al. 2005). Subpopulation-only designs rely on a strong assumption that the selected biomarker is truly predictive of treatment response and the treatment is only effective in the biomarker-positive subgroup. No treatment benefit is expected or the magnitude of this benefit is believed to be substantially reduced in the biomarker-negative subgroup and thus it is difficult to justify exposing biomarker-negative patients to the treatment.

Multi-population designs, also known as biomarker-stratified designs, serve as a viable alternative to subpopulation-only designs. With the multi-population approach, all patients with the condition of interest are enrolled in a trial but the analysis strategy is set up in such a way that the treatment effect is evaluated in the broad overall population as well as in one or more subpopulations based on pre-defined classifiers. These classifiers are set up using biomarkers with strong predictive properties and thus the efficacy signal is expected to be greater in these subpopulations compared to the trial's overall population. An advantage of multi-population trials is that they enable the trial's sponsor to provide a comprehensive characterization of the efficacy and safety properties of an experimental treatment across the entire population of interest.

The SATURN trial (Cappuzzo et al. 2010) mentioned above serves as an example of a multi-population tailoring approach. This trial was conducted to investigate the advantages of erlotinib in patients with advanced non-small-cell lung cancer. A twofold efficacy objective was pursued in this trial, namely, the effect of erlotinib on progression-free survival was evaluated in the overall population of patients and a subset of biomarker-positive patients, i.e., patients with an EGFR-positive status. This biomarker-positive subpopulation was incorporated into the primary analysis to evaluate improved potential benefit from erlotinib compared to the subpopulation of biomarker-negative patients. Other examples of multi-population designs will be given later in this chapter.

In what follows we will go over key challenges arising in multi-population trials, including control of the Type I error rate (Sect. 3.4.2), decision-making considerations (Sect. 3.4.3) and trial design considerations (Sect. 3.4.4).

3.4.2 Multiplicity Issues in Multi-Population Trials

One of the most important objectives of confirmatory subgroup evaluations in pivotal Phase III trials is control of the Type I error rate since multiplicity corrections are mandatory in pivotal trials with multiple clinical objectives (FDA 2017b; EMA 2017). In the context of multi-population designs, there are multiple opportunities to claim treatment effectiveness, namely, the trial can demonstrate beneficial treatment effect in the overall patient population (this claim is often referred to as a *broad effectiveness claim*) or in one of several target subpopulations (these claims are known as *restricted or tailored effectiveness claims*). These claims are independent of each other and a claim in any patient population can form the basis for a regulatory submission for the experimental treatment. In particular, if a beneficial treatment effect is established in a subpopulation, the trial's sponsor can pursue a claim in this subpopulation even if the treatment provides no benefit in the overall population. A multiplicity adjustment must be pre-defined to preserve the overall Type I error rate across the null hypotheses associated with the individual claims. For more information on multiplicity corrections used in pivotal trials and key principles that guarantee error rate control, e.g., the *closure principle*, see Dmitrienko and D'Agostino (2013, 2018).

Dozens of multiplicity adjustments, also known as multiple tests, have been developed in the literature and it is helpful to organize information on the available options by considering the following classification scheme that incorporates information on *logical relationships* among the individual clinical objectives in a multi-population trial and *distributional information*.

The first dimension of this classification scheme deals with the available information on logical restrictions among the null hypotheses corresponding to the overall population and subpopulations. The most basic multiplicity adjustments, e.g., the Bonferroni test, examine each null hypothesis independently of the other null hypotheses and are known as *single-step tests*. More efficient tests that result in a higher overall probability of success belong to the class of *stepwise tests*. These tests rely on sequential testing algorithms. For example, the order in which the null hypotheses are to be tested may be pre-defined and a stepwise test is applied to carry out the hypothesis tests according to this pre-defined hypothesis ordering. Examples of stepwise tests with a pre-defined hypothesis ordering include the fixed-sequence and fallback tests. On the other hand, the null hypotheses may be tested in the order determined by the significance of the hypothesis test statistics. The Holm and Hochberg tests serve as examples of stepwise tests that are based on a data-driven hypothesis ordering. As will be shown below, stepwise tests of this kind are more flexible than, say, the fixed-sequence test and are generally recommended in multi-population clinical trials.

Secondly, when selecting the most efficient method for handling multiplicity in a trial, it is important to fully utilize the available information on the joint distribution of the hypothesis test statistics associated with the hypotheses of interest. Some of popular multiple tests such as the Bonferroni test, fixed-sequence test and a family of chain tests, also known as graphical methods, do not make any assumptions on the joint distribution of the test statistics. These tests, known as *nonparametric tests*, are uniformly less powerful than *semiparametric tests* (e.g., Hochberg test) or *parametric tests* (e.g., Dunnett test). The reason for this is that semiparametric tests make some distributional assumptions, e.g., they control the overall Type I error rate under the assumption that the test statistics follow a multivariate normal distribution with non-negative pairwise correlations, and parametric tests require a full specification of the joint distribution. The joint distribution of the hypothesis test statistics is typically well characterized in multi-population trials and therefore it is advisable to employ semiparametric or parametric tests in confirmatory subgroup analysis.

To review the recommended multiplicity adjustment strategies, we will begin with the discussion of traditional multiplicity problems which can be thought of as "univariate" multiplicity problems, e.g., problems where multiplicity is induced by the evaluation of treatment effects in several patient populations and a single family of null hypotheses of no effect is set up. Advanced settings with "multivariate" multiplicity problems, i.e., problems with several families of null hypotheses, will be considered later in this section. In addition, we will discuss a structured approach to identifying the multiplicity corrections that perform best in a given multi-population trial and are also robust against deviations from the (potentially optimistic) original treatment effect assumptions.

3.4.2.1   Traditional Multiplicity Problems

Two important approaches to handling multiplicity in a traditional setting with a single family of null hypotheses will be compared and contrasted below. Consider, for simplicity, a trial with a single pre-specified subgroup and let H_0 and H_+ denote the null hypotheses of no treatment effect in the overall population and biomarker-positive subpopulation, respectively. These hypotheses will be evaluated as part of the trial's primary analysis. The hypothesis of no effect in the biomarker-negative subpopulation, denoted by H_-, can be considered as well but it will be treated as a secondary hypothesis. The raw or marginal treatment effect p-values in the overall population and biomarker-positive subpopulation will be denoted by p_0 and p_+, respectively, and a multiplicity correction will be applied to control the overall Type I error rate at a two-sided $\alpha = 0.05$ in the strong sense.

Even though basic nonparametric multiple tests are known to be inefficient, they are still often employed in multi-population clinical trials. Most commonly, the fixed-sequence test is applied which requires that the hypothesis ordering should be pre-defined. Depending on the testing sequence, the following two strategies can be considered

- Strategy 1. The overall population effect is evaluated first followed by the subpopulation effect. In this case, the hypothesis H_0 is rejected if $p_0 \leq \alpha$. If H_0 is rejected, the hypothesis H_+ is tested and it is rejected provided $p_+ \leq \alpha$.
- Strategy 2. The subpopulation effect is evaluated first followed by the overall population effect. The hypothesis H_+ is rejected if $p_+ \leq \alpha$. If this hypothesis is rejected, H_0 is rejected if $p_0 \leq \alpha$.

The fixed-sequence testing approach is easily extended to clinical trials with two or more pre-defined subpopulations.

An important feature of Strategy 1 is that it implicitly assumes a fairly strong treatment effect in the biomarker-negative subpopulation. If this effect is weak, i.e., the selected biomarker is a strong predictor of treatment benefit, the overall effect will be attenuated and the probability of passing the gatekeeper, i.e., the probability of rejecting H_0, will likely be low. On the other hand, Strategy 2 relies on an assumption of a pronounced treatment effect in the biomarker-positive subpopulation. This assumption will not be satisfied if the selected biomarker is a weak predictor of treatment response or even completely non-informative. The following two examples illustrate potential weaknesses of fixed-sequence testing strategies.

The PRIME trial (Douillard et al. 2014) was conducted to evaluate the efficacy of panitumumab in combination with FOLFOX4 chemotherapy as a first-line treatment for patients with metastatic colorectal cancer. Patients in the control arm received FOLFOX4 alone. The primary efficacy evaluation in the trial was performed using progression-free survival. A binary classifier based on an oncogene known as KRAS played a key role in this trial. The KRAS status (wild-type versus mutated) has been used in multiple metastatic colorectal cancer trials as a strong predictor of anti-EGFR monoclonal antibody efficacy. A multi-population design stratified by this biomarker was utilized in the PRIME trial because panitumumab is an anti-EGFR therapy and a greater effect on progression-free survival was expected in the subset of patients with wild-type KRAS status. The biomarker-positive and biomarker-negative subsets were defined as follows:

- Biomarker-positive subpopulation: Patients with wild-type KRAS status.
- Biomarker-negative subpopulation: Patients with KRAS mutations.

Table 3.4 PRIME trial results

Population	Hazard ratio for PFS (95% CI)	Two-sided p-value
Biomarker-positive subpopulation	0.80 (0.67, 0.95)	0.01
Biomarker-negative subpopulation	1.27 (1.04, 1.55)	0.02

A summary of PFS findings in the biomarker-positive and biomarker-negative subsets is provided in Table 3.4. The table shows the subpopulation-specific hazard ratios along with two-sided treatment effect p-values computed from the log-rank test. The experimental treatment significantly improved progression-free survival for patients with a biomarker-positive status but no improvement was observed in the complementary subpopulation. In fact, it follows from Table 3.4 that the two-sided p-value in the biomarker-negative subset was significant at a 0.05 level, which means that the experimental treatment was significantly worse than the control in patients with KRAS mutations.

Note that about 40% of the patients in the PRIME trial were biomarker-negative. If a multiplicity adjustment based on Strategy 1 had been used in this trial, it would have most likely led to a negative outcome since the treatment effect in the overall population was reduced due to a strong negative trend in biomarker-negative patients. Since the probability of establishing a significant overall effect in this and similar settings is low, the null hypothesis H_+ simply will not be tested most of the time (recall that H_+ cannot be tested and is automatically accepted unless H_0 is rejected) and a promising treatment effect in the biomarker-positive subpopulation will be missed.

The next example is based on the APEX trial (Cohen et al. 2016) which was conducted in the population of patients at risk for venous thrombosis to evaluate the efficacy and safety of a novel treatment (betrixaban) compared to an active control. The primary efficacy endpoint was a composite of several binary events (deep-vein thrombosis, nonfatal pulmonary embolism or death from venous thromboembolism). The APEX trial employed a multi-population design with two pre-defined subpopulations based on the D-dimer level at baseline and patient's age:

- Biomarker-positive subpopulation 1: Patients with an elevated D-dimer level.
- Biomarker-positive subpopulation 2: Patients with an elevated D-dimer level or patients who are older than 75 years.

The two biomarkers were believed to be predictive of treatment benefit; however, both biomarkers turned out to be non-informative. Table 3.5 presents a summary of the trial results and shows that the relative risk of primary events compared to placebo (betrixaban was expected to reduce the risk of vein thrombosis, pulmonary embolism or death) was virtually constant across the patient populations. The table also presents the two-sided p-values within each patient population that were computed using the Cochran-Mantel-Haenszel test. Strategy 2 was employed in this trial to address multiplicity induced by the analysis of the three populations.

Table 3.5 APEX trial results

Population	Relative risk (95% CI)	Two-sided p-value
Biomarker-positive subpopulation 1	0.81 (0.65, 1.00)	0.054
Biomarker-positive subpopulation 2	0.80 (0.66, 0.98)	0.030
Overall population	0.76 (0.63, 0.92)	0.006

Testing started with the first test in the sequence, i.e., with Biomarker-positive subpopulation 1, but there was not enough evidence to reject the corresponding null hypothesis of no treatment benefit since the p-value was greater than 0.05. Consequently, all three null hypotheses were accepted even though the overall effect was both clinically and statistically significant ($p = 0.006$).

The two examples based on the PRIME and APEX trials show that both fixed-sequence testing strategies (Strategy 1 and Strategy 2) make difficult-to-justify assumptions. If these assumptions are violated, the trial's overall outcome could be negative despite the fact that a subpopulation effect or even the overall effect is significant. An alternative approach to handling multiplicity in a multi-population trial is to apply an efficient multiple test with a data-driven testing sequence, e.g., the Hochberg test. This test is a member of the semiparametric class and is uniformly more powerful than nonparametric tests such as the Bonferroni and chain tests. The Hochberg test treats the null hypotheses of interest, e.g., the hypotheses H_0 and H_+ in a trial with two patient populations, as interchangeable rather than hierarchically ordered.

In a general setting with m null hypotheses denoted by H_1 through H_m, let $p_{(1)} < \ldots < p_{(m)}$ denote the ordered p-values and let $H_{(1)}, \ldots, H_{(m)}$ denote the corresponding ordered null hypotheses. The Hochberg test is carried out using the following algorithm:

- Step 1. Accept $H_{(m)}$ if $p_{(m)} > \alpha$. If $H_{(m)}$ is accepted, proceed to Step 2. Stop testing and reject all null hypotheses otherwise.
- Step 2. Accept $H_{(m-1)}$ if $p_{(m-1)} > \alpha/2$. If $H_{(m-1)}$ is accepted, proceed to Step 3. Stop testing and reject the remaining null hypotheses otherwise.
- Steps $i = 3, \ldots, m - 1$. Accept $H_{(m-i+1)}$ if $p_{(m-i+1)} > \alpha/i$. If $H_{(m-i+1)}$ is accepted, proceed to Step $i + 1$. Stop testing and reject the remaining null hypotheses otherwise.
- Step m. Reject $H_{(1)}$ if $p_{(1)} \leq \alpha/m$.

It is easy to verify that, if the Hochberg test had been applied to the family of three null hypotheses in the APEX trial, the null hypothesis corresponding to the overall population would have been rejected and thus a significant treatment effect would have been established in the all-comers population.

The clinical trial examples presented above demonstrate the strengths and weaknesses of popular multiplicity corrections in the evaluation of treatment effects in multi-population trials. A formal framework for assessing the performance of candidate multiple tests and identifying the best performing multiplicity adjustment in a particular trial will be presented later in this section.

3.4.2.2   Advanced Multiplicity Problems

A class of more complex multiplicity problems, commonly referred to as advanced multiplicity problems, will be briefly discussed in this subsection. This class is characterized by the fact that several sources of multiplicity are present. In addition to the analysis of treatment effects in several patient populations, which defines the first source of multiplicity, multiplicity may also be induced by the evaluation of several efficacy endpoints or assessment of several doses. Multiple testing procedures known as *gatekeeping procedures* are commonly used to address multiplicity in these more challenging settings. Gatekeeping procedures enable trial sponsors to protect the overall Type I error rate in complex multiplicity problems where the null hypotheses of interest are arranged into families to account for the relative importance of the clinical goals corresponding to the individual sources of multiplicity.

Applications of gatekeeping procedures to multivariate multiplicity problems arising in multi-population trials have been discussed in several publications. As an example, Dmitrienko et al. (2011) discussed the principles that govern the selection of gatekeeping procedures in clinical trials with two dose-placebo comparisons (Dose 1 versus placebo and Dose 2 versus placebo) and two patient populations (overall population and subpopulation of biomarker-positive patients). The resulting null hypotheses of no effect are grouped into two families as follows:

- Family 1: H_1 (null hypothesis of no difference between Dose 1 and placebo in the overall population) and H_2 (null hypothesis of no difference between Dose 2 and placebo in the overall population).
- Family 2: H_3 (null hypothesis of no difference between Dose 1 and placebo in the subpopulation) and H_4 (null hypothesis of no difference between Dose 2 and placebo in the subpopulation).

As shown in Fig. 3.5, by creating these two families of null hypotheses, the trial's sponsor can take into account the hierarchical structure of effectiveness claims in the overall population and pre-defined subpopulation. The family of null hypotheses

Fig. 3.5 Two families of null hypotheses in the clinical trial with two dose-placebo comparisons and two patient populations

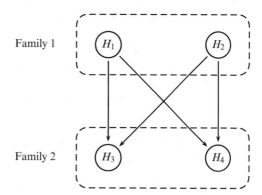

of no treatment benefit in the overall population is examined first and thus it is treated as the primary family. The family of null hypotheses in the subpopulation is positioned as a secondary family. The treatment effect will be evaluated in Family 2 only if there is evidence of a significant effect in the overall population at either Dose 1 or Dose 2. If neither hypothesis is rejected in Family 2, the sponsor will be limited to an effectiveness claim in the overall patient population. However, if a significant effect is established in the subpopulation at either dose level, the sponsor can consider a composite claim of treatment effectiveness in the overall population as well as subpopulation of biomarker-positive patients that will be discussed in more detail in Sect. 3.4.3.

An additional important aspect of this two-family multiplicity problem is the specification of logical relationships among the four null hypotheses. In this particular case, the secondary null hypotheses (H_3 or H_4) will be tested if at least one primary null hypothesis (H_1 or H_2) is rejected. This means that Family 1 serves as a *parallel gatekeeper* for Family 2. More complex relationships among the null hypotheses of interest can be considered to ensure that these relationships are fully aligned with the trial's clinical objectives.

The mixture method (Dmitrienko and Tamhane 2011, 2013) has been successfully applied to multiple clinical trials to develop powerful and flexible gatekeeping procedures in multiplicity problems with several families of null hypotheses. This general method supports arbitrary logical relationships among the null hypotheses and, unlike simple Bonferroni-based tests such as chain tests, it enables trial sponsors to efficiently incorporate available distributional information by utilizing semiparametric or fully parametric tests within the individual families. The mixture method was applied in Dmitrienko et al. (2011) to build several candidate gatekeeping procedures for the two-family problem displayed in Fig. 3.5. The authors performed a simulation-based assessment of the candidate procedures and concluded that the most efficient multiplicity adjustment strategy is the one that relies on Hochberg-based gatekeeping procedures. A general Hochberg-based gatekeeping procedure is constructed as follows:

- Powerful Hochberg-type tests are applied to the null hypotheses within each family. Hochberg-type tests account for positive correlations between the hypothesis test statistics within each family, which is induced by the fact that the subpopulation is nested within the overall population. This results in a uniform power gain compared to gatekeeping procedures that rely on nonparametric tests, e.g., Bonferroni-based gatekeeping procedures.
- A Hochberg-type test, known as the truncated Hochberg test, is applied to the null hypotheses in Family 1 since this family serves as a gatekeeper for the other family. The regular Hochberg test is applied in Family 2 due to the fact it is the last family in the sequence. Since a truncated Hochberg test is used in Family 1, this gatekeeping procedure can proceed to Family 2 even if only one overall population test is significant.

It needs to be noted that Hochberg-based gatekeeping procedures control the overall Type I error rate only if the test statistics within each family follow a multivariate normal distribution with non-negative pairwise correlation coefficients (Dmitrienko and D'Agostino 2013). This condition is satisfied in this clinical trial example as well as in advanced multiplicity problems arising in other multi-population trials since nested subpopulations induce non-negative correlations.

3.4.2.3   Clinical Scenario Evaluation Approach

It was pointed earlier in this section that numerous classes of multiple tests or gatekeeping procedures have been proposed in the literature. To choose the best multiplicity adjustment strategy for a particular multiplicity problem, it is important to review the applicable strategies as well as relevant information on logical relationships among the null hypotheses of interest and joint distribution of the hypothesis test statistics. This process can be facilitated by utilizing the Clinical Scenario Evaluation (CSE) framework that was introduced in Benda et al. (2010).

The general CSE framework encourages trial sponsors to perform a structured quantitative assessment of candidate trial designs and analysis methods in clinical drug development. The following components play a central role within this framework:

- *Data models* define the assumptions for the process of generating trial data.
- *Analysis models* define the analysis strategies applied to the trial data generated from the data models.
- *Evaluation models* define the criteria/metrics for evaluating the performance of the analysis strategies applied to the trial data.

By carefully building plausible data models, applicable analysis models and evaluation models that are based on clinically relevant metrics, a trial's sponsor can perform a comprehensive evaluation of the advantages and disadvantages of selected analysis methods and assess their sensitivity to potential deviations from the original treatment effect assumptions.

This disciplined approach could be used to identify powerful multiplicity adjustments that exhibit a robust performance across multiple sets of assumptions unlike suboptimal tests such as the fixed-sequence test employed in the APEX trial. The emphasis on optimality is not a coincidence since the CSE framework provides a foundation for developing optimal approaches to selecting trial designs and analysis methods in clinical trials. Clinical trial optimization strategies based on the CSE principles have been studied in several recent publications, e.g., an overview of CSE-based approaches to finding multiplicity adjustments that optimize the clinically relevant evaluation criterion in the context of confirmatory subgroup analysis can be found in Dmitrienko and Paux (2017). The following recommendations were formulated in this publication:

- Data models should reflect a broad range of assumptions, including optimistic and pessimistic sets of treatment effect assumptions. For example, scenarios corresponding to strong and weak differential treatment effects in the biomarker-positive and biomarker-negative subpopulations should be considered.

- Analysis models should incorporate several applicable multiple tests, including basic non-parametric tests such as the fixed-sequence test and more powerful semiparametric and parametric tests.
- Evaluation models should include commonly used evaluation criteria such as disjunctive power, i.e., the probability that at least one null hypothesis is rejected, as well other criteria, e.g., weighted power based on a weighted sum of marginal power functions. Note that disjunctive power tends to be driven mostly by the probability of success in the overall population and thus this evaluation criterion may not be particularly meaningful in a multi-population setting.
- Both qualitative and quantitative sensitivity assessments should be performed to assess the robustness of the optimal multiple test against deviations from the most likely set of treatment effect assumptions, known as the main data model. The qualitative approach focuses on sensitivity assessments based on a small number of data models that correspond to clinically distinct treatment effect scenarios. With the quantitative approach, the "robustness profile" of the selected multiple test is examined using a large set of data models that are only quantitatively different from the main model (these models are obtained using random perturbations of the main data model).

These principles provide a useful template for performing CSE-based assessments of multiplicity corrections in multi-population clinical trials.

As an application of the recommendations presented above, Paux and Dmitrienko (2018) performed a comparison of candidate multiple tests in late-stage clinical trials using novel evaluation criteria. The proposed approach relied on penalty-based criteria that facilitate "head-to-head" comparisons of candidate tests to identify the best performing multiplicity adjustment in a particular trial. Using a case study based on the APEX trial, the authors evaluated the performance of the fixed-sequence and Hochberg procedures in the problem of testing the three null hypothesis of interest, i.e.,

- H_1: Null hypothesis of no treatment effect in Biomarker-positive subpopulation 1.
- H_2: Null hypothesis of no treatment effect in Biomarker-positive subpopulation 2.
- H_3: Null hypothesis of no treatment effect in Overall population.

This evaluation included a qualitative sensitivity assessment based on four data models or treatment effect scenarios that corresponded to a range of assumptions about the predictive strength of the two biomarkers employed in the trial. For example, the underlying biomarkers were assumed to be strongly predictive of treatment response under Scenario 1, weak predictive effects were assumed under Scenarios 2 and 3 and lastly the biomarkers were non-informative under Scenario 4, i.e., a homogeneous treatment effect was assumed across the subgroups defined by these biomarkers in this case. The probability that each test is superior to the other test was computed under the selected data models (a multiple test is superior to another test if it rejects all null hypotheses rejected by the other test and may also reject additional null hypotheses). Under Scenario 1, the probability that the Hochberg procedure would be superior to the fixed-sequence procedure was 15% and the probability that the fixed-sequence procedure would be superior to the Hochberg procedure was only 1%. When less optimistic assumptions about the predictive strength of the biomarkers were considered, e.g., Scenarios 2, 3 and 4, the gap between the two procedures widened. In particular, the probability that the Hochberg procedure would be superior to the fixed-sequence procedure increased to 25% and the probability that the fixed-sequence procedure would be superior to the Hochberg procedure was less than 1% under Scenario 4. Note that the most meaningful hypothesis to reject under this scenario is H_3 and the Hochberg procedure correctly rejected the hypothesis most of the time while the fixed-sequence procedure failed to reject H_3. This was due to the fact that the fixed-sequence procedure stopped at the very first hypothesis in the sequence (H_1), which means that the hypotheses H_2 and H_3 were accepted without testing, as explained above. The fixed-sequence test was clearly inferior to the Hochberg test across the selected data models in this CSE exercise, which suggests that multiple tests with a pre-defined hypothesis ordering should be avoided in multi-population clinical trials.

3.4.3   Decision-Making Framework in Multi-Population Trials

In general, multi-population tailoring trials provide an important advantage over single-population trials, e.g., trials that are restricted to biomarker-positive patients or trials that enroll all patients but focus on the evaluation of the overall effect. However, since multiple objectives are pursued within a single trial, the trial's sponsor needs to carefully interpret the outcome of each population test to ensure logical decision making in a multi-population setting.

The importance of creating a statistical framework for decision making in clinical trials that investigate treatment effects in several patient populations has been stressed in recent publications, see, for example, Rothmann et al. (2012) and Millen et al. (2012). This framework addresses additional possibilities to commit errors that go beyond the multiplicity framework that emphasizes Type I error rate control. These errors are related to the broad and restricted effectiveness claims defined at the beginning of Sect. 3.4.2. Consider, for simplicity, a clinical trial with a single pre-defined subpopulation, then the following logical errors could occur:

- If a trial is conducted to pursue the broad claim, an *influence error* will be committed if the trial's sponsor claims effectiveness in the overall population but this conclusion is purely driven by a strong treatment effect in the subpopulation.
- If the sponsor is interested in a composite claim (the broad and restricted claims could be pursued simultaneously), an *interaction error* will be committed if

Table 3.6 Lapatinib trial results

Population	Hazard ratio for PFS (95% CI)	Two-sided p-value
Overall population	0.86 (0.76, 0.98)	0.026
Biomarker-positive subpopulation	0.71 (0.53, 0.96)	0.019
Biomarker-negative subpopulation	0.90 (0.77, 1.05)	0.188

the sponsor concludes that both claims are valid when in reality there is no differential treatment effect and the treatment is equally effective in biomarker-positive and biomarker-negative patients.

We will use the Phase III trial that was conducted to develop a first-line therapy for postmenopausal hormone receptor (HR)-positive metastatic breast cancer (Johnston et al. 2009) to discuss the relevance of the influence error in multi-population trials. In this trial, the combination of lapatinib and letrozole was tested versus letrozole and placebo with the primary endpoint based on progression-free survival. Within the overall population of HR-positive patients, the following two subsets were prospectively defined:

- Biomarker-positive subset: Patients with HER2-positive tumors.
- Biomarker-negative subset: Patients with HER2-negative tumors.

A summary of the PFS assessments in the overall trial population as well as the pre-defined subsets is presented in Table 3.6. The table shows that a significant reduction in the risk of disease progression was observed within the biomarker-positive subpopulation with the hazard ratio of 0.71 (in fact, the median PFS increased from 3 months to 8.2 months). However, the improvement in progression-free survival within the subset of biomarker-negative patients was not considered clinically meaningful. The observed hazard ratio was 0.9 with a rather wide 95% confidence interval. Formally, the overall effect of the experimental treatment was significant with a two-sided p-value of 0.026 but the trial's sponsor could have committed an influence error if the broad effectiveness claim had been pursued in this trial. It could not be appropriate to recommend the experimental treatment for all HR-positive patients since the treatment does not appear to be effective in patients with HER2-negative tumors.

The concept of the interaction error can be illustrated using the KEYNOTE-010 trial in patients with advanced non-small-cell lung cancer (Herbst et al. 2016). Patients in this trial were randomly allocated to two regimens of an experimental treatment (pembrolizumab) or control (docetaxel). The primary analysis focused on overall survival benefits in the all-comers population and efficacy assessments were also performed in the following two subsets:

- Biomarker-positive subset: Patients with the tumor proportion score greater than or equal to 50%.
- Biomarker-negative subset: Patients with the tumor proportion score between 1% and 49%.

Table 3.7 KEYNOTE-010 results

Population	Hazard ratio for OS (95% CI)
Overall population	0.67 (0.56, 0.80)
Biomarker-positive subpopulation	0.53 (0.40, 0.70)
Biomarker-negative subpopulation	0.76 (0.60, 0.96)

The tumor proportion score was used to quantify PD-L1 protein expression. PD-L1 (programmed death-ligand 1) is an immune-related biomarker that can be expressed on tumor cells.

The results of the pooled analysis of overall survival data in the two pembrolizumab regimens versus docetaxel are presented in Table 3.7. We can see from the table that the experimental treatment was highly beneficial within the biomarker-positive subpopulation with the hazard ratio of 0.53. A slightly weaker effect was achieved in the complementary subpopulation but the treatment could still be viewed as beneficial. As a consequence, this setting could be used as an example of a potential interaction error. If the sponsor were to use the trial's results to justify a composite claim of effectiveness in the overall population and biomarker-positive subpopulation, one could argue that this conclusion is not logical since the efficacy signal may in fact be quite similar in the biomarker-positive and biomarker-negative subpopulations.

To address these considerations, Millen et al. (2012) developed a general inferential framework for multi-population trials that relies on the influence and interaction conditions. The first of these conditions states that, to achieve the broad claim, the sponsor should demonstrate that the beneficial effect of a treatment is not limited to the biomarker-positive subpopulation and a meaningful benefit is present in the complementary subpopulation of biomarker-negative patients. The interaction condition states that, to support the composite claim, the sponsor should establish a differential treatment effect in the sense that the treatment effect in the biomarker-positive subpopulation should be appreciably greater than that in the biomarker-negative subpopulation. We will introduce the influence and interaction conditions in the context of fixed-design trials with a pre-defined number of patients or events. This decision-making framework will also play an important role in clinical trials that support adaptive population selection (Sect. 3.4.4).

The influence and interaction conditions could be formulated using either frequentist or Bayesian principles. In a more straightforward frequentist setting, these conditions are defined as follows. Let $\widehat{\theta}_-$ and $\widehat{\theta}_+$ denote the estimated effect sizes in the biomarker-negative and biomarker-positive subpopulations, respectively. Effect sizes are defined as standardized treatment effects, e.g., the effect size is equal to the negative log-hazard ratio in a trial with a time-to-event endpoint.

The influence condition is met and thus the broad effectiveness claim is justified if

$$\widehat{\theta}_- \geq \lambda_{\text{INF}},$$

where λ_{INF} is a pre-specified threshold which could be equal to the minimal clinically important difference. For example, if a clinically relevant effect corresponds to hazard ratios below 0.9, the influence condition threshold can be set to $\lambda_{INF} = -\log 0.9 = 0.105$. If the influence condition is not met, inference from the trial will be limited to the biomarker-positive subpopulation. Returning to the lapatinib trial (see Table 3.6), it is easy to see that the influence condition with $\lambda_{INF} > 0.015$ would have guided the sponsor to the conclusion that it is most meaningful to claim treatment effectiveness in the subset of biomarker-positive patients.

If the trial's sponsor plans to purse the composite claim and the possibility of an influence error has been ruled out, the next step is to check the interaction condition. This condition is met if

$$\widehat{\theta}_+/\widehat{\theta}_- \geq \lambda_{INT},$$

where λ_{INT} is a pre-specified interaction condition threshold. This threshold is greater than 1 and defines a lower bound on the magnitude of desirable differential treatment effect. As a quick example, if $\lambda_{INT} = 1.4$, the effect size in the biomarker-positive subpopulation should be 40% greater than that in the biomarker-negative subpopulation to conclude that a meaningful differential effect is present in the trial. If the interaction condition with $\lambda_{INT} = 1.4$ had been applied to the KEYNOTE-010 trial, this condition would have been met since

$$\widehat{\theta}_+/\widehat{\theta}_- = (-\log 0.53)/(-\log 0.76) = 2.3$$

is clearly greater than 1.4. This indicates that the differential treatment effect in the KEYNOTE-010 trial is quite strong and the composite claim of treatment effectiveness in the overall population and biomarker-positive subpopulation could be justified.

The frequentist formulation of the influence and interaction conditions is easily extended using Bayesian arguments (Millen et al. 2014). For example, within the Bayesian framework, the influence condition is satisfied if the posterior probability of the event of interest, i.e., $\widehat{\theta}_- \geq \lambda_{INF}$, is sufficiently high.

It is helpful to compare the decision rules based on the influence condition to those employed in the marker sequential test (MaST) proposed by Freidlin et al. (2014). This test is closely related to the fallback test but its main focus is the formulation of decision-making criteria, i.e., criteria for selecting an appropriate patient population for a tailored therapy. The corresponding decision rules are formulated in terms of p-values and incorporate a simple nonparametric multiplicity adjustment. Let $0 < \alpha_1 < \alpha$ denote the fraction of the Type I error rate assigned to the treatment effect test in the biomarker-positive subpopulation ($\alpha = 0.05$). The remaining error rate, i.e., $\alpha_2 = \alpha - \alpha_1$, is allocated to the overall population test. Also, as in Sect. 3.4.2, let p_0, p_+ and p_- denote the treatment effect p-values in the overall population, biomarker-positive subpopulation and biomarker-negative subpopulation, respectively. The MaST-based decision rules are set up as follows:

- Broad effectiveness claim is considered if either of the following two conditions is met:

  - Condition 1. The treatment effect is simultaneously significant in the biomarker-positive and biomarker-negative subpopulations, namely, if $p_+ \leq \alpha_1$ and $p_- \leq \alpha$.
  - Condition 2. The treatment effect is significant in the biomarker-positive subpopulation ($p_+ \leq \alpha_1$) and overall populations ($p_0 \leq \alpha_2$).

- Restricted effectiveness claim is considered if

  - Condition 3. The treatment effect is significant in the biomarker-positive subpopulation ($p_+ \leq \alpha_1$) but insignificant in the biomarker-negative subpopulation ($p_- > \alpha$).

It is easy to see that the resulting decision rules are not consistent with the rules based on the influence condition. Beginning with Condition 1, it is logical to recommend the broad claim if there is evidence of a significant treatment effect within the biomarker-negative subpopulation, i.e., if $p_- \leq \alpha$. This condition appears to be extremely stringent since establishing significance within a subset where the treatment effect is not expected to be strong is a very high hurdle. To illustrate this point, let us return to the summary of the lapatinib trial results in Table 3.6. When we discussed these results, we pointed out that the treatment effect in biomarker-negative patients is weak since the hazard ratio is close to 1. Suppose that the minimal clinically important effect in this trial corresponds to the hazard ratio of 0.85 and the influence condition threshold is defined based on this hazard ratio. According to the influence condition, the broad effectiveness claim would be recommended if the hazard ratio in the biomarker-negative subpopulation was less than 0.85. Switching to Condition 1, the broad effectiveness claim would be considered only if the treatment effect was significant at a two-sided 0.05 level within the biomarker-negative subpopulation, i.e., if the hazard ratio was less than 0.75, which is a much higher hurdle. If the MaST-based decision rules had been applied to the lapatinib trial, the efficacy signal in the biomarker-negative subpopulation would have to be almost as strong as that in the biomarker-positive subpopulation to support a claim in the overall population. Next, Condition 2 simply relies on the significance of the treatment effect tests within the overall population and biomarker-positive subpopulation and it does not directly address the problem at hand. The fact that $p_+ \leq \alpha_1$ and $p_0 \leq \alpha_2$ does not provide any evidence to support the broad claim since the significance of the overall p-value is likely to be driven by the significance of the treatment effect p-value in the biomarker-positive subpopulation. As a result, the probability of incorrect recommendations may be quite high if the MaST-based decision rules are applied.

By contrast, the approach defined earlier in this section is easily combined with a multiple test to create a set of rules that protect the Type I error rate and support logical decision making in a multi-population setting:

- Step 1. Apply an efficient and robust multiplicity adjustment which can be found using the CSE approach outlined in Sect. 3.4.2.
- Step 2. If only one test (overall population test or subpopulation test) is significant, the effectiveness claim is selected in a straightforward way. If both tests are significant, apply the influence condition and, if appropriate, the interaction condition to formulate an appropriate effectiveness claim.

3.4.4 Adaptive Designs in Multi-Population Trials

So far we have discussed multi-population trials with a fixed design where the total sample size or target number of events is pre-specified. The multi-population framework has been successfully extended to trial designs with data-driven decision rules that enable the trial's sponsor to select the most promising patient population at an interim look. These designs are known as *adaptive population selection designs* or *adaptive enrichment designs* (FDA 2018). In this section we will discuss well-established adaptive designs aimed at evaluating treatment benefits in a set of subpopulations that are defined at the trial planning stage as well as more advanced designs that are built around subpopulations selected at an interim analysis.

3.4.4.1 Adaptive Trials with Pre-planned Subpopulations

This section provides an overview of adaptive approaches to designing clinical trials aimed at evaluating the efficacy and safety of new treatments in several pre-planned patient populations. Each population is set up using a prospectively defined binary classifier, which means that a cut point must be pre-specified for every continuous biomarker. A multi-population adaptive trial can be designed to evaluate the treatment effect within each pre-planned population at an interim look and identify the most promising populations that will be examined at the final analysis. Furthermore, the number of patients or events can be appropriately increased in the overall population or within a subpopulation if the treatment effect at the final analysis is projected to be borderline non-significant. An important feature of adaptive population selection trials is that they are *inferentially seamless*, i.e., the final analysis is conducted within the selected populations using all relevant data (data collected before and after the decision point at which these patient populations were chosen).

This adaptive approach is appealing in settings where there is enough evidence to justify an assumption of a differential treatment effect, i.e., to assume that the treatment would be more effective in biomarker-positive patients compared to the biomarker-negative subpopulation, and it may be premature to rule out a clinically relevant treatment benefit in the biomarker-negative subpopulation. In a sense, adaptive designs with a provision to choose the most appropriate patient population provide a compromise between subpopulation-only designs that restrict patients enrollment to a pre-defined subset of the overall population and multi-population designs that are open to all comers.

Increasingly more sophisticated adaptive designs are available to support the investigation of subgroup effects in pivotal trials. These designs incorporate flexible data-driven decision rules and enable valid statistical inferences that guarantee Type I error rate control. Adaptive population selection designs have been applied to dozens of pivotal trials and a detailed description of key methodological and operational considerations can be found in Bretz et al. (2006) and Wassmer and Brannath (2016).

When considering the available options for selecting decision rules in adaptive population selection trials, it is important to differentiate between population selection rules and hypothesis selection rules. Suppose, for example, that an adaptive trial needs to be designed to investigate the effects of a novel treatment in two patient populations (overall population and a subpopulation of biomarker-positive patients). As in Sect. 3.4.2, the null hypotheses of no effect in the overall population and biomarker-positive subpopulation will be denoted by H_0 and H_+, respectively. The following population selection rule can be applied at an interim look in this trial:

- Decision 1. Select the overall population, i.e., continue enrolling all patients after the interim analysis.
- Decision 2. Select the biomarker-positive subpopulation, i.e., continue enrolling biomarker-positive patients only after the interim analysis (biomarker-negative patients will no longer be enrolled).
- Decision 3. Select no population, i.e., terminate the trial at the interim analysis.

Depending on the selected patient population, the trial's sponsor has the following options for choosing the hypotheses that will be tested at the final look:

- Decision 1A. If the overall population is selected, evaluate the overall effect as well as the subpopulation effect, i.e., test both H_0 and H_+ at the final analysis.
- Decision 1B. If the overall population is selected, evaluate the overall effect only, i.e., test H_0 at the final analysis.
- Decision 2A. If the biomarker-positive subpopulation is selected, evaluate the subpopulation effect, i.e., test H_+ at the final analysis.

It follows from this hypothesis selection rule that, if a decision is made to continue enrolling all patients, the sponsor may still examine the effectiveness of the experimental treatment within the biomarker-positive subset, in which case an appropriate multiplicity correction will need to be employed. Also, under Decision 1B, the hypothesis H_+ can be tested at the end of the trial but as a secondary hypothesis that does not require a formal multiplicity adjustment.

To construct adaptive designs that guarantee Type I error rate control in multi-population trials, *combination function methods* are commonly applied in conjunction with the closure principle. Assuming a trial with a single interim look, two stages are easily defined, e.g., the first stage includes all patients who complete the trial prior to the interim analysis and the second stage includes all patients who complete the trial after the interim analysis if the primary efficacy endpoint is continuous or binary. The stagewise p-values are computed using the data collected in each trial stage and, to control the Type I error rate, the p-values from the first and second stages need to be pooled using a pre-specified combination function with appropriate adjustments for hypothesis selection. This general approach was used in Brannath et al. (2009) to define a class of adaptive designs with two pre-defined patient populations. Alternatively, *conditional error rate methods* can be applied, see, for example, Friede et al. (2012). For more information on the combination function and conditional error rate methods, see Wassmer and Brannath (2016). The combination function principle is mathematically equivalent to the conditional error rate principle but, as shown in Friede et al. (2012), the former may provide certain advantages over the latter when dealing with the hypothesis test statistics in multi-population settings.

In addition to addressing multiplicity issues arising in clinical trials with adaptive population selection, it is important to define meaningful rules for identifying the most promising populations and hypotheses to examine at the final analysis. First of all, the choice of population or hypothesis selection rules is not driven by multiplicity considerations. In fact, adaptive designs based on the combination function principle or conditional error rate principle can be set up with any population or hypothesis selection rules. The rules should be chosen to ensure logical decision making and it is advisable to apply the decision-making framework introduced in Sect. 3.4.3.

As a quick illustration of the guiding principles, consider again an adaptive design in a multi-population trial with a single pre-defined population. To be consistent with the notation introduced in Sect. 3.4.3, the interim effect sizes in the biomarker-negative subpopulation, biomarker-positive subpopulation and overall population will be denoted by $\widehat{\theta}_-$, $\widehat{\theta}_+$ and $\widehat{\theta}_0$, respectively.

Suppose that there is a strong efficacy trend in the overall as well as biomarker-positive analyses, i.e., both $\widehat{\theta}_+$ and $\widehat{\theta}_0$ are sufficiently large. While it may be tempting to select Decision 1A, i.e., enroll both biomarker-negative and biomarker-positive patients in the second stage of the trial and test the hypotheses H_0 and H_+ at the final analysis, it will be beneficial to carefully examine the evidence in support of this decision. It was explained in Sect. 3.4.3 that a positive signal in the overall analysis does not immediately imply that the treatment is truly effective across the trial's population. This positive signal may simply be caused by a very strong efficacy signal in the biomarker-positive subpopulation. This consideration is related to the concept of an influence error and, according to the influence condition, it will be sensible to consider Decisions 1A or 1B only if there is evidence of a meaningful benefit is the subpopulation of biomarker-negative patients, i.e., if $\widehat{\theta}_- \geq \lambda_{\text{INF}}$, where λ_{INF} is a pre-specified threshold. If this influence condition is not satisfied, the most reasonable course of action would be to consider Decision 2A, i.e., to restrict patient enrollment to the biomarker-positive subset since biomarker-negative patients do not appear to benefit from the treatment. As the next step, the interaction condition needs to be checked to determine whether or not a differential treatment effect is present. This condition is met and thus Decisions 1A is justified if $\widehat{\theta}_+/\widehat{\theta}_- \geq$

λ_{INT} with a pre-specified $\lambda_{INT} > 1$. Otherwise, there is no clinically meaningful difference between the biomarker-negative and biomarker-positive subpopulations and Decisions 1B with the overall population test at the final look needs to be selected.

To summarize, the following population and hypothesis selection rules would be generally recommended if the experimental treatment appears to be effective in both patient populations at the interim look:

- Decision 1A. Select the overall population and test both H_0 and H_+ at the final analysis if the influence and interaction conditions are both met.
- Decision 1B. Select the overall population but test H_0 only at the final analysis if the influence condition is satisfied but the interaction conditions is not.
- Decision 2A. Select the biomarker-positive subpopulation and test H_+ at the final analysis if the influence condition is not satisfied.

The decision-making framework presented above depends of the free parameters known as the influence and interaction thresholds (λ_{INF} and λ_{INT}). To select reasonable values of these parameters, it is advisable to perform a comprehensive evaluation of the operating characteristics of the adaptive design using the Clinical Scenario Evaluation approach presented at the end of Sect. 3.4.2. The evaluation may be aimed at maximizing the probability of success in the trial under clinically meaningful constraints on the influence and interaction thresholds.

3.4.4.2   Adaptive Trials with Data-Driven Subpopulations

A key assumption made in the adaptive designs described above is that the subsets of the overall population must be prospectively defined, i.e., the corresponding binary classifiers must be known at the planning stage. This means that the underlying biomarkers must be selected upfront and, if these biomarkers are measured on interval or ordinal scales, the cut points that define the biomarker-low and biomarker-high subgroups must be selected as well. It is natural to consider extensions of this adaptive design framework that support more flexible decision rules. Examples include adaptive trials where patient subpopulations are partially defined at the onset of a trial, i.e., candidate biomarkers are pre-selected but the rules for defining patient subgroups are not specified, and the subpopulations of interest are fully defined at an interim look using the data from the first stage of the trial. To understand the advantages of this approach over standard adaptive designs with pre-set subpopulations, note that the rules for defining these subpopulations, e.g., cut points for continuous biomarkers, are typically estimated from rather small Phase II trials. With data-driven subpopulations, a larger data set from the first stage of a Phase III trial will be used to identify an optimal cut point for a biomarker. The subpopulation corresponding to this cut point could then be analyzed at the final look as if it was pre-planned.

One of the most straightforward approaches to designing adaptive trials with data-driven subpopulations, often referred to as *signature designs*, was proposed by Freidlin and Simon (2005). Very briefly, consider a two-stage trial with a single interim look. The interim data set is carefully examined to find a subset with a strong efficacy signal and the treatment effect is evaluated at the final analysis in the overall population as well as the subset identified at the interim analysis. The data collected before and after the interim look are used in the overall test and only the second-stage data are utilized in the subpopulation test. To protect the Type I error rate, the Bonferroni test is applied to split the alpha between the two tests. An obvious limitation of the signature designs is that the data from the first stage of the trial are discarded when the subpopulation effect is assessed. In addition, it is important to bear in mind that the null hypothesis of no effect within the subpopulation is not pre-specified. Therefore, it is not clear how to define the Type I error rate in this problem with data-driven subsets and, even though the multiplicity adjustment based on the Bonferroni test is conservative, the probability of an incorrect decision may not be controlled in this setting.

More efficient designs have been developed to support inferences in data-driven subsets in an inferentially seamless way, i.e., by pooling the first-stage and second-stage data within a subset, using the combination function principle, see, for example, Graf et al. (2019, 2020). This general framework can be combined with any of the principled subgroup identification methods discussed in Sect. 3.3. Briefly, suppose that a set of continuous biomarkers is pre-specified in a two-stage Phase III trial and an appropriate subgroup identification method is applied at the interim analysis to choose the strongest predictor of treatment benefit and a corresponding binary classifier that defines a target subpopulation. The trial proceeds to the final analysis and the treatment effect is assessed within the overall trial population as well as the selected subpopulation using an appropriate multiplicity adjustment. This multiplicity adjustment is derived using the combination function principle and takes into account a rich family of null hypotheses corresponding to all cut points for the candidate biomarkers and, as a result, the adjustment guarantees Type I error rate control in the strong sense. This flexible approach to identifying relevant patient subgroups in pivotal trials has the potential to provide a more comprehensive characterization of the efficacy of novel treatments compared to adaptive trials that rely on pre-planned patient subpopulations.

3.5   Discussion

The main goal of this chapter is to provide an overview of key considerations in the evaluation of subgroup effects in late-stage clinical trials, including an overview of applicable trial designs, statistical methods used in exploratory and confirmatory settings, and regulatory considerations.

We started this chapter with a review of applicable regulatory guidance documents, including one of the earliest guidelines (ICH E9). Regional health authorities have carefully considered the topic of subgroup analysis in their guidance documents, as reflected in the special issue of Journal of Biopharmaceutical Statistics on subgroup analysis in clinical trials guest-edited by Dmitrienko and Wang (2014). All of the guidance documents we reviewed agree that a homogeneity of the treatment effect can not be assumed a priori in late-stage clinical trials. Consequently, a comprehensive evaluation of subgroup effects needs to be considered in all late-stage clinical trials when the objective is to achieve corresponding licensing for an investigational treatment. In multiple settings, including innovative therapeutic fields with completely new medical entities and sparse historical data, it remains challenging to specify a homogeneous trial population, define consistency and pre-specify factors that define relevant patient subgroup. The interpretation of trial results in these and similar settings will strongly depend on benefit-risk considerations. To summarize, it follows from multiple guidance documents that subgroup findings should take into consideration biological plausibility, internal and/or external consistency, the strength of evidence, clinical relevance and statistical uncertainty.

Putting these principles to practice may be challenging and we considered many challenges and pitfalls of subgroup investigation in both exploratory and confirmatory settings in separate sections. Numerous patient subgroups are examined in all Phase II and Phase III clinical trials and it is critical to identify appropriate statistical methods that are aligned with the goals of subgroup analysis, e.g., data-driven or confirmatory subgroup analysis.

In the context of data-driven subgroup analysis we followed the established clinical practice and distinguished among the traditional exploratory analysis in clinical trials, post-hoc analysis and subgroup/biomarker discovery, arguing that a unified set of statistical principles should be applied across different types of subgroup evaluations. These principles were derived from the literature on statistical methods that have been proposed recently for data-driven subgroup analysis, as a result of a cross-fertilization of efforts from machine learning, causal inference and multiple testing. Several criteria of principled data-driven subgroup evaluation were established. In particular, complexity and false positive error rate control should be implemented using, for example, resampling methods to account for the uncertainty in complex biomarker/subgroup search strategies. Treatment effects within the identified subgroups or, more broadly, gains from estimated individual treatment assignment rules should be evaluated by methods that account for optimism (selection) bias when applied in the future trials/patient populations.

Multiplicity adjustments play a central role in confirmatory subgroup analysis settings. As a large number of options for multiplicity adjustment exist in a multi-population trial, it is important to identify a procedure that performs best in the context of a given trial while being robust against deviations from the assumptions about the treatment effect. Recently several sophisticated adaptive designs have been developed to support the investigation of subgroup effects in pivotal trials. These designs rely on flexible data-driven decision rules, e.g., support options to select treatment arms or subpopulations, while enabling valid statistical inferences in terms of controlling the overall Type I error rate.

Acknowledgement We are grateful to Lei Xu, Anthony Zagar, Lanju Zhang and the book's editors for their insightful comments.

References

Alosh M, Huque MF (2013) Multiplicity considerations for subgroup analysis subject to consistency constraint. Biom J 55:444–462

Alosh M, Fritsch K, Huque M, Mahjoob K, Pennello G, Rothmann M, Russek-Cohen E, Smith F, Wilson S, Yue L (2015) Statistical considerations on subgroup analysis in clinical trials. Stat Biopharm Res 7(4):286–303

Alosh M, Huque MF, Bretz F, D'Agostino RB (2016) Tutorial on statistical considerations on subgroup analysis in confirmatory clinical trials. Stat Med 36:1334–1360

Athey S, Imbens G (2016) Recursive partitioning for heterogeneous causal effects. Proc Natl Acad Sci 113:7353–7360

Ballarini NM, Rosenkranz GK, Jaki T, König F, Posch M (2018) Subgroup identification in clinical trials via the predicted individual treatment effect. PLoS One 13:e0205971

Battioui C, Denton B, Shen L (2018) TSDT: treatment-specific subgroup detection tool. R package version 1.0.0. https://CRAN.R-project.org/package=TSDT

Benda N, Branson M, Maurer W, Friede T (2010) Aspects of modernizing drug development using clinical scenario planning and evaluation. Drug Inf J 44:299–315

Berger J, Wang X, Shen L (2014) A Bayesian approach to subgroup identification. J Biopharm Stat 24:110–129

Bornkamp B, Ohlssen D, Magnusson BP, Schmidli H (2017) Model averaging for treatment effect estimation in subgroups. Pharm Stat 16:133–142

Bonetti M, Gelber R (2000) A graphical method to assess treatment–covariate interactions using the cox model on subsets of the data. Stat Med 19:2595–2609

Brannath W, Zuber E, Branson M, Bretz F, Gallo P, Posch M, Racine-Poon A (2009) Confirmatory adaptive designs with Bayesian decision tools for a targeted therapy on oncology. Stat Med 28:1445–1463

Breiman L (2001) Statistical modeling: the two cultures. Stat Sci 16:199–231

Breiman L, Friedman JH, Olshen RA, Stone CJ (1984) Classification and regression trees. Wadsworth, Belmont

Bretz F, Schmidli H, Koenig F, Racine A, Maurer W (2006) Confirmatory seamless phase II/III clinical trials with hypotheses selection at interim: general concepts. Biom J 48:623–634

Brookes ST, Whitley E, Peters TJ, Mulheran PA, Egger M, Davey Smith G (2001) Subgroup analyses in randomised controlled trials: quantifying the risks of false-positives and false-negatives. Health Technol Assess 5:1–56

Cappuzzo F et al (2010) Erlotinib as maintenance treatment in advanced non-small-cell lung cancer: a multicentre, randomised, placebo-controlled Phase 3 study. Lancet Oncol 11:521–529

Carroll KJ, Fleming TR (2013) Statistical evaluation and analysis of regional interactions: the PLATO trial case study. Stat Biopharm Res 5(2):91–101

Carroll KJ, Le Maulf F (2011) Japanese guideline on global clinical trials: statistical implications and alternative criteria for assessing consistency. Drug Inf J 45:657–667

CFDA (China Food and Drug Administration) (2007) Provisions for drug registration. State Food and Drug Administration Order No. 28

Chen G, Zhong H, Belousov A, Viswanath D (2015) PRIM approach to predictive-signature development for patient stratification. Stat Med 34:317–342

Chipman HA, George EI, McCulloch RE (2010) BART: Bayesian additive regression trees. Ann Appl Stat 4:266–298

Cohen AT et al (2016) Extended thromboprophylaxis with betrixaban in acutely ill medical patients. N Engl J Med 375:534–544

Dane A, Spencer A, Rosenkranz G, Lipkovich I, Parke T on behalf of the PSI/EFSPI Working Group on Subgroup Analysis (2019) Subgroup analysis and interpretation for phase 3 confirmatory trials: white paper of the EFSPI/PSI working group on subgroup analysis. Pharm Stat 18:126–139. https://doi.org/10.1002/pst.1919

Dixon DO, Simon R (1991) Bayesian subset analysis. Biometrics 47:871–882

Dmitrienko A, D'Agostino RB (2013) Tutorial in biostatistics: traditional multiplicity adjustment methods in clinical trials. Stat Med 32:5172–5218

Dmitrienko A, D'Agostino RB (2018) Multiplicity considerations in clinical trials. N Engl J Med 378:2115–2122

Dmitrienko A, Paux G (2017) Subgroup analysis in clinical trials. In: Dmitrienko A, Pulkstenis E (eds) Clinical trial optimization using R. Chapman and Hall/CRC Press, New York

Dmitrienko A, Tamhane AC (2011) Mixtures of multiple testing procedures for gatekeeping applications in clinical trials. Stat Med 30:1473–1488

Dmitrienko A, Tamhane AC (2013) General theory of mixture procedures for gatekeeping. Biom J 55:402–419

Dmitrienko A, Soulakova JN, Millen B (2011) Three methods for constructing parallel gatekeeping procedures in clinical trials. J Biopharm Stat 53:768–786

Dmitrienko A, Muysers C, Fritsch A, Lipkovich I (2016) General guidance on exploratory and confirmatory subgroup analysis in late-stage clinical trials. J Biopharm Stat 26:71–98

Douillard JY et al (2014) Final results from PRIME: randomized phase 3 study of panitumumab with FOLFOX4 for first-line treatment of metastatic colorectal cancer. Ann Oncol 25:1346–1355

Dusseldorp E, Van Mechelen I (2014) Qualitative interaction trees: a tool to identify qualitative treatment-subgroup interactions. Stat Med 33:219–237

Dusseldorp E, Conversano C, Van Os BJ (2010) Combining an additive and tree-based regression model simultaneously: STIMA. J Comput Graph Stat 19:514–530

EMA (European Medicines Agency) (2007) Reflection paper on methodological issues in confirmatory clinical trials planned with an adaptive design. European Medicines Agency/Committee for Medicinal Products for Human Use. CHMP/EWP/2459/02

EMA (European Medicines Agency) (2014) Guideline on the investigation of subgroups in confirmatory clinical trials. Draft. European Medicines Agency/Committee for Medicinal Products for Human Use. EMA/CHMP/539146/2013

EMA (European Medicines Agency) (2015) Guideline on adjustment for baseline covariates in clinical trials. European Medicines Agency/Committee for Medicinal Products for Human Use. EMA/CHMP/295050/2013

EMA (European Medicines Agency) (2017) Guideline on multiplicity issues in clinical trials. Draft. European Medicines Agency/Committee for Medicinal Products for Human Use. EMA/CHMP/44762/2017

FDA (U.S. Food and Drug Administration) (2014) Guidance: evaluation of sex-specific data in medical device clinical studies. https://www.fda.gov/media/82005/download

FDA (U.S. Food and Drug Administration) (2017a) Guidance for industry: evaluation and reporting of age-, race-, and ethnicity-specific data in medical device clinical studies; doc number 1500626, pp 1–36. https://www.fda.gov/media/98686/download

FDA (U.S. Food and Drug Administration) (2017b) Guidance for industry: multiple endpoints in clinical trials. https://www.fda.gov/media/102657/download

FDA (U.S. Food and Drug Administration) (2018) Guidance for industry: adaptive design clinical trials for drugs and biologics. https://www.fda.gov/media/78495/download

FDA (U.S. Food and Drug Administration) (2019) Guidance for industry: enrichment strategies for clinical trials to support determination of effectiveness of human drugs and biological products. https://www.fda.gov/media/121320/download

Foster JC, Taylor JMC, Ruberg SJ (2011) Subgroup identification from randomized clinical trial data. Stat Med 30:2867–2880

Freidlin B, Simon R (2005) Adaptive signature design: an adaptive clinical trial design for generating and prospectively testing a gene expression signature for sensitive patients. Clin Cancer Res 21:7872–7878

Freidlin B, McShane LM, Korn EL (2010) Randomized clinical trials with biomarkers: design issues. J Natl Cancer Inst 102:152–160

Freidlin B, Korn EL, Gray R (2014) Marker sequential test (MaST) design. Clin Trials 11:19–27

Friede T, Parsons N, Stallard N (2012) A conditional error function approach for subgroup selection in adaptive clinical trials. Stat Med 31:4309–4320

Friedman JH (2002) Stochastic gradient boosting. Comput Stat Data Anal 38:367–378

Friedman JH, Fisher NI (1999) Bump hunting in high-dimensional data. Stat Comput 9:123–143

Fu H (2018) Individualized treatment recommendation (ITR) for survival outcomes. Presentation at conference on statistical learning and data science/nonparametric statistics. Columbia University, New York. https://publish.illinois.edu/sldsc2018/2018/05/20/session-35-machine-learning-and-precision-medicine/

Fu H, Zhou J, Faries DE (2016) Estimating optimal treatment regimes via subgroup identification in randomized control trials and observational studies. Stat Med 35:3285–3302

Graf AC, Wassmer G, Friede T, Gera RG, Posch M (2019) Robustness of testing procedures for confirmatory subpopulation analyses based on a continuous biomarker. Stat Methods Med Res 28:1879–1892

Graf AC, Magirr D, Dmitrienko A, Posch M (2020) Optimized multiple testing procedures for nested subpopulations based on a continuous biomarker. Stat Med. To appear

Gunter L, Zhu J, Murphy S (2011) Variable selection for qualitative interactions in personalized medicine while controlling the familywise error rate. J Biopharm Stat 21:1063–1078

Hemmings R (2014) An overview of statistical and regulatory issues in the planning, analysis, and interpretation of subgroup analyses in confirmatory clinical trials. J Biopharm Stat 24:4–18

Hemmings R (2015) Comment. Stat Biopharm Res 7:305–308

Henderson NC, Louis TA, Rosner G, Varadhan R (2020) Individualized treatment effects with censored data via fully nonparametric Bayesian accelerated failure time models. Biostatistics 21:50–68. https://doi.org/10.1093/biostatistics/kxy02

Herbst RS et al (2016) Pembrolizumab versus docetaxel for previously treated, PD-L1-positive, advanced non-small-cell lung cancer (KEYNOTE-010): a randomised controlled trial. Lancet 387:1540–1550

Hill JL (2016) Bayesian nonparametric modeling for causal inference. J Comput Graph Stat 20:217–240

Hirakawa A, Kinoshita F (2017) An analysis of Japanese patients enrolled in multiregional clinical trials in oncology. Ther Innov Regul Sci 51:207–211

Hodges JS, Cui Y, Sargent DJ, Carlin BP (2007) Smoothing balanced single-error-term analysis of variance. Technometrics 49:12–25

Huang Y, Fong Y (2014) Identifying optimal biomarker combinations for treatment selection via a robust kernel method. Biometrics 70:891–901

Huang X, Sun Y, Trow P, Chatterjee S, Chakravatty A, Tian L, Devanarayan V (2017) Patient subgroup identification for clinical drug development. Stat Med 36:1414–1428

ICH (1998) Ethnic factor in the acceptability of foreign data. ICH E5 Expert Working Group. The US Federal Register, vol 83, pp 31790–31796

ICH (International Council for Harmonisation of Technical Requirements for Pharmaceuticals for Human Use) (1999) Topic E9 Statistical principles for clinical trials. CPMP/ICH/363/96

ICH (2014) Final Concept Paper E9 (R1): Addendum to Statistical principles for clinical trials on choosing appropriate estimands and defining sensitivity analyses in clinical trials. ICH Steering Committee

ICH (International Council for Harmonisation of Technical Requirements for Pharmaceuticals for Human Use) (2017) Guideline E17 on general principles for planning and design of multi-regional clinical trials. EMA/CHMP/ICH/453276/2016

Ikeda K, Bretz F (2010) Sample size and proportion of Japanese patients in multi-regional trials. Pharm Stat 9:207–216

Imai K, Ratkovic M (2013) Estimating treatment effect heterogeneity in randomized program evaluation. Ann Appl Stat 7:443–470

Johnston S et al (2009) Lapatinib combined with letrozole versus letrozole and placebo as first-line therapy for postmenopausal hormone receptor-positive metastatic breast cancer. J Clin Oncol 33:5538–5546

Kehl V, Ulm K (2006) Responder identification in clinical trials with censored data. Comput Stat Data An 50:1338–1355

Koch A, Framke T (2014) Reliably basing conclusions on subgroups of randomized clinical trials. J Biopharm Stat 24:42–57

Koch G, Schwartz TA (2014) An overview of statistical planning to address subgroups in confirmatory clinical trials. J Biopharm Stat 24:72–93

Laber EB, Zhao YQ (2015) Tree-based methods for individualized treatment regimes. Biometrika 102:501–514

Lamont AE, Lyons M, Jaki TF, Stuart E, Feaster D, Ishwaran H, Tharmaratnam K, Van Horn ML (2018) Identification of predicted individual treatment effects in randomized clinical trials. Stat Methods Med Res 27:142–157

Leblanc M, Crowley J (1993) Survival trees by goodness of split. J Am Stat Assoc 88:457–467

Lee JD, Sun DL, Sun Y, Taylor JE (2016) Exact post-selection inference, with application to the lasso. Ann Stat 44:907–927

Lipkovich I, Dmitrienko A (2014a) Strategies for identifying predictive biomarkers and subgroups with enhanced treatment effect clinical trials using SIDES. J Biopharm Stat 24:130–153

Lipkovich I, Dmitrienko A (2014b) Biomarker identification in clinical trials. In: Carini C, Menon S, Chang M (eds) Clinical and statistical considerations in personalized medicine. Chapman and Hall/CRC Press, New York

Lipkovich I, Dmitrienko A, Denne J, Enas G (2011) Subgroup identification based on differential effect search (SIDES): a recursive partitioning method for establishing response to treatment in patient subpopulations. Stat Med 30:2601–2621

Lipkovich I, Dmitrienko A, D'Agostino BR (2017a) Tutorial in biostatistics: data-driven subgroup identification and analysis in clinical trials. Stat Med 36:136–196

Lipkovich I, Dmitrienko A, Patra K, Ratitch B, Pulkstenis E (2017b) Subgroup identification in clinical trials by Stochastic SIDEScreen methods. Stat Biopharm Res 9:368–378

Lipkovich I, Dmitrienko A, Muysers C, Ratitch B (2018) Multiplicity issues in exploratory subgroup analysis. J Biopharm Stat 28:63–81

Liu JT, Tsou HH, Gordon Lan KK et al (2016) Assessing the consistency of the treatment effect under the discrete random effects model in multiregional clinical trials. Stat Med 35:2301–2314

Loh WY, He X, Man M (2015) A regression tree approach to identifying subgroups with differential treatment effects. Stat Med 34:1818–1833

Loh WY, Fu H, Man M, Champion V, Yu M (2016) Identification of subgroups with differential treatment effects for longitudinal and multiresponse variables. Stat Med 35:4837–4855

Lu M, Sadiq S, Feaster DJ, Ishwaran H (2018) Estimating individual treatment effect in observational data using random Forest methods. J Comput Graph Stat 27:209–219

Mahaffey KW, Wojdyla DM, Carroll K, Becker RC, Storey RF, Angiolillo DJ, Held C, Cannon CP, James S, Pieper KS, Horrow J, Harrington RA, Wallentin L (2011) Ticagrelor compared with clopidogrel by geographic region in the Platelet Inhibition and Patient Outcomes (PLATO) Trial. Circulation 124:544–554

Mayer C, Lipkovich I, Dmitrienko A (2015) Survey results on industry practices and challenges in subgroup analysis in clinical trials. Stat Biopharm Res 7:272–282

Millen B, Dmitrienko A, Ruberg S, Shen L (2012) A statistical framework for decision making in confirmatory multi-population tailoring clinical trials. Drug Inf J 46:647–656

Millen B, Dmitrienko A, Song G (2014) Bayesian assessment of the influence and interaction conditions in multi-population tailoring clinical trials. J Biopharm Stat 24:94–109

NMPA (China National Medical Product Administration) (2015) Guidance for international multicenter clinical trials (IMCT). Issued final trial implementation, as of March 2015

Paux G, Dmitrienko A (2018) Penalty-based approaches to evaluating multiplicity adjustments in clinical trials: traditional multiplicity problems. J Biopharm Stat 28:146–168

Piccart-Gebhart MJ et al (2005) Trastuzumab after adjuvant chemotherapy in HER2-positive breast cancer. N Engl J Med 353:1659–1672

PMDA (Pharmaceuticals and Medical Devices Agency) (2007) Ministry of Health, Labour and Welfare. Basic principles on global clinical trials. https://www.pmda.go.jp/files/000153265.pdf

PMDA (Pharmaceuticals and Medical Devices Agency) (2012) Ministry of Health, Labour and Welfare. Basic principles on global clinical trials (Reference cases). https://www.pmda.go.jp/files/000157520.pdf

Qian M, Murphy SA (2011) Performance guarantees for individualized treatment rules. Ann Stat 39:1180–1210

Ridgeway G (1999) The state of boosting. Comput Sci Stat 31:172–181

Rosenkranz GK (2016) Exploratory subgroup analysis in clinical trials by model selection. Biom J 58:1217–1228

Rothmann MD, Zhang JJ, Lu L, Fleming TR (2012) Testing in a pre-specified subgroup and the intent-to-treat population. Drug Inf J 46:175–179

Rothwell PM (2005) Subgroup analysis in randomized controlled trials: importance, indications, and interpretation. Lancet 365:176–86

Royston P, Sauerbrei W (2004) A new approach to modelling interaction between treatment and continuous covariates in clinical trials by using fractional polynomials. Stat Med 23:2509–2525

Royston P, Sauerbrei W (2013) Interaction of treatment with a continuous variable: simulation study of power for several methods of analysis. Stat Med 32:3788–3803

Russek-Cohen E (2014) EMA workshop on the investigation of subgroups in confirmatory clinical trials. Presentation available online: https://www.ema.europa.eu/documents/presentation/presentation-comments-us-food-drug-administration-fdaworking-group-subgroup-analyses-estelle-russek_en.pdf

Seibold H, Zeileis A, Hothorn T (2016) Model-based recursive partitioning for subgroup analyses. Int J Biostat 12:45–63

Seibold H, Zeileis A, Hothorn T (2018) Individual treatment effect prediction for ALS patients. Stat Methods Med Res 10:3104–3125

Su X, Zhou T, Yan X, Fan J, Yang S (2008) Interaction trees with censored survival data. Int J Biostat 4(1):Article 2

Su X, Tsai CL, Wang H, Nickerson DM, Li B (2009) Subgroup analysis via recursive partitioning. J Mach Learn Res 10:141–158

Su X, Peña AT, Liu L, Levine RA (2018) Random forests of interaction trees for estimating individualized treatment effects in randomized trials. Stat Med 37:2547–2560

Sun X, Briel M, Walter SD, Guyatt GH (2010) Is a subgroup effect believable? Updating criteria to evaluate the credibility of subgroup analyses. BMJ 340:c117

Taylor J, Tibshirani R (2017) Post-selection inference for l1-penalized likelihood models. Can J Stat 1:1–21

Tian L, Alizaden AA, Gentles AJ, Tibshirani R (2014) A simple method for detecting interactions between a treatment and a large number of covariates. J Am Stat Assoc 109:1517–1532

Wager S, Athey S (2018) Estimation and inference of heterogeneous treatment effects using random forests. J Am Stat Assoc 113:1228–1242

Wager S, Hastie T, Efron B (2014) Confidence intervals for random forests: the jackknife and the infinitesimal jackknife. J Mach Learn Res 15:1625–1651

Wang SJ, Dmitrienko A (2014) Guest Editors' Note: Special issue on subgroup analysis in clinical trials. J Biopharm Stat 24:1–3

Wang SJ, Hung HMJ (2014) A regulatory perspective on essential considerations in design and analysis of subgroups when correctly classified. J Biopharm Stat 24:19–41

Wang SJ, O'Neill RT, Hung HMJ (2007) Approaches to evaluation of treatment effect in randomized clinical trials with genomic subset. Pharm Stat 6:227–244

Wassmer G, Brannath W (2016) Group sequential and confirmatory adaptive designs in clinical trials. Springer, New York

Xu Y, Yu M, Zhao YQ, Li Q, Wang S, Shao J (2015) Regularized outcome weighted subgroup identification for differential treatment effects. Biometrics 71:645–653

Zhang B, Tsiatis AA, Davidian M, Zhang M, Laber EB (2012) Estimating optimal treatment regimes from a classification perspective. Statistics 1:103–114

Zhang Y, Laber EB, Tsiatis A, Davidian M (2016) Interpretable dynamic treatment regimes. Preprint. arXiv:1606.01472

Zhao Y, Zheng D, Rush AJ, Kosorok MR (2012) Estimating individualized treatment rules using outcome weighted learning. J Am Stat Assoc 107:1106–1118

Zhao Y, Zheng W, Zhuo DY, Lu Y, Ma X, Liu H, Zeng Z, Laird G (2018) Bayesian additive decision trees of biomarker-by treatment interactions for predictive biomarkers detection and subgroup identification. J Biopharm Stat 28:534–549

Chapter 4
Considerations on Subgroup Analysis in Design and Analysis of Multi-Regional Clinical Trials

Hui Quan, Xuezhou Mao, Jun Wang, and Ji Zhang

Abstract The application of multi-regional clinical trials (MRCTs) is a preferred strategy for rapid global new drug development. In a MRCT, besides the other subgroup factors that are in general well defined, region is a special subgroup factor which can be a surrogate of many intrinsic and extrinsic factors. The definition of a region for a MRCT may be trial specific. It depends on where the MRCT will be conducted and how the sample sizes will be allocated across the regions. As a regional health authority will carefully review the regional treatment effect before the approval of the drug for the patients of the region, special attention should be paid to the regional subgroup analysis. In this chapter, we will discuss subgroup analysis in design and analysis of multi-regional clinical trials focusing on regional subgroup analysis. These include the considerations on region definition, analysis model, consistency assessment of regional treatment effects, regional sample size allocation and trial result interpretation. Numerical and real trial examples will be used to illustrate the applications of the methods.

Keywords Assurance probability · Between-region variance · Consistency assessment · Data monitoring committee · Discrete random effect model · Empirical shrinkage estimate · Fixed effects model · Heterogeneity · Intent-to-treat analysis · Interaction test · Intrinsic/extrinsic factors · James-Stein shrinkage estimate · Multi-center clinical trials · Multinomial distribution · Multiple hypotheses · Multiplicity adjustment · Multi-regional clinical trial · Non-inferiority margin · Per-protocol analysis · PK/PD model · Power · Predictive factor · Primary endpoint · Random effects model · Region

H. Quan (✉) · X. Mao · J. Zhang
Biostatistics and Programming, Sanofi, Bridgewater, NJ, USA
e-mail: hui.quan@sanofi.com; Xuezhou.mao@sanofi.com; ji.zhang@sanofi.com

J. Wang
Office of Biostatistics and Clinical Pharmacology, Center for Drug Evaluation, National Medical Products Administration, Beijing, China
e-mail: wangj@cde.org.cn

© Springer Nature Switzerland AG 2020
N. Ting et al. (eds.), *Design and Analysis of Subgroups with Biopharmaceutical Applications*, Emerging Topics in Statistics and Biostatistics,
https://doi.org/10.1007/978-3-030-40105-4_4

definition · Regional treatment effect · Sample size · Statistical analysis plan · Subgroup analysis · Stratification factors · Type I error rate

4.1 Introduction

With the growing trend of globalization and the need for rapid availability of new medicines to patients worldwide, the use of multi-regional clinical trials (MRCTs) is a preferred strategy applied by sponsors for global drug development. A MRCT is clearly much more complex than a trial conducted in a single region. To address issues raised from MRCTs, extensive effort from all parties involved in MRCTs has been made in the related research field. The Japanese health authority issued the initial guidance for international clinical trials in 2007. The European Medicines Agency (EMA) published a reflection paper on the extrapolation of results from trials conducted outside of Europe in 2008. The International Council for Harmonisation (ICH) Expert Working Group recommended the adoption of E17: General principle on planning/designing Multi-Regional Clinical Trials in 2017. Statisticians from academia, regulatory and industry have also contributed by conducting independent research (Kawai et al. 2007; Uesaka 2009; Quan et al. 2010a, b, 2017; Hung et al. 2010; Chen et al. 2010; Khin et al. 2013).

As in any clinical trials, subgroup analyses are important components of data analysis in MRCTs. Besides the regular subgroups defined by baseline characteristics such as gender, age, disease severity and biomarkers, region is a special subgroup factor in a MRCT. Consistency of treatment effects across all subgroups is essential for the drug to be applicable to all patients. However, different from a regular subgroup that has no representatives who impose trial requirements from the perspective of that subgroup, regional health authorities may have specific trial demands that the sponsor should meet. Therefore, particular attention should be paid to the regional subgroup analysis in MRCTs.

In this chapter, we will discuss subgroup analysis in a MRCT setting focusing on regional subgroup analysis. MRCT design related issues will be tackled in Sect. 4.2. These include considerations of region definitions and harmonization of the regional requirements. Models for data analysis with region as a factor for the convenience of regional treatment effect assessment along with the corresponding overall sample size calculations will be specified in Sect. 4.3. Methods for regional treatment effect assessments will be discussed in Sect. 4.4. Given the operational complexity of MRCTs, special care for the conduct and monitoring of a MRCT to ensure the trial quality will be pointed out in Sect. 4.5. Interpretation of trial results is the focus of Sect. 4.6. Further regular subgroup analysis in a MRCT setting will be reviewed in Sect. 4.7 and brief discussion is provided in Sect. 4.8 to conclude the chapter.

4.2 Trial Design Considerations

Clinical trial design depends on the trial objective(s), the primary endpoint(s), the primary population and the method for the primary analyses. One objective of a multi-regional clinical trial is to assess treatment effects simultaneously across regions. An understanding of the disease and the impact of many intrinsic/extrinsic factors on the treatment outcomes across regions is the key for determining whether a MRCT is an appropriate choice for meeting the trial objectives.

The definitions of the regular subgroups which rely on specific factors (e.g., gender) are well understood and are basically the same across studies. However, the definition of regions for a MRCT may be trial dependent. It depends on where the MRCT will be conducted and how the sample sizes will be allocated across the regions. Furthermore, upon the completion of the trial, while interpreting the MRCT results, a country or regional health authority has the obligation to protect the public health of the patients under its jurisdiction and will examine the regional result before marketing approval of the drug in the regions. Thus, special care for handling region related issues should be initiated from MRCT design.

4.2.1 Region Definition

A "region" may be defined as a single country or as a combination of several countries resting on many considerations. The definition of region may impact many aspects of the design and data analysis. For example, as will be seen in Sect. 4.3, if a random effects model is applied to data analysis, a larger number of regions (e.g., region is defined at the country level particularly for large countries) will associate with a smaller overall sample size for the MRCT for the desired power. On the other hand, if regional treatment effects will be quantified based on solely regional data, countries should be combined to form regions so that the number of regions is small and the regional sample sizes are relatively large for producing robust estimates of regional treatment effects. Documenting the definition of region and the analysis plan for the evaluation of regional treatment effects at the time of study design will provide appropriate perspective and integrity for anticipated and unanticipated regional findings at study conclusion.

A typical approach for combining countries to form a region is based on the proximity of geographic locations. For instance, the common practice is to combine Japan, South Korea and China into a single geographic region denoted as "Asia" in a MRCT. However, considering that region is a surrogate for many intrinsic and extrinsic factors (Wittes 2013), the definition of region may not be limited to geographic boundaries. Adequate justification of any definition of region should take into consideration such factors as race or ethnicity, disease epidemiology, background therapies, and medical practice among others. If economic situation or income along with the availability of medications could potentially influence treatment outcome, then combining those three countries with different Human Development Index levels to form an 'Asia' region may not make sense and in fact may increase heterogeneity (Tanaka et al. 2011).

4.2.2 Regional Requirements

As discussed in Girman et al. (2011), for certain disease areas, regulatory agencies in different parts of the world may have different requirements for the approval of a new drug for their regions. These differences could be about the primary endpoints, key secondary endpoints, the minimum treatment duration, a key time point of measurement, a non-inferiority margin, an analysis population, the amount of overall data (e.g., safety data) or amount of regional data, and even the approach for data analysis (e.g., method for handling missing data). Such differences create significant challenges for trial design and data analysis when a MRCT strategy is applied for global drug development. To address these challenges, the trial sponsor should first negotiate with regulatory agencies to harmonize or minimize regional requirement differences and obtain agreement to use a more consistent standard across regions. If the differences are unavoidable due to, for example, different scientific interpretations of the disease, the sponsor needs to carefully address the disparities in the protocol with key design components capable of meeting most of the needs of all the regions in order to maintain the efficiency of the MRCT to enable a successful global new drug development program. Hemmings and Lewis (2013) even propose to "Develop multiple SAPs (Statistical Analysis Plans) for a single trial, conducted under a single trial protocol, so that the same trial data separately analyzed will address the different standards of each region."

In the case of different agencies requiring different primary endpoints (see Girman et al. 2011 for examples), data for all these endpoints should be collected in the same MRCT. However, it should be comprehensively stated in the study protocol that the primary endpoints for different regions are unique and not co-primary endpoints for all the regions. The pre-specified region specific primary endpoints could still be used as supportive endpoints for the other regions. The needs for multiplicity adjustments are only for the endpoints within a region (see also ICH E17). In addition, the overall sample size for the MRCT should be the largest sample size such that there is the desired power for demonstrating the treatment effects on each of the separate region specific primary endpoints, utilizing the total trial information from all regions.

Non-inferiority trials are often conducted to establish that the experimental treatment is not worse than a reference treatment based on a pre-specified non-inferiority margin. There may be a well-established consensus for defining the margin for certain therapeutic areas. For other therapeutic areas, a conservative method requested by an agency will result in a smaller margin. The requests of different non-inferiority margins from different regions will create a situation where there are multiple hypotheses for the non-inferiority assessment for the different regions/margins within the same MRCT. Since the hypothesis of a smaller margin is nested in the hypothesis of a larger margin, all tests for these hypotheses form a closed procedure and multiplicity adjustment is not necessary. That is, even when the nominal significance level is applied to test each hypothesis, the overall type I error rate is still strongly controlled. If no multiplicity adjustment will be considered across regions, there will be no issue of type I error rate inflation anyway. For non-inferiority assessment, some agencies may request an Intent-To-Treat analysis while the other agencies may prefer a Per-Protocol analysis as the primary analysis (see Girman et al. 2011). The point estimates and sample sizes for these two analyses are slightly different. Even with all these issues specified in the protocol, results in the analysis reports for different regions and different populations should be viewed with caution and in proper context.

When different agencies request different time points for the primary endpoint or different amounts of data (e.g., safety data), the MRCT should have sufficient duration for the longest time point and large enough sample size for the largest requested amount of data. This means that the regulatory agency(s) with the most extreme requests for the amount and/or duration of data will drive these study design characteristics. Depending on the specific circumstances, it may not be wise to perform an interim analysis and submit the interim results to the agency that requests relatively shorter study duration in order to maintain the integrity of the whole MRCT. If different active controls, different doses of the experimental treatment or different rescue medications are used by different regions, the complexity of using one MRCT for all the regions will significantly increase. The sponsor will need to address all these complications due to different study design elements, and the resulting trial conduct and data analysis.

The analyses performed to meet the different requirements of different regional health authorities are not really the traditional subgroup analyses. They will use all available data from the whole study across all regions. The results are not specific for individual regions but for the full population as a whole.

4.2.3 Randomization

The number of pre-specified subgroup analyses for a clinical trial is usually not small. Thus, it is not practical to incorporate all the factors which define these subgroups as stratification factors for randomization. As a relatively more formal analysis will be performed for the region subgroup to meet the needs of regional health authorities and region is a surrogate for many intrinsic and extrinsic factors, use region as a stratification factor in randomization and in analysis model may increase the efficiency of the analysis. Stratified randomization can also be conducted at the country level in case later on small countries will be combined to form regions in data analysis.

A subgroup factor may be a predictive factor. That is, the treatment effect is relatively larger in one subgroup than another subgroup defined by the factor. If there is imbalance in terms of the prevalence of patients in these subgroups across regions, the factor itself should also be treated as a stratified randomization factor and be included in the analysis model to avoid heterogeneity of treatment effects across regions (Chen et al. 2012).

4.3 Models and the Overall Sample Sizes

The primary objective of a MRCT is to demonstrate the overall treatment effect based on data from the whole trial. Before the discussion of subgroup analyses, we first need to lay out the approaches for the overall analysis. The calculation of the required overall sample size to have desired power for this primary objective is also an important component of the trial design. For ease of explanation, we assume that the endpoint follows a normal distribution. The methodology can be applied to other types of endpoints through the asymptotic normality of the distributions of the estimators of treatment effects. Suppose δ_i is the true treatment effect, N_i is the sample size per treatment group for region i and s is the number of regions in a MRCT. When stratified randomization and analysis are performed with region as a stratification factor, the estimator $\hat{\delta}_i$ of the regional treatment effect based on data of region i is the between-treatment difference of sample means and follows

$$\left(\hat{\delta}_i | \delta_i\right) \sim N\left(\delta_i, \frac{2\sigma^2}{N_i}\right), \tag{4.1}$$

where σ^2 is the variance of the endpoint assumed to be a constant across regions. For unbalanced design or in the case of heterogeneity of variance across regions, modification on the variance in (4.1) is needed. Basically, $\frac{2\sigma^2}{N_i}$ is replaced by the variance of the estimator of the regional treatment effect. Denoting $N = \sum_{i=1}^{s} N_i$ as the total number of patients in each treatment group in the trial, $f_i = N_i/N$ will be the fraction of patients from region i and $\sum_{i=1}^{s} f_i = 1$.

To address possible regional treatment effect differences in a MRCT, we focus on and compare three major models: fixed effects, random effects and the discrete random effects models, in a MRCT setting. With a fixed effects model (FEM), δ_i's are treated as fixed unknown parameters and can be different across regions. The overall treatment effect can be estimated as a weighted average of those regional treatment effect estimators $\hat{\delta}_i$'s

$$\hat{\delta} = \sum_{i=1}^{s} f_i \hat{\delta}_i. \tag{4.2}$$

Given δ_i, $i = 1, 2, \ldots, s$,

$$\left(\hat{\delta} | \delta_i, i = 1, 2, \ldots, s\right) \sim N\left(\bar{\delta}, \frac{2\sigma^2}{N}\right), \tag{4.3}$$

where the expectation $\bar{\delta} = E\hat{\delta} = \sum_{i=1}^{s} f_i \delta_i$ which relies on the regional sample size configuration and the values of δ_i's is the overall treatment effect under the FEM.

A random effects model (REM) has been applied to multi-center clinical trials (Senn 1998; Fedorov and Jones 2005) which have similar setting of multi-regional clinical trials. Hung et al. (2010) also proposed to use a random effects model for sample size calculation for a MRCT. With a REM, the δ_i's are treated as random variables, assumed to be exchangeable (perhaps after the adjustment for key covariates) and follow the same distribution

$$\delta_i \sim N\left(\delta, \tau^2\right) \tag{4.4}$$

where δ is the overall population mean or treatment effect and τ^2 is the between-region variance. Under the REM, from (4.2),

$$\hat{\delta} \sim N\left(\delta, \frac{2\sigma^2 N + \tau^2 \sum_{i=1}^{s} N_i^2}{N^2}\right). \tag{4.5}$$

With (4.5), to detect an overall treatment effect δ at a one-sided level α with power $1 - \beta$, the overall sample size should satisfy

$$\hat{N} = \left[\delta^2 / \left(2\sigma^2(z_\alpha + z_\beta)^2\right) - \left(\tau^2 / \left(2\sigma^2\right)\right) \sum_{i=1}^{s} f_i^2\right]^{-1} \tag{4.6}$$

where z_a is the $1 - a$ quantile of the standard normal distribution. If $\tau^2 = 0$ or no between-region variation, (4.6) becomes the usual sample size formula based on a fixed effects model

$$N_0 = 2\sigma^2(z_\alpha + z_\beta)^2 / \delta^2 \tag{4.7}$$

(with $\bar{\delta} = \delta$) which is smaller than $\hat{N}$ unless $\tau^2 = 0$. Under the REM, combining (4.1) and (4.4),

$$\hat{\delta}_i \sim N\left(\delta, \tau^2 + 2\sigma^2 / N_i\right). \tag{4.8}$$

Rather than $\hat{\delta}$, an alternative estimator of the overall treatment effect with a smaller variance compared to the one in (4.5) is

$$\tilde{\delta} = \sum_{i=1}^{s} w_i \hat{\delta}_i / \sum_{i=1}^{s} w_i = \sum_{i=1}^{s} w_i \hat{\delta}_i / w \sim N(\delta, 1/w) \tag{4.9}$$

where $w_i = 1/(\tau^2 + 2\sigma^2/N_i)$ is the inverse of the variance of $\hat{\delta}_i$ in (4.8) and $w = \sum_{i=1}^{s} w_i$. Using (4.9) for the inference of the overall treatment effect, the required sample size $\tilde{N}$ for detecting δ is the solution of (Quan et al. 2013)

$$\sum_{i=1}^{s} \frac{1}{\tau^2 + 2\sigma^2/\tilde{N} f_i} = \frac{(z_\alpha + z_\beta)^2}{\delta^2} \tag{4.10}$$

which will become (4.7) when $\tau^2 = 0$. One criticism for the use of $\tilde{\delta}$ is that when τ^2 is not small and sample sizes for individual regions are large, $w_i \approx 1/\tau^2$ so that a relatively smaller region (but still with a large regional sample size) will have a weight similar to that of a larger region and $\tilde{\delta}$ will be approximately the simple average of $\hat{\delta}_i$'s.

The discrete random effects model (DREM) proposed by Lan and Pinheiro (2012) assumes that patients are randomly drawn from the s regions with probability p_i from region i. Therefore, the regional sample size vector $(N_1, \ldots, N_s)$ is random with a multinomial distribution $MN(N, p_1, \ldots, p_s)$. Under the DREM,

$$\hat{\delta} \sim N\left(\bar{\delta}^0, \frac{2\left(\sigma^2 + \sum_{i=1}^{s} p_i \left(\delta_i - \bar{\delta}^0\right)^2\right)}{N}\right) \tag{4.11}$$

where $\bar{\delta}^0 = \sum_{i=1}^{s} p_i \delta_i$ as the overall treatment effect under the DREM in a real MRCT will be approximated by $\bar{\delta} = \sum_{i=1}^{s} f_i \delta_i$ when p_i is estimated by f_i. The additional variance component in (4.11) compared to (4.3) is due to the randomness of the regional sample sizes. The required overall sample size for the MRCT for detecting an overall treatment effect $\bar{\delta}^0 = \delta$ based on (4.11) is

$$N_D = 2\left(\sigma^2 + \sum_{i=1}^{s} p_i (\delta_i - \delta)^2\right)(z_\alpha + z_\beta)^2/\delta^2. \tag{4.12}$$

Ideally, p_i represents the proportion of patients in region i among patients worldwide. However, in a real MRCT setting, the sample size of a region may rely more on the other considerations rather than the proportion of patients in the region. For example, for most diseases, even though China has the largest number of patients in the world, in many MRCTs, it may have a much smaller sample size than the United States. In addition, some regions (or countries) may have their own specific requirements for their own regional sample sizes. Thus, in practice, like the overall sample size for the whole trial which depends on the pre-specified design parameters and is a fixed value, the sample size for a region is also more like a fixed value rather than a random variable.

Besides the assumption on the overall treatment effect δ, the use of (4.6), (4.10) and (4.12) for an overall sample size calculation also need the assumption about the between-region variability which can be estimated based on data of previous multi-regional trials of the same drug or the other drugs in the same therapeutic area. Quan et al. (2014) discussed the consideration for the selection of τ^2 for the use of (4.10). Comparing (4.10) and (4.12), we see that the random components for REM and DREM are different. REM treats the true regional treatment effects as random variables. A robust estimate of the associated between-region variability τ^2 needs a reasonably large number of regions and cannot be obtained by solely increasing the overall sample size. That is the reason why a REM is not suitable for a regular subgroup analysis where the number of subgroups is small. On the other hand, DREM treats regional sample sizes as random variables. The corresponding additional variability for the estimate of the overall treatment effect can be managed by increasing the overall sample size (see (4.11)).

When regional treatment effects are homogeneous, (4.11) theoretically should become (4.3) regardless of the sample sizes across the regions. Nonetheless, since δ_i's are unknown parameters, for data analysis, $\sum_{i=1}^{s} p_i \left(\delta_i - \bar{\delta}^0 \right)^2$ will be estimated by $\sum_{i=1}^{s} f_i \left(\hat{\delta}_i - \hat{\delta} \right)^2$ which is almost always positive, where $\hat{\delta} = \sum_{i=1}^{s} f_i \hat{\delta}_i$. For example, the expectation of $\sum_{i=1}^{s} f_i \left(\hat{\delta}_i - \hat{\delta} \right)^2$ is a positive value $2\sigma^2(s-1)/N$ even for the special case of equal δ_i's and $f_i = 1/s$. When δ_i's are different across regions, the expectation will be larger than $2\sigma^2(s-1)/N$.

Under a fixed or discrete random effects model, $\hat{\delta}_i$ converges strongly to the true regional treatment effect δ_i as N_i tends to infinity while $\hat{\delta} = \frac{\sum_{i=1}^{s} N_i \hat{\delta}_i}{N}$ converges to $\delta^0 = \sum_{i=1}^{s} f_i^0 \delta_i$ where $f_i = N_i/N \to f_i^0$ when all regional sample sizes tend to infinity. In other words, $\hat{\delta}_i$ and $\hat{\delta}$ are consistent estimators of δ_i and δ^0, respectively.

Under a regular random effects model, as seen from (4.5), (4.8) and (4.9), when all regional sample sizes tend to infinity, the distribution of $\hat{\delta}_i$ converges to $N(\delta, \tau^2)$, the distribution of $\hat{\delta}$ converges to $N\left(\delta, \tau^2 \sum_{i=1}^{s} \left(f_i^0 \right)^2 \right)$ while the distribution of $\tilde{\delta}$ converges to $N(\delta, \tau^2/s)$. All the asymptotic variances are greater than zero and therefore, the estimators are not consistent estimators of δ (unless the number of regions also tends to infinity which is not realistic with MRCTs). Moreover, the variance of $\hat{\delta}_i$ is never less than τ^2 (the variance of δ_i in (4.4)). In addition, because $w_i = 1/(\tau^2 + 2\sigma^2/N_i) < 1/\tau^2$, $1/w = 1/\sum_{i=1}^{s} w_i > \tau^2/s$ holds regardless of the values of the N_i's. Therefore, no matter whether $\hat{\delta}$ or $\tilde{\delta}$ is used for the inference and no matter how large the N_i's are, there is the possibility that the overall treatment effect may not be significant under a random effects model if τ^2/s is not small (relative to δ). Thus, extreme care should be exercised when a random effects model is selected for the design and analysis of a MRCT. No matter which of these models is chosen, the model for trial design and sample size calculation should be consistent with the model of the pre-planned data analysis. That is, if a fixed effects model will be used for data analysis, the sample size calculation for the trial should also be based on a fixed rather than a random effects model.

4.4   Considerations for Regional Subgroup Analysis

Besides the sample size for the overall treatment effect assessment, one of the other considerations for MRCT design is the allocation of patients across the regions. Some regional (or country) health authorities have specific requirements on their regional sample sizes for regional safety evaluations and/or reasonable assurance probabilities (Uesaka 2009) for consistency assessment of regional treatment effects (e.g., see the Japanese Guidance 2007). Calculation of the assurance probabilities heavily depends on the model for data analysis, the approach for quantifying regional treatment effects and the criteria of consistency. Besides $\hat{\delta}_i$ based on data of only region i, that we have already discussed, to quantify regional treatment effects, another consideration for alternative approaches is to reduce variability through borrowing information among the regions. Quan et al. (2013, 2014) propose the use of the empirical shrinkage estimate

$$\tilde{\delta}_i \approx \frac{\tau^2}{\tau^2 + 2\sigma^2/N_i}\hat{\delta}_i + \frac{2\sigma^2/N_i}{\tau^2 + 2\sigma^2/N_i}\tilde{\delta}$$

and the James-Stein shrinkage estimate

$$\check{\delta}_i = c\hat{\delta}_i + (1-c)\,\hat{\delta}$$

where $c = \dfrac{\sum_{i=1}^{s}\left(\delta_i - \bar{\delta}\right)^2/s}{\sum_{i=1}^{s}\left(\delta_i - \bar{\delta}\right)^2/s + 2s\sigma^2/N}$ and $\bar{\delta} = \dfrac{\sum_{i=1}^{s} N_i \delta_i}{N}$. The parameters in the formulae can be replaced by their corresponding estimates. Note that $\tilde{\delta}_i$ is defined only under a random effects model while both $\hat{\delta}_i$ and $\check{\delta}_i$ are available under both the fixed and random effects models. Discussion and comparison among these estimates can be found in Quan et al. (2014).

In a MRCT study protocol that assumes a low enough level of heterogeneity among regions such that the MRCT is valid, all regions should be treated equally even when some countries have specific requirements. One commonly used approach for a consistency assessment is a treatment by region quantitative (or the test of $\tau^2 = 0$) or qualitative interaction test which is also the common approach for the regular subgroup analysis. With this approach, there will be no strong evidence for inconsistency unless the p-value for the interaction test is small (e.g., less than 0.1). The concern for using the interaction test approach is the low power for the test. Another simple and straightforward approach that is free of formal statistical inference testing is to require all estimated regional treatment effects to be at least π fraction of the observed overall treatment effect (Quan et al. 2010b). That is

$$D_1 > \pi D_{all}, \ldots, D_s > \pi D_{all} \tag{4.13}$$

where D_i can be $\hat{\delta}_i$, $\tilde{\delta}_i$ or $\check{\delta}_i$ and D_{all} can be $\hat{\delta}$ or $\tilde{\delta}$. Minimally, an agency would hope to see positive observed treatment effect for the region of its interest. Thus, the minimum value for π in (4.13) is zero that is the value suggested in the Japanese guidance (2007). It is important to note that this consistency assessment is meaningful only if the overall treatment effect is significant. Therefore, for trial design, we need to evaluate

$$\Pr(D_1 \geq \pi D_{all}, \ldots, D_s \geq \pi D_{all} | \text{significant overall treatment effect}). \qquad (4.14)$$

Clearly, (4.14) depends on the methods for quantifying the regional treatment effects, the number of regions and sample size configuration across the regions.

Example 4.1 A multi-regional diabetes trial was designed to assess the treatment effect of an experimental drug. The primary endpoint was change from baseline in HbA1c at Month 6. To detect an overall between-treatment difference of $\delta = 0.4\%$ with a standard deviation of 1.1% and 95% power at a one-sided significant level of 0.025, a total of $N_0 = 197$ patients per group would be needed when a fixed effects model was used for trial design and data analysis. Suppose a total of 12 countries from 4 geographic locations would participate in the study. A $\tau^2 = 0.02$ would imply a 95% confidence range for δ_i of $\delta \pm 1.96\sqrt{0.02}\%$ or $(0.4 - 1.96\sqrt{0.02})\% = 0.123\%$ to $(0.4 + 1.96\sqrt{0.02})\% = 0.677\%$. The corresponding required overall sample sizes for the random effects model can be derived using the methods discussed above. We examine (4.14) under different scenarios based on the random effects model where the case of 12 regions implies that regions are defined at country level and the case of 4 regions implies that countries are combined to form regions. Results are presented in Table 4.1. In the table, $\hat{P}$, $\tilde{P}$ and $\check{P}$ denote the values of (4.14) when $\hat{\delta}_i$, $\tilde{\delta}_i$ and $\check{\delta}_i$ are used for D_i, respectively; while $\hat{\delta}$ ($\tilde{\delta}$) is used for D_{all} when $\hat{\delta}_i$ or $\check{\delta}_i$ ($\tilde{\delta}_i$) is used for D_i. Clearly, $\hat{P}$ is too small for $s = 12$ while $\tilde{P}$ may be too large with too much information borrowed from the other regions for quantifying regional treatment effects. The R program that implements Table 4.1 can be found in Mao and Li (2016). Those designing MRCTs have also to evaluate the chance of demonstrating consistency under the selected setting at the design stage.

Besides the overall consistency assessment, a regional health authority can request specific regional treatment effect assessment. The Japanese health authority specifies a method as an example for local consistency assessment in their guidance for global new drug development. Based on the method, consistency of treatment effect for region 1 (e.g., Japan) can be declared if

$$D_1/D_{all} > \pi, \qquad (4.15)$$

where different from the π for (4.13), the π for (4.15) has to be at least 0.5. The guidance further requests a large enough sample size for region 1 when designing the MRCT to ensure at least a $1 - \beta'$ assurance probability to demonstrate consistency, i.e.,

$$\Pr(D_1/D_{all} > \pi) \geq 1 - \beta' \qquad (4.16)$$

and $1 - \beta' = 0.8$.

Table 4.1 Probabilities ($\hat{P}$, $\tilde{P}$ and $\breve{P}$)* (%) of demonstrating overall consistency based on (4.14) for different π values and methods for quantifying regional treatment effects

	s = 4			s = 12		
	$\hat{P}$	$\tilde{P}$	$\breve{P}$	$\hat{P}$	$\tilde{P}$	$\breve{P}$
$\pi = 0$						
f_i case 1 (Equal)	77	99	96	12	100	67
f_i case 2*	77	99	96	12	100	67
f_i case 3*	73	99	95	10	100	59
$\pi = 0.15$						
f_i case 1 (Equal)	68	99	94	6	99	53
f_i case 2*	68	99	94	6	99	52
f_i case 3*	64	99	92	5	99	45
$\pi = 0.3$						
f_i case 1 (Equal)	56	98	89	2	99	35
f_i case 2*	55	98	89	2	99	34
f_i case 3*	53	98	87	2	99	29
*The 2 unequal f_i cases:						
	f_i case 2			f_i case 3		
s = 4	$f_1{:}f_2{:}f_3{:}f_4 = 3{:}3{:}4{:}4$			$f_1{:}f_2{:}f_3{:}f_4 = 5{:}5{:}2{:}2$		
s = 12	$f_1{:}\ldots{:}f_6{:}\ldots{:}f_7{:}\ldots{:}f_{12}{=}3{:}\ldots{:}3{:}\ldots{:}4{:}\ldots{:}4$			$f_1{:}\ldots{:}f_6{:}\ldots{:}f_7{:}\ldots{:}f_{12}{=}5{:}\ldots{:}5{:}\ldots{:}2{:}\ldots{:}2$		

* $\hat{P}$, $\tilde{P}$ and $\breve{P}$ represent the probabilities of (4.14) when $\hat{\delta}_i$ (the regular estimator), $\tilde{\delta}_i$ (empirical shrinkage estimator) and $\breve{\delta}_i$ (James-Stein shrinkage estimator) are used for D_i, respectively; while $\hat{\delta}$ ($\tilde{\delta}$) is used for D_{all} when $\hat{\delta}_i$ or $\breve{\delta}_i$ ($\tilde{\delta}_i$) is used for D_i

Under a fixed effects model, Quan et al. (2010a) derive closed form sample size formulae for region 1 to satisfy (4.16) when estimators $\hat{\delta}_1$ and $\hat{\delta}$ are used for D_1 and D_{all}, respectively. Suppose δ_{1c} is the true treatment effect for all regions other than region 1 and $\delta_1 = u\delta_{1c}$, or the true treatment effect for region 1 is a u factor of the true treatment effect of the other regions. Let f_u ($N_1^u = f_u N$) be the corresponding minimum fraction of sample size for region 1 to satisfy (4.16). Then f_u satisfies (Quan et al. 2010a)

$$\frac{\left(z_{1-\alpha} + z_{1-\beta}\right)\sqrt{f_u}\left(u - \pi - \pi\left(u - 1\right)f_u\right)}{\left(1 + (u - 1)f_u\right)\sqrt{1 + \left(\pi^2 - 2\pi\right)f_u}} = z_{1-\beta'},$$

which can be transformed into a cubic equation of f_u giving a closed form solution for f_u. More specifically, if $u = 1$, i.e., $\delta_1 = \delta_{1c} = \delta$, f_u has a simple expression

$$f_1 = \frac{z_{1-\beta'}^2}{\left(z_{1-\alpha} + z_{1-\beta}\right)^2(1 - \pi)^2 + z_{1-\beta'}^2\left(2\pi - \pi^2\right)}.$$

Again, consistency assessment is meaningful only if the overall treatment effect is significant. The conditional probability of local treatment effect consistency given the significance of the overall treatment effect is

$$\Pr\left(\hat{\delta}_1 > \pi\hat{\delta}|\delta_1, \delta_{1c}, \hat{\delta} - z_{1-\alpha}\sigma/\sqrt{N/2} > 0\right).$$

Quan et al. (2010a) provide values of f_u for various u, π, $1 - \beta$ and $1 - \beta'$ when $\alpha = 0.025$ (one-sided). For $\pi=0.5$, $1 - \beta=0.9$ and $1 - \beta'=0.8$, $f_1 = 22.4\%$. That is, region 1 should have at least 22.4% of the overall sample size in order to satisfy (4.16) even when the true treatment effect of region 1 is exactly the same as the true overall treatment effect. Clearly, such a requirement is not practical when there are more than 5 regions since we should treat all regions equally in a study protocol. As for the overall consistency assessment via (4.14), we can also use shrinkage estimates for local consistency assessment.

If another consistency criterion is used for a MRCT, the impact of the configuration of the regional sample sizes on the assurance probability should also be evaluated. Different approaches can be compared under different scenarios before finalizing the MRCT protocol. The overall sample size and regional sample sizes should be considered simultaneously to satisfy the requirements of all parties. Moreover, due to limited availability of data from early phase studies for dose selection, some Phase III confirmatory trials may have two doses of the experimental treatment. Taking into account both efficacy and safety, it is possible that the optimal dose may be different for different regions due to differences in intrinsic and/or extrinsic factors. For efficient dose justification, it will be ideal to collect PK data during the trial. Then a PK/PD model can be built with the incorporation of key baseline patient characteristics or some intrinsic and extrinsic factors as covariates. With such a PK/PD model, regional treatment effects of different doses including doses not studied in the MRCT can be predicted. Trial simulation can also be conducted to obtain more reasonable variability for statistical inference.

In addition, if an Intent-To-Treat (ITT) design is used for a superiority trial, all patients including those who discontinue study medication will be followed and endpoint values will be measured until the pre-planned end of the study. A region with more treatment discontinuations may have a smaller observed regional treatment effect due to the reduction in treatment effect after treatment discontinuation. This should be taken into consideration for consistency assessment when there are unbalanced treatment discontinuations across regions. For a time to event endpoint, treatment may have delayed effect. This lag in treatment effect needs to be accounted for such that enough treatment duration for each patient should be considered when determining the sample size and study duration. Careful planning and even trial simulation may be needed to get the optimal balance of all these design specifications from the MRCT perspective.

4.5 Trial Conduct

A large scale confirmatory MRCT may institute a Data Monitoring Committee (DMC) to monitor cumulative safety and pre-planned interim efficacy data. If a large enough number of patients (e.g., more than 20%) are to be enrolled from a region, the regional agency may suggest having a DMC representative from the region to provide input from the regional perspective (Japanese PMDA 2013). The regional DMC representative could pay close attention to

- Regional enrollment rate or number of events for an event driven trial
- Strict applications of the inclusion and exclusion criteria in the regional sites
- Any major differences in baseline characteristics across the regions
- Consistency of the interim observed regional efficacy and safety profiles compared to the other regions

Example 4.2a The MERIT-HF (Metoprolol Controlled-Release Randomized Intervention Trial in Heart Failure) (The MERIT-HF Study Group 1999) was a MRCT conducted to evaluate the treatment effect of once-daily metoprolol controlled/extended release in patients with Congestive Heart Failure (CHF). One of the primary endpoints was total mortality. A total of 3991 patients were enrolled from 14 countries. Results for individual countries can be found in (Wedel et al. 2001). Two countries had zero events in the active treatment group. They were separately combined with another two separate countries to form a total of 12 regions (Wedel et al. 2001). The trial used an adaptive design with 4 pre-specified interim analyses. Based on the recommendation of the DMC, the trial was stopped early at the second pre-planned interim analysis with an observed overall hazard ratio on total mortality via the fixed effects model of 0.66 (95% confidence interval (0.53, 0.81), and nominal $p = 0.00015$) (The MERIT-HF Study Group 1999). However, a post hoc regional analysis on the final data found that the observed hazard ratio for the United States (sample size: $1071/3991 = 31.58\%$) was 1.06 associated with an observed negative treatment effect. This created a concern among regulators that, perhaps, could have been avoided had the DMC noted this and not stopped the trial early until more information was collected in the United States.

For a MRCT trial with an adaptive design, even when the interim overall result has crossed the pre-specified early stopping boundary, the DMC should still carefully review consistency of treatment effects across different regions and different subgroups to check whether additional data are needed before making the recommendation of early trial stopping. This type of study 'extension' should not be viewed as a new study adaptation in terms of Type I error rate control since the trial results have already met the criterion for declaring significant treatment effect and the Type I error rate has already been appropriately controlled via the adaptive design procedure. To help in the situations, the sponsor should provide guidance in the DMC charter regarding the expectations of what the DMC should look for in subgroup including regional results, and state preferences as to early stopping (or not) of the trial when there is qualitative interaction of treatment effect across regions, especially if that region is viewed as "key" for the success of the drug approval and subsequent labeling. As noted above, the DMC should not focus only on efficacy, but also confirm that there are sufficient safety data to address any safety concern before early trial stopping.

The study sponsor can further improve the quality of data from all the regions by making effort to reduce protocol violations, treatment discontinuation and missing data based on blinded data reviews. For event type endpoints including adverse events, data monitoring can also detect whether consistent criteria of event definition are applied to all regions and potential over/under reporting from some of the regions.

For trials with a time to event endpoint as the primary endpoint, as discussed earlier, the number of events rather than the number of patients is the measure of the amount of information. A more efficient trial design for this type of trial is the common study stopping time design: all patients will stop the study at the common calendar time once the total number of events for the whole study reaches the target. Note that times for site initiation for different regions may be different particularly when some regions have a delayed protocol approval resulting in differences in the regional mean exposure times. In such a scenario, regional sample sizes may need certain adjustment during the conduct of the trial in order to get the appropriate configuration of the regional number of events for consistency assessment of regional treatment effects. There are methods to connect enrollment rate, study duration, sample size and the number of events (Quan et al. 2010a).

4.6 Interpretation of Results

Interpretation of results from a MRCT depends on many factors. One of them is the model applied to derive the trial results. As discussed in Sect. 4.2, compared to a fixed effects model, the application of a random effects model to a MRCT demands a larger sample size for demonstrating the overall treatment effect for the same power and significance level. Several authors (e.g., Fedorov and Jones 2005; Senn 1998) state that with a random effects model for a multi-center trial, we can make probabilistic statements about patients in general, including those from centers we did not include. Since multi-center trials have a similar setting of multi-regional trials, this statement may also apply to MRCTs.

For a fixed effects model, even though region definition is important for regional treatment effect quantification, the overall estimate of treatment effect derived from such a model is more or less independent with whether and how the regions are combined in the analysis. To see this, the overall estimate of treatment effect based on a fixed effects model is

$$\hat{\delta} = \frac{\sum_{i=1}^{s} N_i \hat{\delta}_i}{N} = \frac{N_1 \hat{\delta}_1 + \left(N_2 \hat{\delta}_2 + \cdots + N_s \hat{\delta}_s \right)}{N_1 + (N_2 + \cdots + N_s)} = \frac{N_1 \hat{\delta}_1 + N_2' \hat{\delta}_2'}{N}$$

where $N_2^{'} = N_2 + \cdots + N_s$ and $\hat{\delta}_2^{'} = \frac{N_2 \hat{\delta}_2 + \cdots + N_s \hat{\delta}_s}{N_2^{'}}$. That is, combining regions 2 to s to form a new region will give the same overall estimate of treatment effect and the corresponding variance based on a fixed effects model. However, for a random effects model, like the required overall sample size, the overall trial result depends on how regions are defined.

Example 4.2b In the MERIT-HF trial example, when a total of 12 regions were considered in a random effects model, the overall estimate of hazard ratio was 0.64. The corresponding 95% confidence interval for treatment effect = (0.48, 0.83) was slightly wider than the one via the fixed effects model provided in Example 4.2a because of the incorporation of the between region variability in the analysis under a random effects model. The estimate of the between region variance $\hat{\tau}^2 = 0.004$ was small and the treatment by region interaction was not significant (p = 0.22). It is not unusual for a health authority to examine the trial result of its jurisdiction compared to those outside its jurisdiction. For example, the US FDA often compares the US versus non-US results in their review. If only two regions, the US and non-US, were considered, the results in the MERIT-HF trial would show a significant treatment by region interaction (p = 0.003). There would also be a large $\hat{\tau}^2 = 0.19$ in this case. The random effects model would provide an overall estimate of the hazard ratio of 0.74 (95% CI = (0.39, 1.41)) which was not statistically significantly smaller than 1 (p = 0.187). Note that the 12-region and 2-region random effects models were two different models. The corresponding parameters including the parameters of between-region variance were also different. These results illustrate the importance of pre-defining regions in the study protocol. When there are a reasonably large number of regions, a random effects model can be appropriately used as it provides a robust estimate of the between region variability. When the number of regions is small, a fixed effects model will be more appropriate. Point estimates and the corresponding 95% confidence intervals of the discrete random effects model with 12 and 2 regions were similar to the fixed effects model for this trial example (see Fig. 4.1).

After the demonstration of significant overall treatment efficacy and good safety profile for the entire MERIT-HF trial, regional results should be carefully reviewed as this is the interest of regional health authorities. By definition, the regions are smaller than the entire trial, and due to the reduced regional sample sizes or numbers of events, demonstrating statistical significance of a regional treatment effect based on solely regional data is not realistic unless a study is specifically designed to do so. The focus of regional treatment effect evaluation should be to assess consistency of regional treatment effects. Three approaches for quantifying regional treatment effects and the corresponding consistency assessment for the MERIT-HF trial example when $s = 12$ have been discussed and compared in Quan et al. (2014). It can be seen that the fixed effect estimate that did not borrow information from the other regions showed negative observed treatment effects for the US and Iceland. Thus, consistency of treatment effects for these two countries based on this approach and (4.13) could not be directly declared. It would be of interest to calculate the

HAZARD RATIO

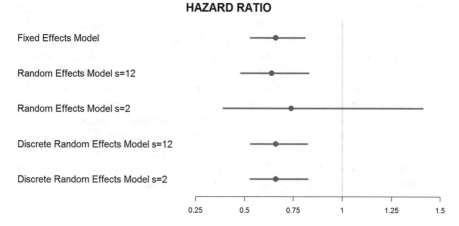

Fig. 4.1 Estimate of hazard ratio (95% CI) for total mortality based on different models (MERIT-HF trial)

probability of observing at least one negative regional treatment effect assuming all regions have the identical true log hazard ratio of $\delta = \log(0.66)$ (note that 0.66 was the observed overall hazard ratio via the fixed effects model). Based on the fixed effects model, the probability is

$$1 - \Pr\left(\hat{\delta}_1 < 0, \ldots, \hat{\delta}_{12} < 0 | \delta\right) = 1 - \prod_{i=1}^{12} | \Pr\left(\hat{\delta}_i < 0 | \delta\right) \tag{4.17}$$

where $\hat{\delta}_i$, $i = 1, \ldots$ 12 are the regional estimates of log hazard ratios of the active treatment versus placebo control that do not borrow information from the other regions and are independent. Keep in mind that a negative $\hat{\delta}_i$ implies a positive observed risk reduction and positive observed treatment effect for region i. Without data of individual patients and with only the regional numbers of events from the publication, we consider asymptotic distribution

$$\hat{\delta}_i \sim N\left(\delta, \hat{\sigma}_i^2\right)$$

where $\hat{\sigma}_i^2$ is the estimate of the variance of $\hat{\delta}_i$ based on data of region i (Quan et al. 2013). For the MERIT-HF trial, the probability (4.17) of observing at least one negative regional treatment effect assuming all regions have the identical true log hazard ratio of $\delta = \log(0.66)$ was 0.892, which was so large that one negative observed regional treatment effect could be expected to be observed by chance. This is in consistent with the other two methods for quantifying regional treatment effects for MERIT-HF, the empirical shrinkage estimate and James Stein shrinkage estimate both showed positive and consistent observed treatment effects for all regions including the US (Quan et al. 2014).

Regional data review should not solely focus on the result of one primary efficacy endpoint. Results of other important endpoints should also be carefully evaluated. For example, in the MERIT-HF trial, even though the observed treatment effect on total mortality (one of the two primary endpoints with a pre-specified alpha of 0.04) for the US patients was negative, the US treatment effects on the second primary endpoint (a composite endpoint of total mortality or all-cause hospitalization with a pre-specified alpha of 0.01) and some of the pre-specified secondary endpoints were all positive. This is referred to as consistency of internal evidence, and is very useful for interpreting inconsistent results of one of the primary endpoints. It is also an informal way of "borrowing" evidence to help interpret the US result for the total mortality primary endpoint. Therefore, utilizing all the evidence, the drug was approved by the FDA with disclosure of the US data in the label.

4.7   The Regular Subgroup Analyses

As any other clinical trials, the regular subgroup analyses should also be of interest in MRCTs. In some occasions, the subgroup analysis results may help us to identify the underlying cause of heterogeneity of regional treatment effects. For example, the heterogeneity in regional treatment effect may be due to the obvious between-region imbalance in a predictive subgroup factor. Diligent blinded data review during the trial may be helpful for us to identify between-region imbalance in a predictive subgroup factor. We can then pre-specify the inclusion of a predictive subgroup factor in the analysis model before data unblinding to reduce the heterogeneity of treatment effects across regions.

The levels for each regular subgroup factor (e.g., only two levels for gender) are generally small. Thus, it is unrealistic to use a random effects model for regular subgroup analysis as the estimate of the between-subgroup variability will not be robust. To borrow information from the alternative subgroups for the assessment of treatment effect for a specific subgroup, we can use James-Stein shrinkage estimate. When the number of subgroup factors is not so small, care should be exercised when interpret the results keeping in mind the multiplicity.

4.8   Discussion

In this chapter, we explore thinking for the design, trial conduct, data analysis and result interpretation of MRCTs treating region as a special subgroup factor. Practitioners can borrow from these experiences to improve future MRCTs. MRCT design components include region definition, endpoint selection, model determination and the others. Agreement from health authorities across regions should first be obtained. When there are disparities in the requirements even after negotiation across agencies, a more conservative approach may be necessary to simultaneously satisfy the needs of all regions with a single MRCT. Separate statistical analysis plans can then be used to analyze the entire data for separate regions (Hemmings and Lewis 2013). Importantly, to avoid being unnecessarily conservative, the familywise Type I error rate should be controlled only within an individual region and not across regions. We have discussed three models for MRCTs. The differences among these three models basically depend on how we treat regional treatment effects as well as regional sample sizes and can be summarized as follows.

- In the fixed effects model, the true regional treatment effects are treated as parameters rather than random variables and could be different across regions. Differing from the fixed effects model, the random effects model treats true regional treatment effects as random variables. With the incorporation of between-region variability, the required overall sample size based on this model can be much larger than that of the fixed effects model particularly if the number of regions is small.
- With the discrete random effects model, patients are assumed to be randomly drawn from different regions. Thus, regional sample sizes are random variables and follow a multinomial distribution. However, in a real MRCT, regional sample sizes have weak relationship with the sizes of the regional patient populations and are more like fixed values.

To be consistent with current practice and for easy interpretation of the trial results, a fixed effects model could be proposed for a MRCT. Then shrinkage estimates could be utilized to quantify regional treatment effects to reduce variability particularly if the number of regions is not small and the regional sample sizes are not large for the assessment of consistency of regional treatment effects.

No matter how well a MRCT is designed and how carefully the trial is conducted, there may be some unexpected regional findings based on pre-planned analyses. Appropriate further analyses should be performed to evaluate whether these findings are real or due to chance. Regional treatment effects on an endpoint should not be interpreted in isolation. Other subgroup analyses should be helpful to address issue of observed heterogeneity of regional treatment effects. For trials with potential early stopping, consistency of treatment effects across subgroups should be confirmed before recommending early trial stopping to avoid potential review issues. The regular subgroup analyses in a MRCT may help us to interpret the potential heterogeneity of regional treatment effects if we observed across region imbalance in some predictive subgroup factors.

Acknowledgements We are grateful for helpful discussion with members of the Society of Clinical Trials QSPI MRCT working group: Yoko Tanaka, Bruce Binkowitz, Gang Li, Josh Chen, Soo Peter Ouyang, Mark Chang Din Chen, Paul Gallo, Qi Jiang, Sandeep Menon, William Wang, Weining Robieson and Sammy Yuan for the preparation of this chapter. Lei Gao, Ming Tan and Ao Yuan also provided valuable comments for improving the manuscript.

References

Chen J, Quan H, Binkowitz B, Ouyang SP, Tanaka Y, Li G, Menjoge S, Ibia E (2010) Assessing consistent treatment effect in a multi-regional clinical trial: a systematic review. Pharm Stat 9:242–253

Chen J, Quan H, Gallo P, Ouyang SP, Binkowitz B (2012) An adaptive strategy for assessing regional consistency in multiregional clinical trials. Clin Trials 9:330–339

European Medicines Agency (2008) Reflection paper on the extrapolation of results from clinical studies conducted outside Europe to the EU-population (draft). EMEA Doc. Ref. CHMP/EWP/692702/2008. Available at https://www.ema.europa.eu/documents/scientific-guideline/reflection-paper-extrapolation-results-clinical-studies-conducted-outside-european-union-eu-eu_en.pdf. Accessed 12 Feb 2019

Fedorov V, Jones B (2005) The design of multicentre trials. Stat Methods Med Res 14:205–248

Girman CJ, Ibia E, Menjoge S, Mak C, Chen J, Agarwal A, Binkowitz B (2011) Impact of different regulatory requirements for trial endpoints in multiregional clinical trials. Drug Inf J 45:587–594

Hemmings RJ, Lewis JA (2013) The implications of current methodological issues for international guidance. Stat Biopharm Res 5:211–222

Hung HMJ, Wang SJ, O'Neill RT (2010) Consideration of regional difference in design and analysis of multi-regional trials. Pharm Stat 9:173–178

ICH E17 (2017) General principle on planning/designing Multi-Regional Clinical Trials. https://www.ich.org/fileadmin/Public_Web_Site/ICH_Products/Guidelines/Efficacy/E17/E17EWG_Step4_2017_1116.pdf

Kawai N, Chuang-Stein C, Komiyama O, Li Y (2007) An approach to rationalize partitioning sample size into individual regions in a multiregional trial. Drug Inf J 42:139–147

Khin NA, Yang P, Hung HMJ, Maung UK, Chen YF, Meeker-O'Connell A, Okwesili P, Yasuda SU, Ball LK, Huang SM, O'Neill RT, Temple R (2013) Regulatory and scientific issues regarding use of foreign data in support of new drug applications in the United States: an FDA perspective. Clin Pharmacol Ther 94:230–242

Lan KKG, Pinheiro J (2012) Combined estimation of treatment effect under a discrete random effects model. Stat Biosci 4(2):235–244

Mao X, Li M (2016) Implementation of multiregional clinical trial design. In: Chen J, Quan H (eds) Multiregional clinical trials for simultaneous global new drug development. Chapman and Hall/CRC, New York, pp 131–147

MERIT-HF Study Group (1999) Effect of metoprolol CR/XL in chronic heart failure: Metoprolol CR/XL Randomised Intervention Trial in Congestive Heart Failure (MERIT-HF). Lancet 353:2001–2007

Ministry of Health, Labour and Welfare of Japan (2007) Basic Concepts for Joint International Clinical Trials

Ministry of Health, Labour and Welfare of Japan (2013) Guideline on Data Monitoring Committee

Quan H, Zhao PL, Zhang J, Roessner M, Aizawa K (2010a) Sample size considerations for Japanese patients in a multi-regional trial based on MHLW Guidance. Pharm Stat 9:100–112

Quan H, Li M, Chen J, Gallo P, Binkowitz B, Ibia E, Tanaka Y, Ouyang SP, Luo X, Li G, Menjoge S, Talerico S, Ikeda K (2010b) Assessment of consistency of treatment effects in multiregional clinical trials. Drug Inf J 44:617–632

Quan H, Li M, Shih JW, Ouyang SP, Chen J, Zhang J, Zhao PL (2013) Empirical shrinkage estimator for consistency assessment of treatment effects in multi-regional clinical trials. Stat Med 32:1691–1706

Quan H, Mao X, Chen J, Shih WJ, Ouyang SP, Zhang J, Zhao PL, Binkowitz B (2014) Multi-regional clinical trial design and consistency assessment of treatment effects. Stat Med 33:2191–2205

Quan H, Mao X, Tanaka Y, Chen J, Zhang J, Zhao PL, Binkowitz B, Li G, Ouyang SP, Chang M (2017) Example-based illustrations of design, conduct, analysis and result interpretation of multi-regional clinical trials. Contemp Clin Trials 58:13–22

Senn SJ (1998) Some controversies in planning and analysing multi-centre trials. Stat Med 17:1753–1765

Tanaka Y, Mak C, Burger B, Lbia EO, Rabbia M, Chen J, Zolotovitski A, Binkowitz B (2011) Points to consider in defining region for a multiregional clinical trial: defining region work stream in PhRMA MRCT key issue team. Drug Inf J 45:575–585

Uesaka H (2009) Sample size allocation to regions in a multiregional trial. J Biopharm Stat 19:580–594

Wedel H, DeMets D, Deedwania P, Fagerberg B, Goldstein S, Gottlieb S, Hjalmarson A, Kjekshus J, Waagstein F, Wikstrand J (2001) Challenges of subgroup analyses in multinational clinical trials: experiences from the MERIT-HF. Am Heart J 142:502–511

Wittes J (2013) Why is this subgroup different from all other subgroups – thoughts on regional differences in randomized clinical trials. In: Fleming TR, Weir BS (eds) Proceedings of the Fourth Seattle Symposium in Biostatistics: Clinical Trials. Springer, New York

Part II
Subgroup Identification and Personalized Medicine

Chapter 5
Practical Subgroup Identification Strategies in Late-Stage Clinical Trials

Pierre Bunouf, Alex Dmitrienko, and Jean-Marie Grouin

Abstract The chapter discusses practical considerations arising in subgroup exploration exercises in late-stage clinical trials. Subgroup identification strategies are commonly applied to characterize the efficacy profile of an experimental treatment based on the results of a failed trial with a non-significant outcome in the overall patient population. Considering this setting, we present a comprehensive overview of relevant considerations related to the selection of clinically candidate biomarkers, choice of statistical models, including the role of covariate adjustment in subgroup investigation, and selection of subgroup search parameters. The subgroup identification methods considered in the chapter rely on the SIDES family of subgroup search algorithms. We discuss applications of this methodology to failed clinical trials and its key features such as biomarker screening, complexity control and Type I error rate control. The statistical methods and considerations discussed in the chapter will be illustrated using a Phase III clinical trial for the treatment of benign prostate hypertrophy.

5.1 Introduction

The broad topic of assessing subgroup effects in late-stage clinical trials has been one of the most important topics from the perspective of both clinical trial sponsors and regulatory authorities (EMA 2019). At a high level, subgroup analysis methods are divided into pre-planned (confirmatory) and post-hoc (exploratory) subgroup

P. Bunouf (✉)
Pierre Fabre, Toulouse, France
e-mail: pierre.bunouf@pierre-fabre.com

A. Dmitrienko
Mediana Inc., Overland Park, KS, USA
e-mail: admitrienko@medianainc.com

J.-M. Grouin
University of Rouen, Rouen, France

© Springer Nature Switzerland AG 2020   117
N. Ting et al. (eds.), *Design and Analysis of Subgroups with Biopharmaceutical Applications*, Emerging Topics in Statistics and Biostatistics,
https://doi.org/10.1007/978-3-030-40105-4_5

analysis. In a confirmatory setting, treatment effects are examined in a small set of prospectively specified subgroups of patients. By contrast, post-hoc subgroup analysis is aimed at examining multiple subgroups in a fairly loose manner. For an overview of exploratory and confirmatory subgroup analysis strategies, including regulatory considerations in subgroup analysis, see, for example, Grouin et al. (2005) and Dmitrienko et al. (2016).

Focusing on post-hoc subgroup analysis, there are several fairly distinct types of post-hoc subgroup investigations carried out in late-stage clinical trials (Lipkovich et al. 2017a). Examples include subgroup analyses aimed at assessing the consistency of the overall treatment effect by examining patient subgroups based on key characteristics such as gender, age, disease severity, etc. Post-hoc subgroup assessments may also be performed to inform future trial designs, especially in the context of failed clinical trials. This is accomplished by reviewing a large set of patient subgroups based on relevant baseline patient characteristics. These characteristics can be collectively referred to as biomarkers and could exhibit prognostic and/or predictive features. Prognostic biomarkers are used in clinical trials for identifying subgroups of patients with different outcomes, e.g., a subgroup of patients with a poor prognosis, irrespective of the treatment. Predictive biomarkers help trial sponsors examine differential treatment effects, e.g., to select a subset of the overall trial population with a marked treatment effect compared to the complementary subset. It is important to know that a biomarker could be neither prognostic nor predictive, purely prognostic without being predictive or simultaneously prognostic and predictive.

Predictive biomarkers have found numerous applications in the development of tailored/targeted therapies (Wang et al. 2007). These therapies are developed to target subgroups based on one or more predictive biomarkers. In this case patients in the subgroup of interest, known as biomarker-positive patients, may derive substantial benefit from the experimental treatment whereas there may be no evidence of treatment benefit in the complementary subgroup (biomarker-negative patients). For this reason, statistical methods for examining predictive properties of biomarkers have attracted much attention in the clinical trial literature (Lipkovich et al. 2017a).

This chapter focuses on practical considerations in an exploratory evaluation of patient subgroups in clinical trials based on a candidate set of potentially predictive biomarkers. Within an exploratory subgroup analysis setting, patient subgroups are not prospectively defined but rather uncovered after the trial ends using a variety of data-driven methods. It is well known that any type of exploratory analysis is subject to optimism bias. For post-hoc subgroup analysis, this means that the trial's sponsor is likely to identify multiple patient subgroups with a strong treatment effect but, in reality, the vast majority of these subgroup findings are false-positive outcomes and ought to be treated with extreme caution. To reduce the likelihood of spurious results in an exploratory setting, it is critical to rely on disciplined approaches to post-hoc subgroup investigations.

Multiple statistical methods have been developed in the literature to support disciplined approaches to post-hoc subgroup analysis, which is known in this context as subgroup identification. Most of these subgroup search procedures, e.g., procedures based on interaction trees or virtual twins, take advantage of recent advances in machine learning and data mining. For a comprehensive review of commonly used subgroup identification procedures, see Lipkovich et al. (2017a) and Dmitrienko et al. (2019).

A popular subgroup identification method, known as SIDES (Subgroup Identification based on Differential Effect Search), was introduced in Lipkovich et al. (2011) and was later extended in several directions to define increasingly more sophisticated subgroup search procedures such as SIDEScreen procedures (Lipkovich and Dmitrienko 2014a) or Stochastic SIDEScreen procedures (Lipkovich et al. 2017b). The resulting family of procedures offers multiple attractive features, e.g., they provide options to control the size of the search space and to filter out non-informative biomarkers, which results in a more efficient subgroup search algorithm. In addition, these procedures come with tools to account for optimism bias, which helps address one of the most important problem plaguing post-hoc subgroup analysis. SIDES-based subgroup search procedures have been successfully applied to dozens of late-stage clinical trials to uncover subgroups with a beneficial treatment effect, see, for example, Hardin et al. (2013) and Dmitrienko et al. (2015). The general SIDES methodology will be utilized in this chapter to investigate patient subgroups in a failed Phase III trial.

The chapter is organized as follows. Section 5.2 introduces a case study based on a Phase III trial for the treatment of benign prostatic hypertrophy (BPH). A high-level summary of SIDES-based subgroup search methods that will be employed in this chapter is provided in Sect. 5.3. Section 5.4 discusses the importance of covariate adjustment in subgroup identification and defines analysis approaches that were utilized in the subgroup search algorithms in the BPH study. Subgroup search results in the BPH trial are presented in Sect. 5.5. Section 5.6 summarizes the results of a simulation study that was conducted to evaluate the performance of subgroup identification methods in the presence of important prognostic variables. The chapter closes with a general discussion in Sect. 5.7.

5.2   Case Study

This section describes a case study that will be used throughout this chapter to illustrate subgroup identification strategies in confirmatory clinical trials. The case study relies on a double-blind Phase III clinical trial that was conducted to compare the effect of an experimental treatment on the irritative and obstructive symptoms in patients suffering from benign prostatic hypertrophy compared to placebo. The primary analysis in the trial was based on the change in the International Prostate Score Symptom (IPSS) total score from randomization to 12 months. A lower value of this endpoint indicates improvement.

The trial was conducted as follows. After the selection visit, patients underwent a 2-month placebo run-in period followed by the randomization visit. If a patient met the eligibility criteria, he or she was expected to take the treatment for 1 year and to undergo visits to the investigator 1, 3, 6, 9, and 12 months after randomization. Patients who discontinued early were asked to undergo a dropout visit at the discontinuation date.

The total sample size in the trial was set to 350 patients (175 patients per arm) to guarantee 90% power based on a mean treatment difference of 2 points, a common standard deviation of 5.5, and a dropout rate of 10%. This treatment difference was regarded as clinically relevant in BPH patients. Of note, the experimental treatment is a plant extract and the reported rate of adverse reactions was quite low. Consequently, a mean treatment difference which is a little lower than 2 is also of clinical interest.

Out of the 364 randomized patients, 359 patients were analyzed in the intention-to-treat (ITT) population, including 177 patients who received the experimental treatment and 182 patients who received placebo. Among them, 18 patients (10.2%) in the experimental arm and 20 patients (11.0%) in the placebo arm did not complete the treatment period mainly due to urinary detention and worsening of the BPH symptoms.

The overall treatment effect was non-significant in this BPH trial and subgroup search methods were applied to better characterize the efficacy profile of the experimental treatment in subsets of the overall population based on information available before randomization. This work focused on selecting biomarkers that help predict treatment response. The chosen biomarkers were utilized to define subgroups of patients who benefitted from the experimental treatment. The subgroup search methodology employed in this trial relied on the SIDES family of subgroup search algorithms (see Sect. 5.3).

5.3 SIDES-Based Subgroup Identification Methods

To prepare for a discussion of subgroup identification strategies in late-stage clinical trials, this section introduces key components of SIDES-based subgroup search algorithms. For more information and technical details, see Lipkovich and Dmitrienko (2014b) and Dmitrienko et al. (2019).

Consider a clinical trial that was conducted to evaluate the efficacy and safety of a novel treatment versus control and suppose that the trial's sponsor is interested in performing a thorough characterization of treatment effects within patient subgroups based on a set of candidate biomarkers. The biomarkers are denoted by $X_1, \ldots, X_m$ and this candidate set includes both continuous and categorical variables (categorical variables can be nominal or ordinal). In what follows, we will define the following key components of the general SIDES method:

1. Subgroup generation algorithm.
2. Subgroup and biomarker selection tools (e.g., complexity control and biomarker screening).
3. Subgroup interpretation tools (e.g., multiplicity adjustment).

5.3.1   Subgroup Generation Algorithm

SIDES-based subgroup search procedures rely on recursive partitioning algorithms to split the trial's population and then split the resulting subsets of the overall population into increasingly smaller subgroups with the ultimate goal of uncovering subgroups of patients who are likely to experience significant treatment benefit. Most commonly, recursive partitioning is carried out using a splitting criterion known as the differential treatment effect criterion. Using the overall population, denoted by S_0, as a starting point, this criterion is applied to each candidate biomarker X. If X is measured on a continuous scale, an optimal split of the parent subgroup S_0 is found by selecting two child subgroups denoted by $S_1(X, c)$ and $S_2(X, c)$. These subgroups are defined as follows

$$S_1(X, c) = \{X \le c\}, \quad S_2(X, c) = \{X > c\}.$$

The two child subgroups can be thought of as biomarker-low and biomarker-high subgroups. The subgroup generation algorithm focuses on evaluating the differential effect between the two child subgroups. In particular, the optimal cutoff c is chosen by minimizing the differential treatment effect criterion, which is given by

$$d(X, c) = 2\left[1 - \Phi\left(\frac{|Z_1(X, c) - Z_2(X, c)|}{\sqrt{2}}\right)\right],$$

where $\Phi(x)$ is the cumulative distribution function of the standard normal distribution and $Z_i(X, c)$ is the appropriately defined test statistics for evaluating the significance of the treatment effect within the subgroup $S_i(X, c)$, $i = 1, 2$. It is worth noting that, unlike the interaction between the treatment effect and biomarker values, the differential effect can be estimated even if the standard deviations are different between the two subgroups.

The choice of the test statistics is determined by the primary endpoint's type, e.g., a log-rank test statistic could be used for survival endpoints. It will be argued in Sect. 5.4.2 that it is important to define test statistics using models with an adjustment for key prognostic covariates. The child subgroup with a stronger treatment effect is retained and is referred to as a promising subgroup. A similar approach is applied to categorical biomarkers. For example, if X is a nominal biomarker with k levels, an optimal split of S_0 is determined by examining all possible ways of partitioning the k levels into two nonempty sets.

As the last step, a pre-set number of promising subgroups with the strongest differential effect is retained (this process is controlled by a parameter known as the subgroup width). The resulting promising subgroups are partitioned further until the maximum number of splits is reached (this process is controlled by a parameter known as the subgroup depth). For example, if the search depth is set to 2, the overall population is split to define Level 1 subgroups and the splitting algorithm is applied to a pre-defined number of most promising subgroups. The selected Level 1 subgroups are now treated as parent subgroups and their children serve as Level 2

Table 5.1 SIDES subgroup search algorithm

Step 1. The parent group representing the overall trial population is defined. The set of promising subgroups is an empty set.

Step 2. Parent subgroups are generated recursively at each pre-defined level. If the maximum level (subgroup depth) is reached, the current parent group is declared a terminal group and the subgroup search algorithm stops. Otherwise:

1. The candidate biomarkers are arranged in terms of the differential treatment effect criterion.
2. For each of the top biomarkers, two child subgroups based on the best split are found. The child subgroup corresponding to the larger positive treatment effect is identified and included in the set of promising subgroups.
3. Each promising subgroup becomes a parent group and Step 2 is repeated.
4. If no child subgroup can be found for any biomarker, the current parent group becomes a terminal group, which means that it is not considered for further partitioning, and Step 2 is repeated for other parent subgroups.

Step 3. A promising subgroup is included in the final set of patient subgroups if the treatment effect p-value within this subgroup is significant at a pre-specified level, e.g., $\alpha = 0.025$.

subgroups. The Level 1 subgroups are defined using a single biomarker whereas the Level 2 subgroups are defined using two biomarkers. A number of restrictions could be imposed, e.g., a sample size restriction is often applied to ensure that the promising subgroups are sufficiently large (smaller promising subgroups are discarded).

A high-level description of the subgroup search algorithm is provided in Table 5.1.

5.3.2 Subgroup and Biomarker Selection Tools

An important feature of any recursive partitioning algorithm is that, without proper constraints, they can easily lead to a very large number of promising subgroups. Roughly speaking, the number of subgroups grows at an exponential rate, which makes it extremely difficult to manage the final set of subgroups. Complexity control tools and biomarker screening tools help reduce the size of the search space, i.e., the set of all promising subgroups, and streamline the process of interpreting the subgroup findings. Complexity control criteria were introduced in the original SIDES procedure, often referred to as the basic SIDES procedure (Lipkovich et al. 2011), and biomarker screens were defined when advanced two-stage SIDES procedures, known as the fixed and adaptive SIDEScreen procedures, were constructed (Lipkovich and Dmitrienko 2014a).

To enable complexity control, the following subgroup pruning tool is typically employed. This tool relies on a penalty parameter denoted by γ, where $0 < \gamma \leq 1$, which is known as the child-to-parent ratio. If this parameter is prospectively defined, a child subgroup is added to the list of promising subgroups only if

$$p_c \leq \gamma p_p,$$

where p_c and p_p are the treatment effect p-values within the child subgroup and its parent group, respectively. In other words, the treatment effect within the selected child subgroup must be much stronger than that within its parent group. It is clear from this definition that, with a smaller child-to-parent ratio, the size of the search space will be reduced due to the fact that the child subgroups that do not provide much improvement over their parents will be discarded.

Next, considering biomarker screening tools, it was shown in Lipkovich and Dmitrienko (2014a) and other papers that the subgroup pruning tool defined above results in efficient complexity control in subgroup identification problems. An application of this tool reduces the burden of multiplicity and ultimately lowers the probability of an incorrect subgroup selection. However, this approach has its limits and, even if most irrelevant child subgroups are discarded, the probability of an incorrect subgroup selection will still be very high if a large number of candidate biomarkers are non-informative. To improve the efficiency of SIDES-based subgroup search, Lipkovich and Dmitrienko (2014a) proposed a family of two-stage algorithms that support an option to screen out those biomarkers that are not good predictors of treatment benefit. Two-stage algorithms are utilized in the fixed and adaptive SIDEScreen procedures that are set up as follows

1. Apply the basic SIDES procedure to the candidate set of biomarkers and select the strongest predictors of treatment benefit using a pre-defined biomarker screening tool.
2. Apply the basic SIDES procedure to the selected set of most promising biomarkers.

A biomarker screening tool is constructed using variable importance scores (Lipkovich and Dmitrienko 2014a) that are computed for the candidate biomarkers in the first stage of this procedure. A biomarker's variable importance score serves as a quantitative measure of its predictive properties. A higher value of the variable importance score indicates that the biomarker modifies treatment effect. On the other hand, if a biomarker is non-informative, e.g., a nuisance variable that is not related to the treatment effect, its variable importance score will be close to 0. Therefore, a biomarker screen can be set up by identifying a fixed number of biomarkers with the highest variable importance score, e.g., the top three biomarkers. The resulting two-stage procedure is termed the fixed SIDEScreen procedure.

A more efficient biomarker screen, which is utilized in the adaptive SIDEScreen procedure, assesses the probability of incorrectly choosing at least one biomarker for the second stage of the procedure under the null case, i.e., under the assumption that all biomarkers are non-informative. To apply the corresponding biomarker screen, the null distribution of the maximum variable importance score is computed using a resampling-based approach and a biomarker is selected for the second stage only if its score is greater than an appropriate threshold computed from the null distribution. Since this null distribution is well approximated by a normal distribution, the adaptive biomarker screen admits the following simple form. Let

E_0 and V_0 denote the mean and variance of the null distribution of the maximum variable importance score, respectively, that are estimated using permutation-based methods. The threshold is defined as follows:

$$E_0 + k\sqrt{V_0},$$

where $k \geq 0$ is a pre-defined multiplier. This multiplier is typically set to 1 or, alternatively, it could be derived from the probability of an incorrect biomarker selection. For example, suppose the trial's sponsor is interested in a stringent biomarker screening tool and this probability is set to 0.1 and thus there is only a 10% chance that at least one non-informative biomarker would be selected for the second stage under the global null hypothesis, i.e., under the assumption that no predictive biomarker is present. In this case, the multiplier needs to be set to $\Phi^{-1}(0.9) = 1.28$.

The resulting SIDEScreen procedures have been shown to efficiently handle nuisance variables and result in superior performance, compared to the basic SIDES procedure, in subgroup identification problems with large sets of candidate biomarkers.

5.3.3 Subgroup Interpretation Tools

It was explained in the Introduction that, to correctly interpret subgroup findings, it is important to quantify the impact of optimism bias. With unconstrained subgroup search, the search space will be broad, which will likely result in substantial multiplicity burden. In other words, the probability of incorrectly selecting at least one patient subgroup without any treatment benefit will be very high and thus treatment effect p-values within the identified subgroups can no longer be trusted.

To estimate this probability and perform appropriate multiplicity adjustments within the final set of promising subgroups, SIDES relies on a resampling-based approach (Lipkovich et al. 2018). To define this approach, consider a subgroup from the final set of promising subgroups. This subgroup is denoted by S and let $p(S)$ denote the treatment effect p-value within this subgroup. By applying a resampling method to the original trial database, a large number of null data sets can be generated. The null data sets are characterized by the fact that there is no treatment effect within any patient subgroup. After that the basic SIDES procedure is applied to each null data set to identify a subgroup with the strongest differential effect. Let q_j denote the treatment effect p-value within the best subgroup selected from the jth null data set, $j = 1, \ldots, m$. A multiplicity-adjusted p-value within the subgroup S is defined as

$$\widetilde{p}(S) = \frac{1}{m} \sum_{j=1}^{m} I\{q_j \leq p(S)\},$$

where $I()$ is the binary indicator. This means that the adjusted p-value associated with the subgroup S is the proportion of the null data sets where the treatment effect is stronger than the effect in S. If the search space is very broad, the adjusted p-value is likely to be much greater than the original p-value. As a consequence, a highly significant p-value may no longer be significant after a resampling-based multiplicity adjustment.

It is worth noting that this resampling-based algorithm assumes a complete null hypothesis (all null hypotheses of no subgroup effect are simultaneously true) and, as a result, the error rate is preserved in the weak sense. An extended version of this resampling-based adjustment can be applied to the two-stage SIDEScreen procedures defined in Sect. 5.3.2. The multiplicity adjusted treatment effect p-values support reliable inferences in the subgroups selected using a SIDES-based subgroup search method.

5.4 Practical Considerations in Subgroup Identification

This section provides a detailed discussion of key practical considerations in subgroup search projects with applications to the BPH trial introduced in Sect. 5.2. This includes important topics such as the selection of candidate biomarkers, choice of the primary analysis model and covariate adjustment in subgroup identification.

5.4.1 Candidate Biomarkers

It was explained in the Introduction that predictive biomarkers play a central role in subgroup identification projects. The choice of biomarkers to be investigated in a particular trial is a crucial aspect of a subgroup identification problem. It is important to specify a set of meaningful candidate biomarkers to ultimately ensure that the chosen subgroups can be interpreted and considered as true clinical population of interest by academic or regulatory bodies rather than an artificial subset. This implies that each biomarker must be clinically relevant and, in addition, the selected combination of biomarker levels should have a clear clinical interpretation.

It is also important to note that the final set of subgroups identified by a SIDES algorithm is driven by the size of the search space. With a large number of biomarkers in the candidate set, the trial's sponsor is likely to face a large set of promising subgroups, which will induce multiplicity and will directly affect the final inferences, as explained in Sect. 5.3.3. Therefore, a project team interesting in carrying out post-hoc subgroup searches must provide clear justifications for the inclusion and exclusion of potentially relevant biomarkers with regard to available patient information.

In our case study, 14 variables grouped into 6 clinical dimensions were selected as candidate predictive biomarkers. The variables and clinical dimensions are defined as follows: patient information available before randomization includes demography, social status, medical history, prior medications, subjective and objective assessments of BPH symptoms, prostatic anatomy based on echography, biological parameters and safety data. Formally, each variable is determined by its assessment time. Some parameters, such as the PSA levels and prostate echography, were measured at the randomization visit only and are referred to as baseline values. Some others, such as objective and subjective assessments of BPH symptoms, were assessed at the selection and randomization visits. As a general rule, we consider the changes from the selection visit to the randomization visit which are referred to as changes during the run-in period.

According to the clinical objectives of subgroup analysis, we focus on the biomarker dimensions related to urology without considering other dimensions possibly involved in prostate inflammation, such as those related to the metabolic syndrome. Totally, 14 variables grouped into 6 dimensions were selected as candidate predictive biomarkers. The variables are defined as follows:

- Demographic characteristics: Body-mass index (BMI) and age (AGE).
- Prior medication: Previous BPH treatment (BPH_TREAT).
- Patient assessments of BPH symptoms:

  - IPSS total score at baseline (IPSSTB) and change during the run-in period (IPSSTR).
  - IPSS obstructive subscore at baseline (IPSSOB) and change during the run-in period (IPSSOR).
  - IPSS irritative subscore at baseline (IPSSIB) and change during the run-in period (IPSSIR).

- Objective assessment of BPH symptoms: Maximum urinary flow at baseline (QmaxB) and change during the run-in period (QmaxR).
- Prostatic anatomy based on echography: Prostate volume (PRVOLB).
- Biology: Total PSA at baseline (PSATB) and free PSA at baseline (PSAFB).

All of these biomarkers are continuous, except for BPH_TREAT, which is a binary variable (Yes/No).

Some variables were not selected since the clinical team could not rationalize possible relationships with IPSS changes in BPH patients. Some other variables cannot formally characterize patient populations. This is the case for the IPSS items (i.e., questions) and the detailed characteristics of patient prostate. It was also stated that all relevant information on medical history and prior medications is summarized in the indicator variable of previous BPH treatment (BPH_TREAT). Lastly, except for the PSA levels, none of the safety variables were selected. The variables available before randomization that were not selected for the candidate set are:

- Demography: Height, weight, smoker, alcohol consumption.
- Social status: Family status, employment.
- Medical history: List of pathologies.

- Prior medications: List of medications.
- Patient assessment of symptoms: IPSS items, including Questions 1 through 7 and Question 8 (Quality of life).
- Prostatic anatomy based on echography: Width, thickness, and height of the prostate and the transition zone and the volume of the transition zone at baseline.
- Safety data: Biological parameters (except for PSA levels) and vital signs.

It is well known that missing observations have a strong impact in subgroup analysis. Missing biomarker values for a particular patient will likely interfere with subgroup search since this patient may not be included in all subgroups based on the selected biomarkers. If biomarker values are missing completely at random, this issue can be overcome by removing the entire patient record from the database; however, the subgroup analysis results may no longer reflect the original trial's population and the sample size will be reduced, which will lead to lower power. In the BPH trial, the original data set was complete for all the candidate biomarkers except for the maximum urinary flow (Qmax) whose values were missing at the selection visit in four patients. These values were missing because the Qmax parameter is calculated from a curve of urinary flow which is corrected from experimental artifacts but the corrected curves were not produced. For the purpose of subgroup analysis, a decision was made to impute the missing Qmax values using the values obtained at an intermediate visit during the placebo run-in period or, if not available, by the Qmax values calculated from the non-corrected curves.

In our case study, all the clinical dimensions are equally relevant. The resulting classes are considered to be equally important and the corresponding biomarkers are to be examined at all levels of the subgroup search algorithm. However, there are many situations wherein some dimensions of candidate biomarkers are less important than others. These could be designated as second-tier or third-tier biomarkers and studied only at lower levels of a subgroup search algorithm. The SIDES-based subgroup search procedures defined in Sect. 5.3 belong to a family of recursive partitioning methods, which means that patient subgroups are constructed sequentially beginning with the overall trial population. Then, targeted "constructs" should be highlighted as much as possible. This can easily be implemented by considering multiple sets of candidate biomarkers that are ordered according to their clinical interest. After that patient subgroups are generated by sequentially introducing the sets of candidate biomarkers by level. This sequential strategy will be illustrated in Sect. 5.5.1.

5.4.2   Primary Analysis Model

Post-hoc subgroup search procedures are often applied using simplified analysis methods that do not match the original primary analysis methodology. For example, a quick review of recent publications that presented the results of SIDES-based subgroup assessments, e.g., Lipkovich and Dmitrienko (2014a,b) or Dmitrienko et al. (2015), reveals that basic methods that relied on the Z-test for the difference in proportions or the log-rank test were utilized whereas the original primary analysis was performed using more complex models that accounted for important covariates.

This section discusses the role of covariate adjustment and selection of the most appropriate primary analysis model in subgroup search procedures. An influential covariate is a variable which has an impact on the outcome values and thus may cause an imbalance among treatment arms which directly affects the evaluation of treatment effect in a trial. As stated in the Introduction, such variables are said to be prognostic of the outcome variable. An adjustment for key prognostic covariates is essential to ensure that subgroup determination is not biased by any imbalance with respect to the prognostic covariates, which is especially important in the context of small patient subgroups. The modeling strategy in post-hoc subgroup assessments should follow the primary analysis in the original trial as much as possible. However, the trial's sponsor could also consider including other covariates in the primary analysis model if it is clear that relevant covariates were not accounted for in the original primary analysis. Additionally, some covariates could be removed from the model, e.g., categorical covariates with too few patients in some of the categories.

Considering the BPH trial, the original primary analysis was performed using an ANCOVA model adjusted for the following covariates (all covariates were included as fixed effects):

- IPSS total score at baseline (IPSSTB).
- Binary variable derived from the IPSS change during the run-in period (i.e., IPSSTR ≤ -3 or IPSSTR > -3).
- Study center.

The results obtained from this model are shown in Table 5.2. The mean treatment difference was 0.30 (the standard error was 0.59) with a one-sided p-value of 0.31.

When subgroup analyses are performed, adjusting for the study center as a fixed effect may not be an optimal strategy since some of the centers will not contribute to the treatment effect evaluation due to empty cells in smaller subgroups.

Table 5.2 Analysis results (ITT population)

	Original primary analysis	Model 0	Model 1	Model 2
Treatment[a]	0.30 (0.59)	0.61 (0.62)	0.51 (0.58)	0.72 (0.58)
	$p = 0.31$	$p = 0.16$	$p = 0.19$	$p = 0.10$
IPSSTB	$p < 0.001$		$p = 0.005$	$p = 0.005$
IPSSTR (binary)	$p < 0.001$			
CENTER	$p = 0.09$			
IPSSTR (continuous)			$p < 0.001$	$p = 0.005$
BPH_TREAT				$p = 0.02$
IPSSIR (IPSSOR)				$p = 0.01$
PRVOLB				$p = 0.02$

[a]Mean treatment difference (standard error) and one-sided p-value

Consequently, the center term was removed from the model. Another modification concerns the IPSSTR variable. A dichotomized version of IPSSTR was a pre-specified covariate in the primary analysis model because the randomization scheme was stratified by this factor. It was natural to replace this binary variable with the original continuous variable to gain precision in the treatment effect estimation.

Based on these considerations, the following models were examined as candidate analysis models for the subgroup identification exercise. Model 0 was based on a simple analysis of variance (ANOVA) model without covariate adjustment, i.e., it included only the treatment term, and Model 1 was defined as analysis of covariance (ANCOVA) model that incorporated two continuous covariates (IPSSTB and IPSSTR). This model can be interpreted as a model which is very similar to the original primary analysis model but optimized for subgroup analysis. These models were fitted to the data in the ITT population and the evaluation results are presented in Table 5.2. Considering Model 0, the mean treatment difference was 0.61 with a one-sided p-value of 0.16. Similarly, the mean treatment difference based on Model 1 was 0.51 with a one-sided p-value of 0.19.

Furthermore, another model, termed Model 2, was set up based on a more informed adjustment strategy wherein several influential covariates were added to the covariates considered in Model 1. The selection of covariates was based on a multi-step procedure and the covariates used in Model 1 (IPSSTB and IPSSTR) were forced into the resulting model in each step. It is important to note here that covariate influence was assessed without adjusting for the treatment effect. First, each candidate covariate was tested separately using a model with the two forced covariates IPSSTB and IPSSTR. Then all covariates that were significant at a 0.1 level were included in a multivariate model. The last step consisted of removing non-significant covariates in a step-down manner, namely, the least significant covariate was removed if the corresponding p-values was greater than 0.1. This operation was repeated until no non-significant covariates were left in the model.

Again, the context of post-hoc subgroup analysis imposes a clear rationale for the selection of candidate covariates which must be based on sound clinical arguments. In this case study, the candidate covariates for the primary analysis model were chosen from the set of 14 candidate biomarkers defined in Sect. 5.4.1. It is important to mention here that IPSSTB and IPSSTR are the sums of the related quantities for obstructive and irritative symptoms, i.e., IPSSTB = IPSSOB + IPSSIB and IPSSTR = IPSSOR + IPSSIR. Consequently, IPSSIB and IPSSOB on the one side and IPSSIR and IPSSOR on the other side yield the same results as IPSSTB and IPSSTR are included in the model. The one-sided p-values corresponding to the biomarker effects within the model used in the first step are presented in Table 5.3. It should also be mentioned that the one-sided p-values of the two covariates that were forced into the models, i.e., IPSSTB and IPSSTR, were $p < 0.001$ and $p = 0.02$, respectively. According to the initial assessments summarized in Table 5.3, BPH_TREAT, IPSSIR (IPSSOR) and PRVOLB were selected for the next step. Since the corresponding p-values in the multivariate model were all below 0.1, the three biomarkers, in addition to IPSSTB and IPSSTR, were chosen as the covariates for Model 2.

Table 5.3 Analysis of each
biomarker using the model
with the two forced covariates
IPSSTB and IPSSTR (ITT
population)

Biomarker	p-value
BMI	$p = 0.35$
AGE	$p = 0.08$
BPH_TREAT	$p = 0.03$
IPSSIB (IPSSOB)	$p = 0.33$
IPSSIR (IPSSOR)	$p = 0.02$
PSATB	$p = 0.06$
PSATB	$p = 0.07$
QmaxB	$p = 0.36$
QmaxR	$p = 0.22$
PRVOLB	$p = 0.02$

Table 5.4 Analysis of
treatment by biomarker
interactions (ITT data set)

Biomarker	Interaction effect p-values		
	Model 0	Model 1	Model 2
BMI	$p = 0.24$	$p = 0.28$	$p = 0.20$
AGE	$p = 0.001$	$p = 0.01$	$p = 0.015$
BPH_TREAT	$p = 0.05$	$p = 0.12$	$p = 0.11$
IPSSTB	$p = 0.001$	$p = 0.002$	$p = 0.004$
IPSSTR	$p = 0.11$	$p = 0.13$	$p = 0.11$
IPSSOB	$p = 0.08$	$p = 0.13$	$p = 0.16$
IPSSIB	$p = 0.001$	$p = 0.002$	$p = 0.002$
IPSSOR	$p = 0.40$	$p = 0.59$	$p = 0.59$
IPSSIR	$p = 0.03$	$p = 0.03$	$p = 0.02$
PSATB	$p = 0.31$	$p = 0.30$	$p = 0.32$
PSAFB	$p = 0.15$	$p = 0.17$	$p = 0.19$
QmaxB	$p = 0.44$	$p = 0.21$	$p = 0.25$
QmaxR	$p = 0.06$	$p = 0.08$	$p = 0.06$
PRVOLB	$p = 0.41$	$p = 0.41$	$p = 0.44$

Table 5.4 summarizes additional modeling results. Note that Models 0 and 1 used in this table refer to the original models with two additional terms (biomarker and treatment-by-biomarker interaction). Focusing on the results obtained from Model 2, the mean treatment difference based on this model was 0.72 and the corresponding one-sided treatment effect p-value was 0.10. Although the adjustment strategy in Model 2 does not overturn the analysis results in terms of statistical significance, treatment effect variability across the selected models emphasizes the impact of covariate adjustment.

Finally, a simple assessment of predictive properties of the selected candidate biomarkers was performed using the models defined above. Under a linearity assumption, the predictive strength of a biomarker can be examined by including this biomarker as a covariate along with the corresponding treatment-by-biomarker interaction in any of the three models. The one-sided p-values associated with the interaction terms are listed in Table 5.4. It follows from this table that AGE,

IPSSTB, IPSSIB, IPSSIR and, to a lesser extent, QmaxR were linearly predictive of treatment response. However, it is possible that other biomarkers could be predictive of treatment benefit but the relationship between a biomarker and the treatment effect could be non-linear. To perform a comprehensive evaluation of predictive properties of the candidate biomarkers, more general methods such as SIDES should be employed. These methods will be discussed in the next section.

5.5 Subgroup Search Strategies in the BPH Trial

The basic SIDES procedure as well as SIDEScreen procedures were applied to the BPH trial to identify subgroups of patients who experienced enhanced treatment benefit. The primary analysis was performed using Model 2 defined in Sect. 5.4.2, i.e., the treatment effect was evaluated using an ANCOVA model adjusted for four continuous covariates (IPSSIR, PRVOLB, IPSSTB and IPSSTR) and one categorical covariate (BPH_TREAT).

It is explained in Sect. 5.3 that several parameters need to be defined before a SIDES-based subgroup search procedure can be applied to a clinical trial database. This includes the maximum number of child subgroups for a given parent subgroup (search width), maximum number of biomarkers to define a patient subgroups (subgroup depth), smallest acceptable size of a promising subgroup, etc. Based on the sample size and number of candidate biomarkers in the BPH trial, the following parameter values were selected:

- The search width was set to 2 or 3 and the search depth was set to 2. These values were believed to facilitate the process of identifying patient subgroups without unduly increasing the complexity of interpretation.
- The smallest sample size per subgroup was fixed at 60 or 120 patients, which is roughly a sixth or a third of the total number of patients in the trial.
- The child-to-parent ratio γ was set to 1 to enable liberal complexity control.
- To help speed up subgroup search, the continuous biomarkers from the candidate set were discretized by converting them into categorical variables with 20 levels based on the 20 percentile groups (unless the number of unique values was already less than 20).

Two approaches to conducting subgroup searches will be compared and contrasted in the context of the BPH trial:

- A less formal subgroup search strategy, which relies on general subgroup exploration without explicitly controlling the Type I error rate, i.e., the probability of incorrectly identifying at least one patient subgroup where there is no treatment benefit within any subset of the overall trial population. To expand the final set of promising subgroups, the search width was set to 3 and the smallest sample size per subgroup was set to 60 patients.

- A more formal subgroup search strategy, which employs a number of tools described in Sect. 5.3 to control the error rate and thus can support hypothesis generation for subsequent clinical trials. To focus on a set of most important subgroups with a larger number of patients, the search width was set to 2 and the smallest sample size per subgroup was set to 120 patients.

The search results based on the two strategies are presented below.

5.5.1 Less Formal Subgroup Search Strategy

As part of a purely exploratory approach to subgroup evaluation, the basic SIDES procedure was applied to the BPH trial with the parameters listed above. The subgroup search results are summarized in Table 5.5. Since the search width and depth were equal to 3 and 2, respectively, the final set of subgroups is expected to consist of 12 subgroups, i.e., 3 first-level subgroups defined using a single biomarker and 9 second-level subgroups defined using two biomarkers. However, one of the second-level subgroups did not satisfy the pre-defined complexity control criterion and the table defines the 11 promising subgroups selected by the basic SIDES procedure. The table also lists the key characteristics, including the total number of patients in a subgroup, mean treatment difference and one-sided treatment

Table 5.5 Promising patient subgroups selected by the basic SIDES procedure (child-to-parent ratio $\gamma = 1$)

Subgroup	Total sample size	Mean treatment difference	Raw p-value	Adjusted p-value
QmaxR > -2.25	281	1.57	0.0074	0.6833
QmaxR > -2.25 and BPH_TREAT = No	165	2.28	0.0027	0.4890
QmaxR > -2.25 and AGE ≤ 68.5	183	2.13	0.0048	0.5964
QmaxR > -2.25 and IPSSIB > 7	145	2.89	0.0015	0.3751
IPSSIB > 7	185	2.25	0.0043	0.5787
IPSSIB > 7 and BMI > 25.7	115	3.43	0.0006	0.2374
IPSSIB > 7 and IPSSIR > -1	121	2.90	0.0032	0.5243
IPSSIB > 7 and AGE ≤ 63.5	77	4.30	0.0009	0.2895
AGE ≤ 63.5	154	2.15	0.0105	0.7426
AGE ≤ 63.5 and IPSSTB > 15.5	92	3.82	0.0013	0.3542
AGE ≤ 63.5 and IPSSIR > -1	87	3.54	0.0022	0.4502

effect p-value (raw p-value). In addition, the multiplicity-adjusted p-value was computed within each of the subgroups. The adjusted p-values were found using the resampling-based algorithm described in Sect. 5.3.3 using 10,000 null data sets.

It follows from Table 5.5 that the mean treatment difference in some of the patient subgroups was well above the desirable level of 2 points. For example, there was strong evidence of treatment benefit in a large subgroup with 145 patients that was defined based on QmaxR and IPSSIB, i.e., Subgroup 4 (QmaxR > −2.25 and IPSSIB > 7). The mean treatment difference within this subgroup was 2.89 and the corresponding one-sided treatment effect p-value was highly significant ($p = 0.0015$). The efficacy signal was much stronger in smaller subgroups, e.g., the mean treatment difference in Subgroup 8 (IPSSIB > 7 and AGE ≤ 63.5) with 77 patients was 4.3.

The first-level partitioning yields three subgroups that are based on biomarkers pertaining to the three-dimensional classification scheme described in Sect. 5.4.1. These biomarkers are AGE (demographic characteristics), QmaxR (objective assessment of BPH symptoms) and IPSSIB (subjective assessment of BPH symptoms). Subgroup 9 (AGE ≤ 63.5) consisted of the 42.8% youngest patients whereas Subgroup 5 (IPSSIB > 7) focused on the 51.5% more severe patients on the irritative symptoms at baseline. Subgroup 1 (QmaxR > −2.25) should be interpreted as an exclusion of a small subset of patients (14.4%) with an abnormal worsening of obstructive symptoms during the run-in period. In the absence of a rationale to justify this finding, this subgroup is cautiously considered as not clinically relevant for interpretation. Within the set of first-level subgroups, the mean treatment difference was increased up to 2.25 in patients with IPSSIB > 7. Next, considering the second-level subgroups, as noted above, the mean treatment difference could reach 4.3, see Subgroup 8 based on IPSSIB and AGE. This subgroup represented 21.4% of the trial's population. An interesting aspect of this subgroup is that, judging by the subgroup sizes and the additivity of treatment effects, IPSSIB and AGE appear to be independent predictors of treatment response. It is also interesting to note that some of the identified biomarkers are predictive of greater treatment effect on one or the other IPSS subscore. To summarize, based on subgroup analyses of the IPSS subscores, higher values of IPSSIR and IPSSIB were predictive of stronger treatment effects on the irritative symptoms whereas lack of previous treatment for BPH (BPH_TREAT = No) and lower patient age were associated with greater treatment effect on the obstructive symptoms.

The discussion presented above focused on a clinical interpretation of subgroup effects. It is broadly recognized that appraising clinical relevance and plausibility of identified patient subgroups plays a key role in understanding how these subgroups can be used to inform the design of future trials. It was explained earlier in the chapter that clinical relevance is insured by the selection of candidate biomarkers which unambiguously characterize the clinical status of patients. However, plausibility of the subgroup analysis findings often remains questionable, especially from the perspective of replicating the subgroup effect in other trials.

The following approach based on the reference treatment effect serves as a useful tool for assessing the relevance of subgroup findings in addition to significance tests and clinical interpretation. Recall that SIDES-based subgroup identification methods rely on the assessment of the differential effect between child subgroups with the same parent groups. In simple terms, the differential effect can be defined as the difference between the mean treatment effects estimated in a pair of subgroups. Assume that a larger value of the endpoint indicates a beneficial effect. Let d_1 denote the mean treatment difference in the promising subgroup (subgroup with a beneficial treatment effect) and let d_2 denote the mean treatment difference in the complementary subgroup.

In general, a strong differential effect can be caused by a higher value of d_1 but also by an unexpectedly low value of d_2. A straightforward method to evaluate a substantial differential effect in a pair of subgroups is to check whether the treatment effect in the complementary subgroup is realistic. However, d_1 and d_2 can be computed based on observations with high variability that mitigate the relevancy of this approach. To address this problem, one may consider instead the effect sizes, denoted by θ_1 and θ_2, within each of the two subgroups and compare them.

An alternative more formal approach is based on the value of θ_2 in the complementary subgroup. This value can be 0 when an active treatment is compared to placebo, as there is often no reason for the treatment effect to be in favor of placebo. By extension, θ_2 can also be represented as a proportion of the effect size in the overall trial population denoted by θ, i.e., $\theta_2 = r\theta$, where $r \leq 1$. Conditional on θ and θ_2, the effect size in the promising subgroup, i.e., θ_1, depends only on the subgroup sizes n_1 and $n_2 = n - n_1$, i.e.,

$$\theta_1 = (n - (n - n_1)r)\theta/n_1.$$

The reference treatment effect (RTE) is defined as the value of θ_1 corresponding to the null hypothesis of no treatment effect in the complementary subgroup, i.e., $r = 0$. In this case, it is easy to see that RTE is equal to $n\theta/n_1$. It is reasonable to consider the effect size in the promising subgroup (θ_1) to be plausible if it does not substantially exceed RTE. In other words, θ_1 should remain in an acceptable range of values under reasonable assumptions on the treatment effect in the complementary subgroup given the effect size in the overall population.

To explain the rationale behind RTE, note that the metric does not represent the maximum possible value of the effect size in the promising subgroup (in fact, the maximum does not exist) since the effect size in the complementary subgroup could be potentially negative. But, since a strong negative effect within the complementary subgroup may not be plausible, the assumption of no effect could serve as a useful reference point when evaluating the magnitude of the treatment effect in the corresponding promising subgroup. To illustrate this concept, consider Subgroup 5 (IPSSIB > 7) and Subgroup 8 (IPSSIB > 7 and AGE $\leq$ 63.5) selected by the basic SIDES procedure (see Table 5.5). There are 185 patients in Subgroup 5 and $\theta_1 = 0.39$. Given that the effect size in the overall trial population is $\theta = 0.13$, it is easy to see that RTE is equal to 0.26 for this promising subgroup and thus the observed effect size greatly exceeds RTE (note that the ratio of the observed effect size to RTE is 1.5). By contrast, when examining Subgroup 8 with 77 patients, we see that $\theta_1 = 0.76$ and RTE is equal to 0.61. The resulting ratio is 1.25, which indicates that the treatment effect in Subgroup 8 is more plausible than that in Subgroup 5 according to the RTE criterion.

Before continuing to the discussion of more formal subgroup evaluation strategies that incorporate Type I error rate control, it is helpful to present subgroup search methods based on ordered biomarker sets. The possibility of defining several biomarker sets and introducing them at different levels of a SIDES-based algorithm was mentioned in Sect. 5.4.1. There are many situations where candidate biomarkers characterize different aspects of the condition of interest and it would be natural to introduce the pre-defined sets of biomarkers sequentially to align the subgroup search process with a trial's clinical objectives.

In what follows we will consider a subgroup search exercise based on the following three-level classification scheme:

- Demography and medical history: AGE, BMI, BPH_TREAT.
- Objective assessment of symptoms: QmaxB, QmaxR, PRVOLB, PSAFB, PSATB.
- Subjective assessment of symptoms: PSSIB, IPSSOB, IPSSIR, IPSSOR, IPSSTB, IPSSTR.

The basic SIDES procedure was applied to the ordered biomarker sets and the resulting patient subgroups are listed in Table 5.6. It is clear that, compared to the standard subgroup search approach with a single biomarker set presented in

Table 5.6 Promising patient subgroups selected by the basic SIDES procedure with three ordered sets of biomarkers

Subgroup	Total sample size	Mean treatment difference	Raw p-value
AGE $\leq$ 63.5	154	2.15	0.0105
AGE $\leq$ 63.5 and PRVOLB $>$ 44.7	75	3.63	0.0021
BPH_TREAT = No	210	1.42	0.0311
BPH_TREAT = No and QMAXR $>$ -1.14	141	2.52	0.0023
BPH_TREAT = No and QMAXR $>$ -1.14 and AGE $\leq$ 64.5	77	4.15	0.0013
BPH_TREAT = No and QMAXR $>$ -1.14 and IPSSTB $>$ 15.5	76	4.34	0.0008
BPH_TREAT = No and QMAXR $>$ -1.14 and IPSSIB $>$ 7	62	4.82	0.0009
BPH_TREAT = No and AGE $\leq$ 63.5	99	3.40	0.0022
BPH_TREAT = No and PSATB $\leq$ 3.37	143	2.36	0.0069
BPH_TREAT = No and PSATB $\leq$ 3.37 and IPSSIR $>$ -1	75	4.33	0.0015
BPH_TREAT = No and PSATB $\leq$ 3.37 and AGE $\leq$ 63.5	77	4.29	0.0011
BPH_TREAT = No and PSATB $\leq$ 3.37 and IPSSTB $>$ 16.5	76	3.97	0.0023
BMI $>$ 28.8	92	1.90	0.0336

Table 5.7 Promising patient subgroups selected by the basic SIDES procedure with two ordered sets of biomarkers

Subgroup	Total sample size	Mean treatment difference	Raw p-value
AGE ≤ 63.5	154	2.15	0.0105
AGE ≤ 63.5 and IPSSTB > 15.5	92	3.82	0.0013
AGE ≤ 63.5 and IPSSIB > 7	77	4.29	0.0009
AGE ≤ 63.5 and IPSSIR > −1	87	3.55	0.0022
BPH_TREAT = No	210	1.42	0.0311
BPH_TREAT = No and IPSSTB > 16.5	99	3.77	0.0008
BPH_TREAT = No and AGE ≤ 63.5	99	3.40	0.0022
BPH_TREAT = No and QmaxR > −1.15	140	2.55	0.0029
PSAFB ≤ 1.07	283	1.00	0.0714
PSAFB ≤ 1.07 and BPH_TREAT = No	163	2.33	0.0050
PSAFB ≤ 1.07 and AGE ≤ 63.5	134	2.77	0.0038
PSAFB ≤ 1.07 and QmaxR > −2.28	221	1.64	0.0163

Table 5.5, the first-tier biomarkers such as BPH_TREAT and BMI play a prominent role in this setting.

Exploratory subgroup analysis can also be aimed at guiding clinicians to examine patients on the basis of available information. Demographic, medical history and biological data are commonly available before clinicians could conduct further investigations on the basis of the IPSS questionnaire, urodynamic exams and prostate volume assessments. We can address these goals using a strategy with the following ordered sets of biomarkers that are applied at the first and second levels of the subgroup search algorithm:

- AGE, BMI, PSAFB, PSATB, BPH_TREAT.
- IPSSIB, IPSSOB, IPSSIR, IPSSOR, IPSSTB, IPSSTR, QmaxB, QmaxR, PRVOLB.

The patient subgroups identified by the basic SIDES procedure are shown in Table 5.7. The results presented in this table can aid in determining whether or not it is worthwhile conducting certain types of exams in some groups of patients to prescribe the active treatment. For example, Table 5.7 confirms that the IPSS questionnaire should be submitted to patients who are younger than 63.5 years.

5.5.2 More Formal Subgroup Search Strategy

When interpreting the subgroup findings in the BPH trial, it is important to keep in mind that many apparent subgroup effects discovered in post-hoc subgroup investigations may be spurious. Using Table 5.5 as an example, the raw p-values presented in this table suggest that there was strong beneficial treatment effect within all of the selected subgroups; however, this is likely to be misleading since hundreds of tests were carried out to arrive at the subgroups presented in this table. After the resampling-based multiplicity adjustment was applied, none of the multiplicity-adjusted p-values was even remotely significant. For example, despite a very strong treatment effect, the multiplicity-adjusted p-value in Subgroup 8 (IPSSIB $>$ 7 and AGE $\leq$ 63.5) was greater than 0.2 ($p = 0.2895$). The same resampling-based algorithm was utilized to estimate the overall Type I error rate (probability of incorrectly identifying at least one subgroup as promising) associated with the basic SIDES procedure. In this particular case, the error rate was 0.87, which clearly shows that the burden of multiplicity was too high to support reliable inferences in this setting.

To enable a more formal framework for post-hoc subgroup investigations, it may be useful to explore the impact of complexity control on the Type I error rate. To lower the burden of multiplicity in this subgroup identification problem, one could try imposing stricter complexity control criteria but, even with a very conservative approach, complexity control does not typically have much impact on the overall Type I error rate. For example, when the child-to-parent ratio was set to 0.1, the basic SIDES procedure identified only four promising subgroups. The other subgroups were discarded since they did not provide meaningful improvement relative to their parents. This resulted in a smaller search space but the associated error rate was still over 0.7. This is due to the important fact that complexity control tools focus on shrinking the set of potential promising subgroups (search space) but cannot lower the level of background noise caused by non-informative biomarkers included in the candidate set. To reduce the influence of non-informative biomarkers and build efficient subgroup search algorithms, SIDEScreen procedures with an appropriate biomarker screen need to be employed.

A more formal approach to subgroup search that set the stage for hypothesis generation will be illustrated using two-stage SIDEScreen procedures. As explained above, these procedures were applied with a stricter set of parameters to reduce the size of the search space and support more reliable inferences (in particular, the search width was set to 2 and the smallest sample size per subgroup was set to 120 patients).

Table 5.8 displays the two promising patient subgroups identified by the fixed SIDEScreen procedure with the biomarker screen that chose the best two biomarkers for the second stage of the procedure. To identify the two biomarkers, the variable importance scores (Lipkovich and Dmitrienko 2014a) were computed for all candidate biomarkers to assess their predictive ability. The highest variable importance scores were 1.49 (IPSSIB) and 1.24 (AGE) and thus IPSSIB and AGE were identified as the biomarkers with the strongest predictive effects. Using these biomarkers, two single-level subgroups were chosen in the second stage of the SIDEScreen procedure (a second-level subgroup based on the two biomarkers was excluded due to the complexity control constraints). It is easy to see that these subgroups correspond to Subgroups 5 and 9 from Table 5.5.

A quick comparison of the adjusted p-values for the first three subgroups listed in Table 5.8 to those derived using the basic SIDES procedure (see Table 5.5) reveals that the fixed SIDEScreen procedure with the biomarker screen based on two biomarkers helped substantially reduce the burden of multiplicity in this subgroup identification problem. Indeed, the adjusted p-values produced by the fixed SIDEScreen procedure were considerably smaller than those produced by the basic SIDES procedure, e.g., the adjusted p-value for Subgroup 1 (IPSSIB $>$ 7) was 0.1053 compared to 0.5787 with the basic procedure. In addition, the overall Type I error rate was lowered to 0.35. Furthermore, when the fixed SIDEScreen procedure that chose a single biomarker for the second stage was applied, only one patient subgroup was identified (IPSSIB $>$ 7). The adjusted treatment effect p-value within this subgroup was 0.0968, which was even lower than the adjusted p-value shown in Table 5.8, and the associated Type I error rate was 0.33.

The exploration of patient subgroups with enhanced treatment effect continued by applying the adaptive SIDEScreen procedure to the BPH trial. The biomarker screen was constructed by setting the multiplier k, defined in Sect. 5.3.2, to 0. This resulted in a threshold for variable importance scores that corresponded to a 50% probability of incorrectly selecting a non-informative biomarker for the second stage of the procedure (the probability is computed under the assumption that no predictive biomarker is present). The mean and standard deviation of the null distribution of the maximum variable importance score were given by:

$$E_0 = 1.348, \quad \sqrt{V_0} = 0.647.$$

These quantities were estimated using 1000 null data sets. Therefore the threshold for variable importance scores used in this subgroup search algorithm was $E_0 + k\sqrt{V_0} = 1.348$. It was noted above that the highest variable importance scores were 1.49 (IPSSIB) and 1.24 (AGE) and thus only IPSSIB passed this biomarker screen. Using this biomarker, the SIDEScreen procedure identified a single subgroup shown in Table 5.9. The same subgroup was found by the fixed SIDEScreen procedure with the biomarker screen that chooses a single biomarker for the second stage

Table 5.8 Promising patient subgroups selected by the fixed SIDEScreen procedure (biomarker screen with two top biomarkers)

Subgroup	Total sample size	Mean treatment difference	Raw p-value	Adjusted p-value
IPSSIB $>$ 7	185	2.25	0.0043	0.1053
AGE $\leq$ 63.5	154	2.15	0.0105	0.2003

Table 5.9 Promising patient subgroups selected by the adaptive SIDEScreen procedure (biomarker screen with the multiplier $k = 0$)

Subgroup	Total sample size	Mean treatment difference	Raw p-value	Adjusted p-value
IPSSIB $>$ 7	185	2.25	0.0043	0.0782

(IPSSIB > 7). However, due to a more efficient approach to handling multiplicity, the adjusted p-value within this subgroup was more significant than that computed using the fixed SIDEScreen procedure, namely, the adjusted p-value approached borderline significance ($p = 0.0782$) and, additionally, the overall Type I error rate was lowered to 0.21.

Another approach to addressing the high burden of multiplicity in subgroup search problems is restricting the subgroup search space by splitting the set of candidate biomarkers into two or more ordered sets. This general approach was illustrated in Table 5.6 using three ordered sets of biomarkers that represented (1) demography and medical history, (2) objective assessment of BPH symptoms and (3) subjective assessment of BPH symptoms.

As an illustration, the adaptive SIDEScreen procedure with $k = 0$ was applied to these three biomarker sets and the results are summarized in Table 5.10. It can be seen from this table that a sequential introduction of biomarkers from ordered candidate sets tends to decrease the adjusted p-values as well as the overall Type I error rate. In this particular case, the Type I error rate was 0.18.

As the final note, even though we recognize the importance of lower multiplicity-adjusted p-values and Type I error rate, we would like to stress that this should not be the only goal to pursue when designing post-hoc subgroup analysis strategies. It is clear that one can easily achieve significant adjusted p-values by artificially restricting the subgroup space via refining the set of candidate biomarkers or by selecting appropriate values of the algorithm search parameters.

To give an example of extreme situations, consider the patient subgroups identified by the basic SIDES procedure in Table 5.5. Using the seven biomarkers shown in this table, let us define a set of binary variables that are derived by dichotomizing the original variables biomarkers, i.e., BPH_TREAT = Yes, AGE $\leq$ 63.5, BMI > 25.7, QmaxR > -2.25, IPSSTB > 15.5, IPSSB > 7 and IPSSIR > -1. As shown in Table 5.11, when the adaptive SIDEScreen procedure was applied to the resulting seven binary biomarkers, a single promising subgroup (IPSSB > 7) was selected. The treatment effect within this subgroup was significant at a one-

Table 5.10 Promising patient subgroups selected by the adaptive SIDEScreen procedure with three ordered sets of biomarkers (biomarker screen with the multiplier $k = 0$)

Subgroup	Total sample size	Mean treatment difference	Raw p-value	Adjusted p-value
AGE $\leq$ 63.5	154	2.15	0.0105	0.0956
BPH_TREAT = No	210	1.42	0.0311	0.2200

Table 5.11 Promising patient subgroups selected by the adaptive SIDEScreen procedure using a set of binary biomarkers (biomarker screen with the multiplier $k = 0$)

Subgroup	Total sample size	Mean treatment difference	Raw p-value	Adjusted p-value
IPSSB > 7	185	2.25	0.0043	0.0244

sided 0.025 level and the corresponding Type I error rate was 0.10. These results
may look appealing even though the approach has limited scientific value.

In general, the trial sponsor's approach should be guided by different types of
analysis. The SIDESscreen procedures help discover important predictive biomarkers which are based on the strongest statistical evidence of a differential treatment
effect whereas subgroup search results based on the basic SIDES procedure provide
a relevant basis for discussions with the clinical team. In our case study, the
partitioning schemes based on ISSIB $>$ 7, AGE $\leq$ 63.5 and BPH_TREAT $=$ No
emerge from the set of formal analyses. When examining the resulting subgroups, a
team of urologists also noted the clinical relevance of another partitioning scheme
obtained from less formal analyses based on ISSIB $>$ 7 and IPSSIR $>$ -1. By
combining these factors, one can define the profile of a super responder who
is not too old, did not receive previous BPH treatment and has severe irritative
symptoms at baseline without improvement during the run-in placebo period. The
mean treatment difference within the resulting subgroup of 39 patients is equal to
6.7!

5.6 Simulation Study

A simulation study was performed to evaluate the key characteristics of subgroup
search strategies that are adjusted for prognostic variables. We have pointed out
multiple times throughout this chapter that failure to account for important variables
that are prognostic of the primary endpoint in a trial is likely to negatively affect the
performance of subgroup search algorithms, especially when dealing with small
subgroups of patients. This simulation study is built around a simple example with
a strong prognostic variable and demonstrates the value of adjusting for this variable
in SIDES-based subgroup search algorithms.

Consider a two-arm clinical trial (experimental treatment versus placebo) with
the total of $n = 200$ patients. Assume that the primary endpoint is normally
distributed and larger values of this endpoint indicate a beneficial effect. Let y_i
denote the endpoint's value and t_i denote the treatment indicator for the ith patient
(i.e., $t_i = 0$ in the placebo arm and $t_i = 1$ in the treatment arm), $i = 1, \ldots, n$. The
model for the primary endpoint is defined as follows

$$y_i = I(x_{1i} > 0.5)\delta t_i + x_{2i} + \sigma \varepsilon_i, \ i = 1, \ldots, n,$$

where ε_i is the error term that follows a standard normal distribution, I is a binary
indicator, i.e., $I(x_{1i} > 0.5) = 1$ if x_{1i} is greater than 0.5 and 0 otherwise, and δ and
σ are model parameters. Furthermore, x_{1i} and x_{2i} are the values of two biomarkers,
denoted by X_1 and X_2. The biomarkers are independent of the error term and are
generated as follows. Let Z_1 and Z_2 denote two random variables that follow a
standard bivariate normal distribution with the correlation coefficient ρ and let $X_i =
\Phi(Z_j), j = 1, 2$, where $\Phi(z)$ is the cumulative distribution function of the standard

normal distribution. This immediately implies that the marginal distribution of X_j is uniform on $[0, 1]$, $j = 1, 2$.

It follows from this model that X_1 is predictive of treatment response in this trial and the value of δ determines the degree of a differential treatment effect. It is clear that the mean treatment difference is equal to 0 for any value of this biomarker if $\delta = 0$. When δ is positive, biomarker-high patients experience treatment benefit while there is no treatment effect in biomarker-low patients. The cutoff that defines a differential effect is 0.5 and thus the true subgroup with an enhanced treatment effect is given by $\{X_1 > 0.5\}$. The other biomarker, i.e., X_2, is a prognostic biomarker since the endpoint's value increases when X_2 is large but this biomarker exhibits no predictive properties.

To identify a subgroup with enhanced treatment effect, a subgroup search was carried out using the basic SIDES procedure with the following parameters:

- The search width and depth parameters were 2 and 1, respectively.
- The smallest sample size per subgroup was 30.
- The child-to-parent ratio was 1.
- The biomarkers were not discretized.

Two primary analysis strategies were considered in the SIDES procedure. Strategy A relied on the two-sample t-test and Strategy B was based on a simple ANCOVA model adjusted for the prognostic variable X_2.

The ability of the SIDES procedure based on Strategies A and B to correctly select the true subgroup was evaluated as follows. First of all, the probability of an incorrect decision was computed. An incorrect decision was defined as (1) a decision to form a subgroup based on X_1 by including biomarker-low patients, e.g., if the subgroup was incorrectly identified as $\{X_1 \leq c\}$, or (2) if no subgroup based on X_1 was found at all, e.g., the subgroup did not meet the complexity control criterion. Secondly, descriptive statistics were obtained for the cutoff estimated by the SIDES procedure, namely, the lower quartile (25th percentile), median and upper quartile (75th percentile), if a subgroup based on X_1 was identified and consisted of biomarker-high patients.

The performance of the two strategies was evaluated using 1000 simulation runs. The simulation results for three selected values of δ are summarized in Table 5.12 (σ was set to 0.03 and ρ was set to 0.9). The table presents the probability of an incorrect decision for each primary analysis strategy as well as the descriptive statistics.

Beginning with the most challenging case when the differential effect parameter δ was 0.02, Table 5.12 shows that the SIDES procedure based on the two-sample t-test (Strategy A) struggled to correctly identify the true subgroup, i.e., $\{X_1 > 0.5\}$. The probability of an incorrect decision was over 40% and, when a biomarker-high subgroup was chosen by the procedure, the estimated cutoff for X_1 was quite unstable. The lower and upper quartiles were 0.3 and 0.7, which resulted in the interquartile range of 0.4. This means that promising subgroups based on X_1 were chosen more or less randomly. By contrast, the SIDES procedure that accounted for the prognostic effect of X_2 (Strategy B) correctly selected a subgroup of patients

Table 5.12 Performance of the SIDES procedure based on the two-sample t-test (Strategy A) and ANCOVA model (Strategy B)

Primary analysis strategy	Probability of an incorrect decision	Estimated cutoff		
		Lower quartile	Median	Upper quartile
$\delta = 0.02$				
Strategy A	0.44	0.30	0.50	0.70
Strategy B	0.05	0.27	0.42	0.50
$\delta = 0.1$				
Strategy A	0.09	0.45	0.54	0.75
Strategy B	0.00	0.49	0.50	0.50
$\delta = 0.2$				
Strategy A	0.01	0.49	0.53	0.72
Strategy B	0.00	0.49	0.50	0.50

with high values of X_1 most of the time (the probability of an incorrect decision was 5%) and the interquartile range of the estimated cutoff was 0.23.

When the differential effect parameter was set to 0.1, the performance of Strategy A clearly improved. The probability of an incorrect decision was below 10% and the interquartile range of the estimated cutoff was down to 0.3. But there was much more improvement if the SIDES procedure with covariate adjustment (Strategy B) was applied. With this primary analysis strategy, no incorrect decisions were made and the cutoff estimates were very tight with the interquartile range of only 0.01.

Considering the last scenario that corresponds to $\delta = 0.2$, we see that, with either primary analysis strategy, the probability of an incorrect decision was very low. However, Strategy A was still inferior to Strategy B because the latter resulted in very consistent cutoff estimates whereas the interquartile range for Strategy A was quite wide (0.23).

The simulation results presented in this section demonstrate the importance of accounting for key prognostic variables in the primary analysis model when running subgroup searches. For example, if the primary endpoint is measured on a continuous scale, the SIDES or SIDEScreen procedures based on ANCOVA models are more likely to correctly identify patient subgroups with a marked differential treatment effect than those based on a simple t-test. Similarly, an adjustment for key prognostic covariates will improve the performance of subgroup search algorithms in trials with binary or time-to-event endpoints.

5.7 Discussion

Post-hoc assessments of patient subgroups in clinical trials are often criticized as being unreliable. Subgroup effects that appear to be highly significant are likely to be spurious and thus it is critical to focus on disciplined approaches to subgroup identification that support more reliable inferences. A disciplined approach based on a family of SIDES subgroup search methods was presented in this chapter and practical considerations arising in subgroup search exercises were illustrated using a case study based on a Phase III clinical trial in patients with benign prostatic hypertrophy.

Beginning with the issue of error rate control in post-hoc subgroup investigation, the goal of lowering the Type I error rate should not generally be the main goal of subgroup exploration, it is more important to ensure that the resulting patient subgroups are clinically relevant. In fact, the SIDES-based subgroup search algorithms with several sets of parameters may lead to the same Type I error rate but completely different sets of promising patient subgroups. It may be counterproductive to pre-define the highest acceptable Type I error rate because the focus on this artificially selected threshold may interfere with the process of fully evaluating treatment effects across clinically meaningful subgroups. In many exploratory settings, the degree of Type I error rate inflation should be considered as a precautionary measure against unconstrained subgroup searches that are likely to substantially increase the burden of multiplicity. To assess the relevance of identified patient subgroups, multiplicity-adjusted p-values and Type I error rate calculations should be combined with other metrics such as the reference treatment effect (RTE) introduced in Sect. 5.5 to evaluate the plausibility of observed subgroup effects.

An important topic discussed in the chapter is the role of complex modeling strategies that account for important covariates in subgroup search exercises. Most commonly, SIDES-based subgroup assessments presented in the literature focus on simple analysis methods such as the t-test that ignore information on prognostic variables. These oversimplified analysis methods are not aligned with the primary analysis methods utilized in the original trial. In addition, they do not recognize an important fact that an adjustment for prognostic covariates plays a key role in ensuring that subgroup assessments are not biased by potential imbalances with respect to these covariates. These considerations are especially important when treatment effects are evaluated within small patient subgroups. Using a series of examples based on the case study, it was shown in this chapter that the modeling strategy in post-hoc subgroup assessments should generally follow the original primary analysis. To help quantify the importance of utilizing covariate-adjusted models in subgroup exploration, a simulation study was presented in Sect. 5.6. A comparison of SIDES-based subgroup search algorithms based on a simple t-test and an ANCOVA model adjusted for an important prognostic variable clearly showed that the likelihood of correctly identifying patient subgroups with a differential treatment effect is reduced if a key prognostic covariate is ignored.

To help define a recommended approach to SIDES-based subgroup search in a late-stage clinical trial, it will be ideal to fix a set of SIDES parameters as well as a covariate adjustment strategy after the trial's database has been cleaned but before breaking the blind. This pre-specification freezes the subgroup search space and enables true weak control of the Type I error rate. In this context, subgroup investigations can address confirmatory issues provided the estimand and analysis strategy are unambiguously specified in the statistical analysis plan (in this particular context, an estimand reflects what is to be estimated to address the scientific question posed by subgroup analysis). It will be especially important to clarify whether or not the investigated subgroups and the measure of treatment effect differ from the primary analysis applied to the overall trial population. The question that needs to be asked is, does "the difference between the trial arms in the most promising subgroups" represent the endpoint of interest? It seems natural to apply the same statistical model but this approach may be inappropriate in the analysis of subgroups. For example, in the case study used in this chapter, the original primary analysis was performed using a certain ANCOVA model with an LOCF-based imputation of missing data caused by patient dropout. The center term was included as a factor in the original model but, with a large number of centers, an ANCOVA model with center effects could not be utilized within smaller subgroups. Similar restrictions would apply in a longitudinal setting. Specifically, the treatment effect in the overall population may be estimated from a longitudinal model which requires numerous parameters. If parameter estimation is not possible due to smaller sample sizes within subgroups, a change in the analysis strategy is quite justifiable if a similar missing data mechanism can be assumed across the subgroups of interest.

Acknowledgements We would like to thank the two reviewers for valuable comments and suggestions that have helped us improve the presentation of the material in this chapter. This chapter is dedicated to the memory of Sylvie Audrain, who took an active part in the realization of this project.

References

Dmitrienko A, Lipkovich I, Hopkins A, Li YP, Wang W (2015) Biomarker evaluation and subgroup identification in a pneumonia development program using SIDES. In: Chen Z, Liu A, Qu Y, Tang L, Ting N, Tsong Y (eds) Applied statistics in biomedicine and clinical trials design. Springer, New York, pp 427–466

Dmitrienko A, Muysers C, Fritsch A, Lipkovich I (2016) General guidance on exploratory and confirmatory subgroup analysis in late-stage clinical trials. J Biopharm Stat 26:71–98

Dmitrienko A, Lipkovich I, Dane A, Muysers C (2019) Data-driven and confirmatory subgroup analysis in clinical trials. In: Ting N et al. (ed.) Design and analysis of subgroups with biopharmaceutical applications. Emerging topics in statistics and biostatistics. https://doi.org/10.1007/978-3-030-40105-4_3

EMA (2019) Guideline on the investigation of subgroups in confirmatory clinical trials. European Medicines Agency/Committee for Medicinal Products for Human Use. EMA/CHMP/539146/2013

Grouin JM, Coste M, Lewis J (2005) Subgroup analyses in randomized clinical trials: statistical and regulatory issues. J Biopharm Stat 15:869–882

Hardin DS, Rohwer RD, Curtis BH, Zagar A, Chen L, Boye KS, Jiang HH, Lipkovich IA (2013) Understanding heterogeneity in response to antidiabetes treatment: a post hoc analysis using SIDES, a subgroup identification algorithm. J Diabetes Sci Technol 7:420–429

Lipkovich I, Dmitrienko A, Denne J, Enas G (2011) Subgroup identification based on differential effect search (SIDES): a recursive partitioning method for establishing response to treatment in patient subpopulations. Stat Med 30:2601–2621

Lipkovich I, Dmitrienko A (2014a) Strategies for identifying predictive biomarkers and subgroups with enhanced treatment effect clinical trials using SIDES. J Biopharm Stat 24:130–153

Lipkovich I, Dmitrienko A (2014b) Biomarker identification in clinical trials. In: Carini C, Menon S, Chang M (eds) Clinical and statistical considerations in personalized medicine. Chapman and Hall/CRC Press, New York

Lipkovich I, Dmitrienko A, D'Agostino BR (2017a) Tutorial in biostatistics: data-driven subgroup identification and analysis in clinical trials. Stat Med 36:136–196

Lipkovich I, Dmitrienko A, Patra K, Ratitch B, Pulkstenis E (2017b) Subgroup identification in clinical trials by stochastic SIDEScreen methods. Stat Biopharm Res 9:368–378

Lipkovich I, Dmitrienko A, Muysers C, Ratitch B (2018) Multiplicity issues in exploratory subgroup analysis. J Biopharm Stat 28:63–81

Wang SJ, O'Neill RT, Hung J (2007) Approaches to evaluation of treatment effect in randomized clinical trials with genomic subset. Pharm Stat 6:227–244

Chapter 6
The GUIDE Approach to Subgroup Identification

Wei-Yin Loh and Peigen Zhou

Abstract GUIDE is a multi-purpose algorithm for classification and regression tree construction with special capabilities for identifying subgroups with differential treatment effects. It is unique among subgroup methods in having all these features: unbiased split variable selection, approximately unbiased estimation of subgroup treatment effects, treatments with two or more levels, allowance for linear effects of prognostic variables within subgroups, and automatic handling of missing predictor variable values without imputation in piecewise-constant models. Predictor variables may be continuous, ordinal, nominal, or cyclical (such as angular measurements, hour of day, day of week, or month of year). Response variables may be univariate, multivariate, longitudinal, or right-censored. This article gives a current account of the main features of the method for subgroup identification and reviews a bootstrap method for conducting post-selection inference on the subgroup treatment effects. A data set pooled from studies of amyotrophic lateral sclerosis is used for illustration.

Keywords Bootstrap · Classification and regression tree · Confidence interval · Missing value · Post selection inference · Recursive partitioning · Variable selection

6.1 Introduction

GUIDE (Loh 2002, 2009) is an algorithm for fitting classification and regression tree models to data. AID (Morgan and Sonquist 1963) was the first algorithm but CART (Breiman et al. 1984) and RPART (Therneau and Atkinson 2018) brought the basic ideas to the mainstream. GUIDE grew out of work on an alternative approach to CART classification (Loh and Vanichsetakul 1988; Loh and Shih 1997; Kim and

W.-Y. Loh (✉) · P. Zhou
Department of Statistics, University of Wisconsin, Madison, WI, USA
e-mail: loh@stat.wisc.edu; pzhou9@wisc.edu

© Springer Nature Switzerland AG 2020
N. Ting et al. (eds.), *Design and Analysis of Subgroups with Biopharmaceutical Applications*, Emerging Topics in Statistics and Biostatistics,
https://doi.org/10.1007/978-3-030-40105-4_6

147

Loh 2001, 2003) and regression (Loh 1991b; Ahn and Loh 1994; Chaudhuri et al. 1994, 1995; Chaudhuri and Loh 2002; Chan and Loh 2004). See Loh (2014) for a recent review of classification and regression trees. Unlike AID and CART that only fit a constant in each node of the tree, GUIDE can fit linear and generalized linear models. This makes GUIDE well suited for subgroup identification—the terminal nodes of the tree are the subgroups and the regression coefficients in the node models give the treatment effects. It is unique among subgroup methods in having properties such as unbiased selection of split variables, approximately unbiased estimation of treatment effects, ability to use treatment variables with more than two levels, optional local adjustment for linear effects of prognostic variables, and automatic handling of missing values without needing prior imputation. Predictor variables may be continuous, ordinal, nominal, or cyclical (such as angles, hour of day, day of week, and month of year). Response variables may be univariate, multivariate, longitudinal, or right censored. Missing values may be coded in more than one way; for example a missing value for age of spouse may be coded as "refuse to answer" if the respondent did not provide an answer and as "valid nonresponse" if the respondent is single, widowed or divorced; see Loh et al. (2019b) for other examples.

This article gives a current account of the GUIDE method for subgroup identification. It uses data combined from several studies of ALS (Amyotrophic Lateral Sclerosis) for illustration. The data were selected because they contained all of the types of response variables that GUIDE can model and because many of the predictor variables had missing values (denoted by "NA" here). ALS is a neurological disease that affects voluntary muscle movement. Death typically occurs within 3–5 years of diagnosis. Only about a quarter of patients survive for more than 5 years after diagnosis. The data were obtained from the Pooled Resource Open-Access ALS Clinical Trials (PRO-ACT) Database (Atassi et al. 2014). In 2011, Prize4Life, in collaboration with the Northeast ALS Consortium, and with funding from the ALS Therapy Alliance, formed the PRO-ACT Consortium. The data in the PRO-ACT Database were provided by the PRO-ACT Consortium members. They were pooled from 23 completed ALS clinical trials and one observational study, and contained information on demographics, family history, and clinical and laboratory test data from more than 10700 ALS patients.

The ALS Functional Rating Scale (ALSFRS) is often used to evaluate the functional status of ALS patients. It is the sum of ten scores (speech, salivation, swallowing, handwriting, cutting food and handling utensils, dressing and hygiene, turning in bed and adjusting bed clothes, walking, climbing stairs, and breathing), with each score measured on a scale of 0–4, with 4 being normal. Seibold et al. (2016) used a subset of the data to study the effectiveness of *riluzole*, a drug approved for treatment of ALS by the US FDA, on ALSFRS at 6 months as well as survival time from trial enrollment. Using the MOB algorithm (Zeileis et al. 2008), they found that for patients with less than 468 days between disease onset and start of treatment, riluzole had a negative treatment effect on ALSFRS at 6 months.

A major difficulty with the PRO-ACT data is that besides riluzole, other medications were also tested in many of the trials (Atassi et al. 2014, Table 1).

Even worse, the additional medications were not identified in the data. To avoid confounding the effects of riluzole and that of other medications, the analysis here is restricted to the subset of 1270 subjects who were assigned to placebo or riluzole only, without other medications. Thirty-six variables were chosen as predictor variables; their names are given in Table 6.1 together with their minimum and maximum values and numbers of missing values. Three additional variables were chosen as dependent variables: (1) change in ALSFRS from baseline at 6 months, (2) monthly change in ALSFRS from baseline at months 1, 2, ..., 6, and (3) survival time in days. ALSFRS scores of subjects who had died by the time the scores were to be measured were set to 0. ALSFRS variables at 0, 1, ..., 6 months are denoted by ALSFRS0, ALSFRS1, ..., ALSFRS6, respectively.

6.2   Univariate Uncensored Response

Figure 6.1 shows a basic GUIDE tree for predicting change in ALSFRS after 6 months (ALSFRS6 minus ALSFRS0), where a linear model (6.1) with treatment as the only predictor variable is fitted in each node. A node of the tree represents a partition of the data, with the root node corresponding to the whole data set. The sample size in each partition is printed beside the node. At each node, a variable X is selected to split the data there into two child nodes. The split, in the form $X \in A$, is printed on the left of the node. The set A is chosen to minimize the sum of the squared residuals in the left and right child nodes. Observations in the node are sent to the left child node if and only if the condition is satisfied. Node labels start with 1 for the root node; for a node with label k, its left and right child nodes are labeled $2k$ and $2k + 1$, respectively.

The root node in Fig. 6.1 is split on Diagnosis_Delta, which is the number of days from clinical diagnosis to the first time the subject was tested in a trial. The 239 subjects with missing values in Diagnosis_Delta go to terminal node 2 and the others go to intermediate node 3. It is unknown why the subjects have missing values in Diagnosis_Delta. One possibility is the variable was not measured in some of the trials, but this cannot be verified because trial ID was not included in the data. Nevertheless, as the barplot for node 2 in Fig. 6.1 shows, subjects in this subgroup deteriorate much more on average with riluzole than without. Subjects in node 3 are split on Hematocrit. Those with Hematocrit missing or ≤ 37.95 (abbreviated in the tree diagrams by the symbol "$\leq_*$" with the asterisk standing for "is missing") go to node 6 where they are split on BP_Diastolic and then on Potassium.

6.2.1   Node Models

Let $\mathbf{X} = (X_1, X_2, \ldots, X_K)$ denote a K-dimensional vector of covariates, Y a univariate response variable, and Z a treatment variable taking values $0, 1, \ldots, G$,

Table 6.1 Predictor variables, minimum and maximum values, numbers of categorical levels, and numbers of missing values for modeling the difference `ALSFRS6-ALSFRS0`

Name	Definition	Min	Max	Miss
Demographics_Delta	Demographic measurement day	−35.00	32.00	19
Age	Subject age at start of trial	18.00	82.00	
Sex	Subject gender (female, male)			
Race	Subject race (5 categories)			3
ALS_History_Delta	Day ALS history reported	0.00	3.00	43
Symptom	Major symptom (10 categories)			1085
Onset_Delta	Day of disease onset, from first test	−1900.00	−84.00	47
Diagnosis_Delta	Day of diagnosis, from first test	−1666.00	0.00	239
Site_of_Onset	Site of disease onset (3 categories)			
Albumin	Albumin in blood (g/L)	31.67	53.00	332
ALT_SGPT	Alanine amino transferase (U/L)	6.00	181.00	259
AST_SGOT	Aspartate amino transferase (U/L)	7.50	116.00	258
Basophil_Count	Amount in white blood cell (×109/L)	0.00	5.56	341
Basophils	Percent in white blood cell count	0.00	3.00	365
Blood_Urea_Nitrogen	Ureas (mmol/L)	0.95	17.34	218
Calcium	Calcium in metabolic panel (mmol/L)	1.55	3.00	333
Creatinine	Creatinine from kidney test	25.00	159.10	216
Eosinophils	Percent in white blood cell count	0.00	15.00	365
Glucose	Glucose in blood (mmol/L)	0.07	18.56	325
Hematocrit	Percent red blood cells	0.00	56.00	326
Hemoglobin	Hemoglobin in blood (g/L)	94.50	181.00	326
Lymphocytes	Percent lymphocyte in blood	8.70	50.00	365
Monocytes	Percent in white blood cell count	0.00	21.40	365
Platelets	Platelets in blood (×109/L)	0.20	552.00	332
Potassium	Potassium in electrolytes (mmol/L)	3.30	5.50	258
Sodium	Sodium in electrolytes (mmol/L)	125.00	150.00	257
Urine_Ph	Acidity of urine	5.00	9.00	355
SVC (Slow_vital_Capacity)	Volume of air exhaled slowly (L)	1.00	7.00	737
Slow_vital_Capacity_Delta	Day of SVC assessment	0.00	14.00	737
BP_Diastolic	Diastolic blood pressure (mmHg)	52.00	125.00	217
BP_Systolic	Systolic blood pressure (mmHg)	90.00	200.00	217
Height	Subject height (in)	131.00	205.00	225
Pulse	Beats per minute	42.00	120.00	218
Weight	Subject weight (kg)	38.33	138.20	178
ALSFRS0	ALSFRS at baseline	10.00	40.00	
ALSFRS_Delta0	Day of ALSFRS0 measurement	−7.00	154.00	

Variables with names containing "_Delta" are days from trial onset to the date that an assessment took place, with negative values for occurrences before trial onset

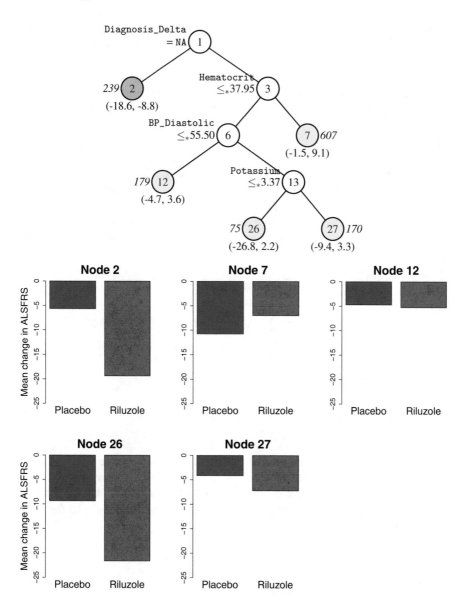

Fig. 6.1 GUIDE tree for change in ALSFRS (ALSFRS6-ALSFRS0) using 1270 observations and node model (6.1). At each split, an observation goes to the left branch if and only if the condition is satisfied. The symbol "$\leq_*$" stands for "$\leq$ or missing." Sample sizes *(in italics)* are printed beside nodes. Bootstrap-calibrated 90% simultaneous intervals of treatment effect are given below nodes. Calibrated alpha is 1.3×10^{-5}. Treatment effect is statistically significant in green node. Barplots show means of change in ALSFRS for placebo and riluzole subjects in the terminal nodes

with 0 being the reference (or placebo) level. Let t denote a node of the tree. A regression tree model is constructed by recursively partitioning a training sample into subsets that are represented by the nodes of a tree. A large majority of regression tree methods for subgroup identification employ stopping rules based on Bonferroni-corrected p-values (Lipkovich et al. 2011; Seibold et al. 2016; Su et al. 2009). Other methods (Dusseldorp and Meulman 2004; Foster et al. 2011), including GUIDE, first grow an overly large tree and then use cross-validation to prune it to a smaller size. We only describe the GUIDE node fitting and splitting steps here because the pruning step is the same as that of CART.

For least-squares regression, GUIDE fits a linear model $Y = f(\mathbf{X}, Z) + \epsilon$ to the data in each node of a tree; ϵ is an independent zero-mean random variable with variance that is constant within each node but may vary between nodes. Four choices of $f(\mathbf{x}, z)$ are available, depending on the number of X variables to be included. Let β_z $(z = 1, 2, \ldots, G)$ denote the effect of treatment level z (versus level 0). The choices are:

$$f(\mathbf{x}, z) = \eta + \beta_z \quad \text{(Treatment only)} \tag{6.1}$$

$$f(\mathbf{x}, z) = \eta + \beta_z + \sum_{j=1}^{p} \gamma_j x_{k*}^j \quad \text{(Polynomial of degree } p) \tag{6.2}$$

$$f(\mathbf{x}, z) = \eta + \beta_z + \sum_{k}^{K} \gamma_k x_k \quad \text{(Multiple linear)} \tag{6.3}$$

$$f(\mathbf{x}, z) = \eta + \beta_z + \sum_{k \in S} \gamma_k x_k \quad \text{(Stepwise linear)} \tag{6.4}$$

In (6.2), p is a user-specified positive integer and k^* is the value of k such that X_k minimizes the sum of squared residuals in the node (k^* may vary from node to node). In (6.4), the set S is the set of indices of the variables X_k that are selected by forward and backward stepwise regression in the node. Thus the model for a tree with terminal nodes $t_1, t_2, \ldots, t_\tau$ may be written as

$$Y = \begin{cases} f_1(\mathbf{X}, Z) + \epsilon_1, \ \mathbf{X} \in t_1 \\ \quad \vdots \\ f_\tau(\mathbf{X}, Z) + \epsilon_\tau, \ \mathbf{X} \in t_\tau \end{cases} \tag{6.5}$$

where $f_1, f_2, \ldots, f_\tau$ take one of the functional forms (6.1)–(6.4) and $\epsilon_1, \ldots, \epsilon_\tau$ are independent random variables with mean zero and variances $\sigma_1^2, \ldots, \sigma_\tau^2$. This is different from the model

$$Y = \sum_{j=1}^{\tau} f_j(\mathbf{X}, Z) I(\mathbf{X} \in t_j) + \epsilon \tag{6.6}$$

which assumes that the error variance is the same in all nodes. The least-squares estimates of the regression coefficients are the same in models (6.5) and (6.6), but not their standard error estimates. In (6.2)–(6.4), missing values in the X variables are imputed by their node means.

Figure 6.1 was constructed using model (6.1) and Fig. 6.2 was constructed using model (6.2) with $p = 1$. The name of the best linear prognostic variable X_{k*} is given beneath each terminal node. The root node splits on "Diagnosis_Delta ≤ -1072 or missing." Of the 245 subjects in this subgroup, 239 are missing Diagnosis_Delta. The best linear prognostic variable in node 2 is Pulse. Plots of the data and regression lines for placebo and riluzole subjects in each node are shown in the lower half of Fig. 6.2. Mean imputation of Sodium is clearly shown by the vertical line of points in the plot of node 13.

6.2.2 Split Variable Selection

To find a variable to split a node t, a test of treatment-covariate interaction is performed for each X_k on the data in t. (This is the default "Gi" method.) Let n_t denote the number of observations in t. The following steps are carried out for each variable X_j, $j = 1, 2, \ldots, K$.

1. If X_j is a categorical variable, define $V = X_j$ and let h denote its number of levels (including a level for NA, if any).
2. If X_j is ordinal and takes only one value (including NA) in the node, do not use it to split the node. Otherwise, let m denote the number of distinct values (including NA) of X_j in t. Transform it to a discrete variable V with h values as follows.

   (a) If $m \leq 4$ or if $m = 5$ and X_j has missing values, define $h = m$. Otherwise, define $h = 3$ if $n_t < 30(G + 1)$ and $h = 4$ otherwise.

   (i) If X_j has missing values in t, define $r_k = k/(h-1)$, $k = 1, 2, \ldots, h-2$.
   (ii) If X_j has no missing values in t, define $r_k = k/h$, $k = 1, 2, \ldots, h - 1$.

   (b) Define $q_0 = -\infty$ and let q_k $(k > 0)$ be the sample r_k-quantile of X_j in t.

   (i) If X_j has missing values in t, define

   $$V = \sum_{k=1}^{h-2} kI(q_{k-1} < X_j \leq q_k) + (h-1)I(X_j > q_{h-2}) + hI(X_j = \text{NA}).$$

   (ii) If X_j has no missing values in t, define

   $$V = \sum_{k=1}^{h-1} kI(q_{k-1} < X_j \leq q_k) + hI(X_j > q_{h-1}).$$

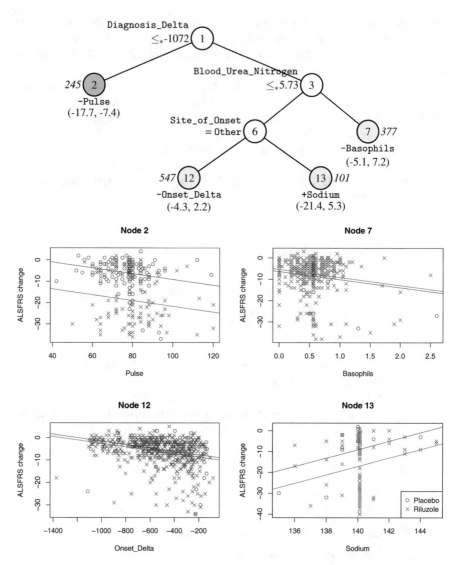

Fig. 6.2 GUIDE tree for `ALSFRS6-ALSFRS0` using 1270 observations and node model (6.2) with polynomials of degree 1. At each split, an observation goes to the left branch if and only if the condition is satisfied. The symbol '$\leq_*$' stands for '$\leq$ or missing'. Sample sizes (*in italics*) are printed beside nodes. Name of best linear prognostic variable (with sign of slope) and bootstrap-calibrated 90% simultaneous confidence interval for treatment effect are below each node. Calibrated alpha is 8.9×10^{-6}. Treatment effect is statistically significant in green node. Plots of change in ALSFRS versus best linear predictor show data points and fitted regression lines in the terminal nodes. Missing values in predictor variables are imputed by the means of their non-missing values in the nodes

3. Test the additive model $E(Y|Z, V) = \eta + \sum_z \beta_z I(Z = z) + \sum_v \gamma_v I(V = v)$, with $\beta_0 = \gamma_1 = 0$, against the full model $E(Y|Z, V) = \sum_z \sum_v \omega_{vz} I(V = v, Z = z)$ and obtain the p-value p_j.

Split node t on the X_j with the smallest value of p_j.

6.2.3 Split Set Selection

After X is selected, a search is carried out for the best split "$X \in A$", where A depends on whether X is ordinal or categorical.

6.2.3.1 Ordinal Variable

If X is ordinal, three types of splits are evaluated.

1. $X = $ NA: an observation goes to the left node if and only if its value is missing.
2. $X = $ NA or $X \leq c$: an observation goes to the left node if and only if its value is missing or if it is less than or equal to c.
3. $X \leq c$: an observation goes to the left node if and only if its value is not missing and it is less than or equal to c.

Candidate values of c are the midpoints between consecutive order statistics of X in t. If X has m order statistics, the maximum number of possible splits is $(m - 1)$ or $\{1 + 2(m - 1)\}$, depending on the absence or presence of missing X values in t. Permissible splits are those that yield two child nodes with each having two or more observations per treatment. The selected split is the one that minimizes the sum of the deviances (or sum of squared residuals in the case of least-squares regression) in the two child nodes.

This method of dealing with missing values is unique to GUIDE. CART uses a hierarchical system of "surrogate splits" on alternative X variables to send observations with missing values to the child nodes. Because the surrogate splits depend on the (missing and non-missing) values of the alternative X variables, observations with missing values do not necessarily go to the same child node. Therefore it is impossible to predict the path of an observation by looking at the tree without knowing the values of its predictor variables. Besides, CART's surrogates splits are biased towards X variables with few missing values (Kim and Loh 2001). Other subgroup methods are typically inapplicable to data with missing values (Dusseldorp and Meulman 2004; Su et al. 2009; Foster et al. 2011; Seibold et al. 2016).

Sometimes, missing-value imputation is illogical, e.g., a prostate-specific antigen test result for a female subject or the age of first cigarette for a subject who never smoked. Other times, imputation erases useful information. For example, if missing values were imputed before application of GUIDE, the large difference in treatment effect between subjects with and without missing values in Diagnosis_Delta in Fig. 6.1 would be undetected.

6.2.3.2 Categorical Variable

If X is a categorical variable, the split has the form $X \in A$, where A is a non-trivial subset of the values (including NA) of X in t. A complete search of all possible values of A can be computationally expensive if the number, m, of distinct values (including NA) of X in t is large, because there are potentially $(2^{m-1} - 1)$ splits (less if some splits yield child nodes with fewer than two observations per treatment). Therefore GUIDE carries out a complete search only if $m \leq 11$. If $m > 11$, it performs an approximate search by means of linear discriminant analysis , based on an idea from Loh and Vanichsetakul (1988), Loh and Shih (1997), and Loh (2009).

1. Let $\bar{y}_z$ denote the sample mean of the Y values in t that belong to treatment $Z = z$ $(z = 0, 1, \ldots, G)$.
2. Define the class variable

$$
C = \begin{cases} 2z - 1, & \text{if } Z = z \text{ and } Y > \bar{y}_z \\ 2z, & \text{if } Z = z \text{ and } Y \leq \bar{y}_z. \end{cases}
$$

3. Let $\{a_1, a_2, \ldots, a_m\}$ denote the categorical values of X in t. Transform X to an m-dimensional 0–1 dummy vector $\mathbf{D} = (D_1, D_2, \ldots, D_m)$, where $D_i = I(X = a_i)$, $i = 1, 2, \ldots, m$.
4. Apply linear discriminant analysis to the data $(\mathbf{D}, C)$ in t to find the discriminant variables $B_j = \sum_{i=1}^{m} b_{ij} D_i$, $j = 1, 2, \ldots$. These variables are also called canonical variates (Gnanadesikan 1997).
5. For each j, find the split $B_j \leq c_j$ that minimizes the sum of the squared residuals of the least-squares models fitted in the child nodes induced by the split.
6. Let j^* be the value of j for which $B_j \leq c_j$ has a smallest sum of squared residuals.
7. Split the node with $B_{j^*} \leq c_{j^*}$. Because $B_{j^*} = \sum_{i=1}^{m} b_{ij^*} D_i = \sum_{i=1}^{m} b_{ij^*} I(X = a_i)$, the split is equivalent to $X \in A$ with $A = \{a_i : b_{ij^*} \leq c_{j^*}\}$.

6.3 Bootstrap Confidence Intervals

The barplots in the lower half of Fig. 6.1 show that the subgroups defined by nodes 2 and 26 have the largest treatment effects. Similarly, the graphs in the lower half of Fig. 6.2 suggest that node 2 has the largest treatment effect. Are the effects statistically significant? This question cannot be answered by means of traditional methods because the subgroups were not specified independently of the data. It is a question of *post-selection inference*.

Given node t and z, let $\hat{\beta}(t, z)$ be the estimated treatment effect for $Z = z$ in t, let $\hat{\sigma}_\beta(t, z)$ denote its usual estimated standard error, and let ν_t be the residual degrees of freedom. Further, let $t_{\nu, \alpha}$ denote the $(1 - \alpha)$-quantile of the t-distribution with ν degrees of freedom and let τ denote the number of terminal nodes of the tree. Let

Table 6.2 90% simultaneous intervals for subgroup treatment effects in Figs. 6.1 and 6.2

Model	Node	$B(0.10, t, z)$	$J(\alpha_{\hat{F}}, t, z)$
Figure 6.1	2	$(-16.0, -11.4)$	$(-18.7, -8.7)$
$\alpha_{\hat{F}} = 1.3 \times 10^{-5}$	7	$(1.3, 6.3)$	$(-1.6, 9.2)$
	12	$(-2.5, 1.4)$	$(-4.7, 3.6)$
	26	$(-18.7, -5.8)$	$(-26.2, 1.7)$
	27	$(-6.0, -0.1)$	$(-9.4, 3.3)$
Figure 6.2	2	$(-15.1, -10.0)$	$(-17.7, -7.4)$
$\alpha_{\hat{F}} = 8.9 \times 10^{-6}$	7	$(-2.1, 4.1)$	$(-5.1, 7.2)$
	12	$(-2.7, 0.6)$	$(-4.3, 2.2)$
	13	$(-14.5, -1.6)$	$(-21.4, 5.3)$

$$B(\alpha, t, z) = \hat{\beta}(t, z) \pm t_{v_t, \alpha/(2\tau)} \hat{\sigma}_\beta(t, z) \tag{6.7}$$

be the Bonferroni-corrected $100(1 - \alpha)\%$ simultaneous t-interval for the treatment effect of $Z = z$ in node t. The middle column of Table 6.2 gives the values of $B(0.10, t, z)$ for the trees in Figs. 6.1 and 6.2. Despite the Bonferroni correction, the standard errors $\hat{\sigma}_\beta(t, z)$ are biased low because they do not account for the uncertainty due to split selection. As a result, the intervals $B(\alpha, t, z)$ tend to be too short and their simultaneous coverage probability is less than $(1 - \alpha)$.

There are two obvious ways to lengthen the interval widths to improve their coverage probabilities. One is to correct the standard error estimates, but this is formidable due to the complexity of the tree algorithm. Another way is to reduce the nominal value of α in (6.7). For example, to obtain 90% simultaneous coverage, we could use $B(\alpha, t, z)$ with a nominal $\alpha < 0.10$. To find the right nominal value of α, we first need to define the *estimand* of $\hat{\beta}(t, z)$, which is the true treatment effect in t. Let $\hat{F}$ denote the training data and F the population from which they are drawn. By definition, $\hat{\beta}(t, z)$ $(z = 1, \ldots, G)$ are the values of the treatment effect coefficients that minimize $\sum_{i \in t}(y_i - f(\mathbf{x}_i, z_i))^2$, where the sum is over the observations in node t. Their estimands, denoted by are $\beta_F(t, z)$, are the values of the treatment effect coefficients that minimize $E\{(Y - f(\mathbf{X}, Z))^2 I(\mathbf{X} \in t)\}$. Clearly, $\beta_F(t, z)$ is a random variable, because it depends on t, which in turn depends on $\hat{F}$. If F is known and t is given, however, $\beta_F(t, z)$ can be computed, by simulation from F if necessary.

Let $J(\alpha, t, z) = \hat{\beta}(t, z) \pm t_{v_t, \alpha/2} \hat{\sigma}_\beta(t, z)$ denote the nominal $100(1 - \alpha)\%$ t-interval, let $\tilde{T}$ be the set of terminal nodes, and let $\gamma_F(\alpha) = P[\cap_{t \in \tilde{T}}\{\beta_F(t, z) \in J(\alpha, t, z)\}]$ denote the simultaneous coverage probability. Clearly, $\gamma_F(\alpha) \uparrow 1$ as $\alpha \downarrow 0$. Given a desired simultaneous coverage probability γ^*, let α_F be the solution of the equation $\gamma_F(\alpha_F) = \gamma^*$. Then the intervals $J(\alpha_F, t, z)$ have exact simultaneous coverage γ^*. We call α_F the "calibrated α." Note that there is no need to work with the Bonferroni-corrected interval (6.7) because $\gamma_F(\alpha)$ is, by definition, a simultaneous coverage probability.

Of course, the value of α_F is not computable if F is unknown. In that case, a natural solution is *bootstrap calibration*, a method proposed in Loh (1987, 1991a) for the simpler problem of estimating a population mean. It was extended to

---

Algorithm 1: Bootstrap calibration of confidence intervals for treatment effects

Data: Given $K > 0$ and $\alpha \in (0, 1)$, $\alpha_1 < \alpha_2 < \ldots < \alpha_K = \alpha$; tree T with nodes
$t_1, t_2, \ldots, t_L$ constructed from $\mathscr{D} = \{(\mathbf{X}_i, Y_i, Z_i), i = 1, 2, \ldots, n\}$; and model M
(one of (6.1), ..., or (6.4)) based on T with estimated treatment effects $\hat{\beta}_{tz}$,
$z = 1, 2, \ldots, G$; $t = t_1, t_2, \ldots, t_L$.

Result: $(1 - \alpha)$ simultaneous t-intervals for $\{\beta_{tz}\}$.

begin

    $\gamma_k \leftarrow 0$ for $k = 1, 2, \ldots, K$;

    for $b \leftarrow 1$ **to** B **do**

        bootstrap data $\mathscr{D}_b^* = \{(\mathbf{X}_i^*, Y_i^*, Z_i^*), i = 1, 2, \ldots, n\}$ from $\mathscr{D}$;

        construct from $\mathscr{D}_b^*$ tree T_b with nodes $t_{b1}^*, t_{b2}^*, \ldots, t_{bL_b}^*$;

        fit model M based on T_b to $\mathscr{D}$ observations to get "true" effects $\beta(t_{bl}^*, z)$;
        $z = 1, \ldots, G$; $l = 1, \ldots, L_b$;

        fit model M based on T_b to $\mathscr{D}_b^*$ observations to get estimates $\hat{\beta}(t_{bl}^*, z)$, residual
        degrees of freedom ν_{bl} and standard errors $\hat{\sigma}_\beta(t_{bl}^*, z)$; $z = 1, \ldots, G$; $l = 1, \ldots, L_b$;

        for $z \leftarrow 1$ **to** G **do**

            for $l \leftarrow 1$ **to** L_b **do**

                for $k \leftarrow 1$ **to** K **do**

                    $J_{klz} \leftarrow (1 - \alpha_k)$ t-interval $\hat{\beta}(t_{bl}^*, z) \pm t_{\nu_{bl}, \alpha_k/2} \hat{\sigma}_\beta(t_{bl}^*, z)$;

                    if $\beta(t_{bl}^*, z) \in J_{klz}$ **then**

                        $c_{klz} \leftarrow 1$;   `/* interval contains true beta */`

                    else

                        $c_{klz} \leftarrow 0$;   `/* interval does not contain true`
                        `beta */`

                    end

                end

            end

        end

        for $k \leftarrow 1$ **to** K **do**

            if $\min_{lz} c_{klz} = 1$ **then**

                $\gamma_k \leftarrow \gamma_k + 1$

            end

        end

    end

    $\gamma_k \leftarrow \gamma_k/B$ for $k = 1, 2, \ldots, K$;

    $q \leftarrow$ smallest k such that $\gamma_k < 1 - \alpha$;

    $g \leftarrow (\gamma_{q-1} - 1 + \alpha)/(\gamma_{q-1} - \gamma_q)$;

    $\alpha' \leftarrow (1 - g)\alpha_{q-1} + g\alpha_q$;

    construct $(1 - \alpha')$ simultaneous t-intervals for β_{tz} for $t = t_1, t_2, \ldots, t_L$; $z = 1, \ldots, G$

end

---

estimation of subgroup treatment effects in Loh et al. (2016, 2019c). The idea is
to replace F with $\hat{F}$ in the calculations. Specifically, use simulation from $\hat{F}$ to find
the solution $\alpha_{\hat{F}}$ of the equation $\gamma_{\hat{F}}(\alpha_{\hat{F}}) = \gamma^*$. The resulting intervals $J(\alpha_{\hat{F}}, t, z)$
are called "bootstrap-calibrated" $100\gamma^*\%$ simultaneous intervals. Algorithm 1 gives
the instructions in pseudo-code, using a grid search to find $\alpha_{\hat{F}}$. The numerical
results here (including those in the last column of Table 6.2) were obtained with
a grid of 200 nominal values of α and 1000 bootstrap iterations. Simultaneous

90% bootstrap-calibrated intervals of treatment effect are given beneath the terminal nodes of the trees in Figs. 6.1 and 6.2. Their respective bootstrap-calibrated alpha values are $\alpha_{\hat{F}} = 1.3 \times 10^{-5}$ and 8.9×10^{-6}. In the tree diagrams, nodes with statistically significant treatment effects are in green color.

6.4 Multivariate Uncensored Responses

GUIDE can construct a least-squares regression tree for data with longitudinal or multivariate response variables as well. Given d response variables $Y_1, Y_2, \ldots, Y_d$, it fits the treatment-only model $E(Y_j|Z) = \eta_j + \sum_{z=1}^{G} \beta_{jz} I(Z = z)$, $j = 1, \ldots, d$, separately to each variable in each node. To find the variable to split a node, the test for treatment-covariate interaction in Sect. 6.2.2 is performed d times for each X_i (once for each Y_j) to obtain the p-value $p_{i1}, p_{i2}, \ldots, p_{id}$. Let $\chi_{\nu,\alpha}^2$ denote the $(1 - \alpha)$-quantile of the chi-squared distribution with ν degrees of freedom. The variable X_i for which $\sum_{j=1}^{d} \chi_{1,p_{ij}}^2$ is maximum is selected to split the node. To allow for correlations in the response variables, GUIDE can optionally apply the treatment-covariate interaction tests to the principal component (PC) or linear discriminant (LD) variates computed from the Y_j values in the node. Specifically, if principal component transformation is desired, the $(Y_1, Y_2, \ldots, Y_d)$ data vectors in the node are transformed to their PCs $(Y_1', Y_2', \ldots, Y_d')$ first; then the treatment-covariate interactions tests are applied to the $(Y_1', Y_2', \ldots, Y_d')$ data vectors. Similarly, if LD is desired, the $(Y_1, Y_2, \ldots, Y_d)$ data vectors in the node are transformed to their linear discriminant variates, using the treatment levels as class labels. The PC and LD transformations are carried out *locally* at each node. After the split variable X_i is selected, its split point (if X_i is ordinal) or split set (if X_i is categorical) is the value that yields the smallest total sum of squared residuals (where the total is over the d models $E(Y_j|Z) = \eta_j + \sum_z \beta_{jz} I(Z = z)$) in the left and right child nodes. See Loh and Zheng (2013) and Loh et al. (2016) for more details.

Using change from baseline of ALSFRS1, ALSFRS2, ..., ALSFRS6 as longitudinal response variables, only the PC option yielded a nontrivial tree, as shown in Fig. 6.3. Subjects who died after 6 months and had missing values in any response variable were omitted, leaving a training sample of 627 observations. The tree has only one split, the same as the split at the root node of Fig. 6.2. The plots below the tree diagram show bootstrap-calibrated 90% simultaneous intervals for the treatment effect for each response variable in each terminal node. The longer lengths of the intervals in the left node are due to its much smaller sample size. Because every interval contains 0, there is no subgroup with statistically significant treatment effect.

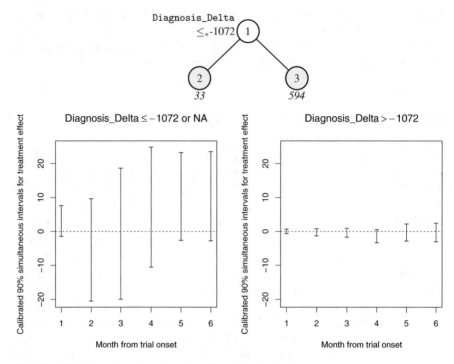

Fig. 6.3 GUIDE tree for change from baseline of longitudinal responses ALSFRS1, ALSFRS2, ..., ALSFRS6, using 627 observations and PCA at each node. At each split, an observation goes to the left branch if and only if the condition is satisfied. The symbol '$\leq_*$' stands for '$\leq$ or missing'. Sample size *(in italics)* printed below nodes. Bootstrap-calibrated 90% simultaneous intervals for treatment effect of each response variable in each node plotted below tree. Calibrated alpha is 0.011

6.5 Time-to-Event Response

Let $(U_1, \mathbf{X}_1)$, $(U_2, \mathbf{X}_2)$, ..., $(U_n, \mathbf{X}_n)$ be the survival times and predictor variable values of n subjects. Let $V_1, V_2, \ldots, V_n$ be independent and identically distributed observations from a censoring distribution and let $\delta_i = I(U_i < V_i)$ be the event indicator. The observed data vector of subject i is $(Y_i, \delta_i, \mathbf{X}_i)$, where $Y_i = \min(U_i, V_i)$. Let $\lambda(y, \mathbf{x}, z)$ denote the hazard function at time y and covariates $\mathbf{x}$ and z. The proportional hazards model stipulates that $\lambda(y, \mathbf{x}, z) = \lambda_0(y) \exp(\eta)$, where $\lambda_0(y)$ is a baseline hazard function independent of $(\mathbf{x}, z)$, and η is a function of $\mathbf{x}$ and z. Many methods fit a proportional hazards model to the data in each node separately (Negassa et al. 2005; Su et al. 2009; Lipkovich et al. 2011; Lipkovich and Dmitrienko 2014; Seibold et al. 2016), giving the tree model

$$\lambda(y, \mathbf{x}, z) = \sum_{j=1}^{\tau} \lambda_{j0}(y) \exp(\eta_j + \beta_{jz}) I(\mathbf{x} \in t_j); \quad \beta_{j0} = 0; \ j = 1, 2, \ldots, \tau.$$

Because the baseline hazard $\lambda_{j0}(y)$ varies from node to node, the model does not have proportional hazards overall. Therefore estimates of regression coefficients cannot be compared between nodes and relative risks are not independent of y.

GUIDE (Loh et al. 2015) fits one of the following three truly proportional hazards models instead.

$$\lambda(y, \mathbf{x}, z) = \lambda_0(y) \exp\left[\sum_{j=1}^{\tau} \{\eta_j + \beta_{jz}\} I(\mathbf{x} \in t_j) \right] \tag{6.8}$$

$$\lambda(y, \mathbf{x}, z) = \lambda_0(y) \exp\left[\sum_{j=1}^{\tau} \left\{ \eta_j + \beta_{jz} + \sum_{i=1}^{p} \gamma_{ji} x_{k*}^i \right\} I(\mathbf{x} \in t_j) \right] \tag{6.9}$$

$$\lambda(y, \mathbf{x}, z) = \lambda_0(y) \exp\left[\sum_{j=1}^{\tau} \left\{ \eta_j + \beta_{jz} + \sum_{k}^{K} \delta_{jk} x_k \right\} I(\mathbf{x} \in t_j) \right] \tag{6.10}$$

where $\beta_{j0} = 0$ ($j = 1, \ldots, \tau$) and the η_j satisfy a constraint such as $\sum_j \eta_j = 0$ to prevent over-parameterization. Model fitting is carried out by means of a well-known connection between proportional hazards regression and Poisson regression (Aitkin and Clayton 1980; Laird and Olivier 1981). Let $\Lambda_0(y) = \int_{-\infty}^{y} \lambda_0(u)\, du$ denote the baseline cumulative hazard function. The regression coefficients in (6.8), (6.9), or (6.10) are estimated by iteratively fitting a GUIDE Poisson regression tree (Chaudhuri et al. 1995; Loh 2006), using the event indicators δ_i as Poisson responses, $\log \Lambda_0(y_i)$ as offset variable, and the Poisson models

$$\log E(\delta|Z) = \log \Lambda_0(y) + \xi_j + \sum_z \beta_{jz} I(Z = z),$$

$$\log E(\delta|Z, X_{k*}) = \log \Lambda_0(y) + \xi_j + \sum_z \beta_{jz} I(Z = z) + \sum_{i=1}^{p} \gamma_{ji} X_{k*}^i,$$

$$\log E(\delta|Z, X_1, X_2, \ldots, X_k) = \log \Lambda_0(y) + \xi_j + \sum_z \beta_{jz} I(Z = z) + \sum_{k}^{K} \delta_{jk} X_k,$$

respectively, in each node t_j. At the first iteration, $\Lambda_0(y_i)$ is estimated by the Nelson-Aalen method (Aalen 1978; Breslow 1972). Then the estimated relative risks of the observations from the tree model are used to update $\Lambda_0(y_i)$ for the next iteration; see, e.g., Lawless (1982, p. 361).

Figure 6.4 gives the result of fitting model (6.8) from the 966 subjects with non-missing censored or observed survival time in the ALS data. The tree splits on Symptom to give two terminal nodes. The left node consists of 815 subjects with Symptom either missing or is speech. The other 151 subjects go to the right node, which has a statistically significant treatment effect based on the bootstrap-

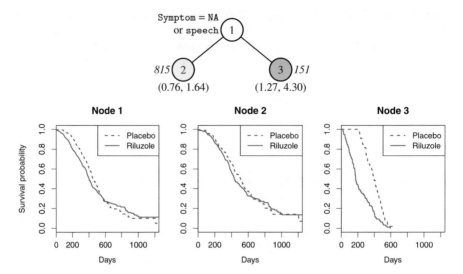

Fig. 6.4 GUIDE proportional hazards regression tree for differential treatment effects using model (6.8). Kaplan-Meier survival curves in each node are shown below the tree. Numbers in italics beside terminal nodes are sample sizes. Bootstrap-calibrated 90% simultaneous confidence intervals of relative risks (riluzole versus placebo) are given below terminal nodes. Calibrated alpha is 0.0003. Treatment effect is statistically significant in green node

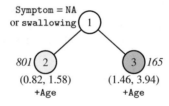

Fig. 6.5 GUIDE proportional hazards regression tree for differential treatment effects using model (6.9). Numbers in italics beside terminal nodes are sample sizes. Bootstrap-calibrated 90% simultaneous confidence intervals of relative risks (riluzole versus placebo) and name of linear prognostic variable (with sign indicating slope) are given below nodes. Calibrated alpha is 0.003. Treatment effect is statistically significant in green node

calibrated 90% simultaneous confidence intervals of relative risks printed below the nodes. Kaplan-Meier survival curves for placebo and riluzole subjects in each node are shown below the tree diagram.

Figure 6.5 gives the result for model (6.9) with polynomial degree $p = 1$. The root node is split into two terminal nodes on Symptom, but now the model in each node includes the effect of the best linear prognostic variable (which turns out to be Age in both child nodes). According to the bootstrap-calibrated 90% simultaneous intervals for relative risk printed below the nodes, the subgroup with significant treatment effect consists of subjects for which Symptom is neither missing nor swallowing.

6.6 Concluding Remarks

We have explained and demonstrated the main features of the GUIDE method for subgroup identification and discussed a bootstrap method of confidence interval construction for subgroup treatment effects. The bootstrap method is quite general and is applicable to algorithms other than GUIDE. Because it expands the traditional t-intervals to account for uncertainty due to split selection, it is more efficient if the estimated subgroup treatment effects are unbiased. The method may still be applicable if the estimates are biased, but the calibrated intervals would be wider as a result. Biased estimates of subgroup treatment effects are common among algorithms that search for splits to maximize the difference in treatment effects in the child nodes. A comparison of methods on this and other criteria is reported in a forthcoming article (Loh et al. 2019a).

Although GUIDE does not impute missing values for split selection, it does impute them in the predictor variables with their node sample means when fitting models (6.2)–(6.4) in the nodes. Therefore these models, e.g., Figs. 6.2 and 6.5, assume that missing values in the X variables are missing at random (MAR). But the MAR assumption is not needed for model (6.1), such as Figs. 6.1, 6.3, and 6.4.

There are two newer GUIDE features that are not discussed here. One is cyclic or periodic predictor variables, such as angle of impact in an automobile crash, day of week of hospital admission, and time of day of medication administration. If GUIDE splits a node on such a variable, the split takes the form of a finite interval of values $a < X \leq b$ instead of a half-line $X \leq c$. Another feature is accommodation of multiple missing-value codes. For example, the result of a lab test may be "missing" for various reasons. It may not have been ordered by the physician because it was risky for the patient, it may be inappropriate (e.g., a mammogram for a male or a prostate-specific antigen test for a female), the patient may have declined the test due to cost, or the result of the test was accidentally or erroneously not reported. If the "missing" values are all recorded as NA, a split would take the form "$X \leq c$ or $X = $ NA" or "$X \leq c$ and $X \neq $ NA". But if the reasons for missingness are known, GUIDE would use the information to produce more specific splits of the form "$X \leq c$ or $X \in $ S", where S is a subset of missing-value codes. Illustrative examples of these two features are given in the GUIDE manual (Loh 2018).

Acknowledgements The authors thank Tao Shen, Yu-Shan Shih and Shijie Tang for their helpful comments. Data used in the preparation of this article were obtained from the Pooled Resource Open-Access ALS Clinical Trials (PRO-ACT) Database. As such, the following organizations and individuals within the PRO-ACT Consortium contributed to the design and implementation of the PRO-ACT Database and/or provided data, but did not participate in the analysis of the data or the writing of this report: Neurological Clinical Research Institute, MGH Northeast ALS Consortium Novartis Prize4Life Israel Regeneron Pharmaceuticals, Inc., and Sanofi Teva Pharmaceutical Industries, Ltd.

References

Aalen O (1978) Nonparametric inference for a family of counting processes. Ann Stat 6:701–726

Ahn H, Loh W-Y (1994) Tree-structured proportional hazards regression modeling. Biometrics 50:471–485

Aitkin M, Clayton D (1980) The fitting of exponential, Weibull and extreme value distributions to complex censored survival data using GLIM. Appl Stat 29:156–163

Atassi N, Berry J, Shui A, Zach N, Sherman A, Sinani E, Walker J, Katsovskiy I, Schoenfeld D, Cudkowicz M, Leitner M (2014) The PRO-ACT database: Design, initial analyses, and predictive features. Neurology 83(19):1719–1725

Breiman L, Friedman JH, Olshen RA, Stone CJ (1984) Classification and regression trees. Wadsworth, Belmont

Breslow N (1972) Contribution to the discussion of regression models and life tables by D. R. Cox. J R Stat Soc Ser B 34:216–217

Chan K-Y, Loh W-Y (2004) LOTUS: An algorithm for building accurate and comprehensible logistic regression trees. J. Comput. Graph. Stat. 13:826–852

Chaudhuri P, Loh W-Y (2002) Nonparametric estimation of conditional quantiles using quantile regression trees. Bernoulli 8:561–576

Chaudhuri P, Huang M-C, Loh W-Y, Yao R (1994) Piecewise-polynomial regression trees. Stat Sin 4:143–167

Chaudhuri P, Lo W, Loh W, Yang C (1995) Generalized regression trees. Stat Sin 5:641–666

Dusseldorp E, Meulman JJ (2004) The regression trunk approach to discover treatment covariate interaction. Psychometrika 69:355–374

Foster JC, Taylor JMG, Ruberg SJ (2011) Subgroup identification from randomized clinical trial data. Stat Med 30:2867–2880

Gnanadesikan R (1997) Methods for statistical data analysis of multivariate observations, 2nd edn. Wiley, New York

Kim H, Loh W-Y (2001) Classification trees with unbiased multiway splits. J Am Stat Assoc 96:589–604

Kim H, Loh W-Y (2003) Classification trees with bivariate linear discriminant node models. J Comput Graph Stat 12:512–530

Laird N, Olivier D (1981) Covariance analysis of censored survival data using log-linear analysis techniques. J Am Stat Assoc 76:231–240

Lawless J (1982) Statistical models and methods for lifetime data. Wiley, New York

Lipkovich I, Dmitrienko A (2014) Strategies for identifying predictive biomarkers and subgroups with enhanced treatment effect in clinical trials using SIDES. J Biol Stand 24:130–153

Lipkovich I, Dmitrienko A, Denne J, Enas G (2011) Subgroup identification based on differential effect search — a recursive partitioning method for establishing response to treatment in patient subpopulations. Stat Med 30:2601–2621

Loh W-Y (1987) Calibrating confidence coefficients. J Am Stat Assoc 82:155–162

Loh W-Y (1991a) Bootstrap calibration for confidence interval construction and selection. Stat Sin 1:477–491

Loh W-Y (1991b) Survival modeling through recursive stratification. Comput Stat Data Anal 12:295–313

Loh W-Y (2002) Regression trees with unbiased variable selection and interaction detection. Stat Sin 12:361–386

Loh W-Y (2006) Regression tree models for designed experiments. In: Rojo J (ed) Second E. L. Lehmann Symposium, vol 49. IMS lecture notes-monograph series. Institute of Mathematical Statistics, pp 210–228

Loh W-Y (2009) Improving the precision of classification trees. Ann Appl Stat 3:1710–1737

Loh W-Y (2014) Fifty years of classification and regression trees (with discussion). Int Stat Rev 34:329–370

Loh W-Y (2018) GUIDE user manual. University of Wisconsin, Madisons

Loh W-Y, Shih Y-S (1997) Split selection methods for classification trees. Stat Sin 7:815–840

Loh W-Y, Vanichsetakul N (1988) Tree-structured classification via generalized discriminant analysis (with discussion). J Am Stat Assoc 83:715–728

Loh W-Y, Zheng W (2013) Regression trees for longitudinal and multiresponse data. Ann Appl Stat 7:495–522

Loh W-Y, He X, Man M (2015) A regression tree approach to identifying subgroups with differential treatment effects. Stat Med 34:1818–1833

Loh W-Y, Fu H, Man M, Champion V, Yu M (2016) Identification of subgroups with differential treatment effects for longitudinal and multiresponse variables. Stat Med 35:4837–4855

Loh W-Y, Cao L, Zhou P (2019a) Subgroup identification for precision medicine: a comparative review of thirteen methods. Data Min Knowl Disc 9(5):e1326

Loh W-Y, Eltinge J, Cho MJ, Li Y (2019b) Classification and regression trees and forests for incomplete data from sample surveys. Stat Sin 29:431–453

Loh W-Y, Man M, Wang S (2019c) Subgroups from regression trees with adjustment for prognostic effects and post-selection inference. Stat Med 38:545–557

Morgan JN, Sonquist JA (1963) Problems in the analysis of survey data, and a proposal. J Am Stat Assoc 58:415–434

Negassa A, Ciampi A, Abrahamowicz M, Shapiro S, Boivin J (2005) Tree-structured subgroup analysis for censored survival data: validation of computationally inexpensive model selection criteria. Stat Comput 15:231–239

Seibold H, Zeileis A, Hothorn T (2016) Model-based recursive partitioning for subgroup analyses. Int J Biostat 12(1):45–63

Su X, Tsai C, Wang H, Nickerson D, Bogong L (2009) Subgroup analysis via recursive partitioning. J Mach Learn Res 10:141–158

Therneau T, Atkinson B (2018) rpart: recursive partitioning and regression trees. R package version 4.1-13

Zeileis A, Hothorn T, Hornik K (2008) Model-based recursive partitioning. J Comput Graph Stat 17:492–514

Chapter 7
A Novel Method of Subgroup Identification by Combining Virtual Twins with GUIDE (VG) for Development of Precision Medicines

Jia Jia, Qi Tang, and Wangang Xie

Abstract A lack of understanding of human biology creates a hurdle for the development of precision medicines. To overcome this hurdle we need to better understand the potential synergy between a given investigational treatment (vs. placebo or active control) and various demographic or genetic factors, disease history and severity, etc., with the goal of identifying those patients at increased "risk" of exhibiting clinically meaningful treatment benefit. For this reason, we propose the VG method, which combines the idea of an individual treatment effect (ITE) from Virtual Twins with the unbiased variable selection and cutoff value determination algorithm from GUIDE. Simulation results show the VG method has less variable selection bias than Virtual Twins and higher statistical power than GUIDE Interaction in the presence of prognostic variables with strong treatment effects. Type I error and predictive performance of Virtual Twins, GUIDE and VG are compared through the use of simulation studies. Results obtained after retrospectively applying VG to data from an Alzheimer's disease clinical trial also are discussed.

7.1 Introduction

The concepts of personalized medicine and precision medicine have "evolved" over time, with precision medicine now viewed as an approach that allows for the treatment of patients while taking their personal signatures, such as genes, environment and lifestyles, into consideration in order to maximize the benefit

J. Jia (✉) · W. Xie
AbbVie Inc., North Chicago, IL, USA
e-mail: jia.jia@abbvie.com; Wangang.xie@abbvie.com

Q. Tang
Sanofi US, Bridgewater, NJ, USA
e-mail: Qi.Tang@sanofi.com

© Springer Nature Switzerland AG 2020   167
N. Ting et al. (eds.), *Design and Analysis of Subgroups with Biopharmaceutical Applications*, Emerging Topics in Statistics and Biostatistics,
https://doi.org/10.1007/978-3-030-40105-4_7

(efficacy) and/or minimize the risk (safety) they receive from the treatment. In other words, treatment effects are often heterogeneous in a given patient population. Thus, it is necessary to improve our understanding of differential treatment effects observed in patients with different signatures.

Although it is difficult to accurately assess and maximize the treatment effect for every patient, it may be possible to categorize patients into subgroups according to some known and pre-defined signatures, for example, demographics, biomarkers and lab values and then assess the treatment effect for those subgroups. The subgroup in which the "best" treatment effect can be observed also could be identified via complicated statistical methods without predefining the variables to be used in the analysis. In this paper, we focus on the latter approach, also known as retrospective (or ad hoc) subgroup identification. Retrospective subgroup identification is a critical approach used to develop precision medicines.

Many subgroup identification methods have been developed. For example, Negassa et al. (2005) developed RECPAM which attempts to maximize the Cox partial likelihood. Su et al. (2008, 2009) developed Interaction trees (IT), which attempts to minimize the p-value for testing the significance of the interaction term between the subgroup indicator and the treatment. Foster et al. (2011) developed Virtual twins (VT) which uses random forests to predict the treatment effect for each patient and then applies Classification and Regression Trees (CART) to identify potential subgroups. Lipkovich et al. (2011) developed the Subgroup Identification based on Differential Effect Search (SIDES), which targets on the treatment effect difference but may lead to selection bias associated with variables having more possible cut-off values. Dusseldorp and Van Mechelen (2014) developed QUalitative INteraction Trees (QUINT) that attempts to optimize a weighted sum of measures of effect size and subgroup size. Loh et al. (2002) developed a method called Generalized, Unbiased, Interaction Detection and Estimation (GUIDE) that is a multi-purpose machine learning algorithm for constructing classification and regression trees. Also, Loh et al. (2015) later developed a new method called GUIDE-Interaction (Gi), which was based on the original GUIDE but targeted on the treatment effect difference. Also, Gi has been compared to other methods regarding the statistical properties in certain scenarios. The results showed that Gi was a preferred solution when the goal is to find the signature based on treatment difference.

In precision medicine, the treatment effect difference forms the basis for subgroup identification. Most of the existing methods can identify subgroups, and some of the methods can obtain unbiased signature selection. However, only a few of these methods can target directly on the treatment effect difference, especially the individual treatment effect (ITE) difference, which in general is the benefit obtained from receiving treatment compared to the benefit obtained from receiving placebo for a particular patient.

During our review of the existing methods, we found that the variable selection process of Gi is partly driven by how well a variable predicts the response variable. Thus, in a case of a strong prognostic effect, Gi may select prognostic variables more often than predictive variables, since prognostic variables may in fact predict the response variable better than predictive variables. In contrast, VT directly targets on the treatment effect difference and may have better variable selection performance than Gi in the scenario of a strong prognostic effect. However, CART is implemented in the variable selection step of VT and may lead to bias in variable selection (Loh et al. 2015).

In this paper, we propose VG, a novel method that targets directly on individual treatment effects using an unbiased variable selection procedure by combining two methods, VT and GUIDE. In VG, the CART part in VT is replaced by GUIDE. The performance of VG, VT and Gi will be compared via simulations and a case study will be presented.

7.2 Methods

The VG method contains two steps: (1) estimate the individual treatment effect (ITE) difference using Virtual Twins; and (2) identify potential subgroup(s) using GUIDE. Benefits inherited from GUIDE include the ability to utilize missing covariate information and simultaneously model multiple endpoints.

Without loss of generality, we illustrate the VG method for the case of binary response variable.

7.2.1 Step I

The first step is to estimate the ITE difference by using Random Forests (need a citation here). Let Y_n and T_n represent the original response variable (continuous or binary) and treatment variable ($0 =$ placebo; $1 =$ investigational treatment), respectively; and let $X_{n,p}$ represents the matrix that contains all the covariates, where n is the sample size, and p is the number of covariates.

Let T'_n represent the flipped (or opposite) treatment variable, where $T'_n = 1_n - T_n$. The purpose of doing this is to estimate the counterfactual response of each patient, in other words, the response of a patient under the treatment that was not received.

Let Y'_n represent the estimated counterfactual response given T'_n and $X_{n,p}$. In the VG method, GUIDE is used to provide nonparametric estimation of Y'_n. In order to do that, similar to Foster et al. (2016), we utilize two additional matrices that contain all possible two $-$ way interactions between (a) T_n and $X_{n,p}$ and (b) T'_n and $X_{n,p}$, respectively.

As a consequence, data used for predicting Y'_n have the structure provided below, which contains five components and a total sample size of 2n:

$$Y_{2n}^* = \begin{pmatrix} Y_n \\ Y_{/n} \end{pmatrix}, \; T_{2n}^* = \begin{pmatrix} T_n \\ T_{/n} \end{pmatrix}, \; X_{2n,p}^* = \begin{pmatrix} X_{n,p} \\ X_{n,p} \end{pmatrix},$$

$$XT_{2n,p} = \begin{pmatrix} T_n X_{n,p} \\ T_{/n} X_{n,p} \end{pmatrix}, \; X(1-T)_{2n,p} = \begin{pmatrix} (1_n - T_n) X_{n,p} \\ (1_n - T_{/n}) X_{n,p} \end{pmatrix}$$

The terms $T_n X_{n,p}$, $T'_n X_{n,p}$, $(1_n - T_n)X_{n,p}$ and $(1_n - T'_n)X_{n,p}$ represent interactions between the treatment indicator, flipped treatment indicator and the covariates. Note that Y'_n is unknown but can be predicted using GUIDE with a weight variable, which assigns 0 weight to Y_n and 1_n weight to Y'_n.

After Y'_n is predicted, we can calculate the individual treatment effect (ITE) difference. Let's assume the i-th patient with individual covariates X_i received treatment ($t_i = 1$) and obtained outcome y_i, and the j − th patient with individual covariates X_j received placebo ($t_j = 0$) obtained outcome y_j. By predicting Y'_n, we have now obtained the 'flipped' outcome for patients i and j, y'_i and y'_j, given their individual covariates X_i and X_j, respectively. Thus, the ITE difference for the i − th patient can be calculated as $ITE_i = y_i - y'_i$; while the ITE difference for the j − th patient can be calculated as $ITE_j = y'_j - y_j$. By doing this, we obtain the vector of length n containing the ITE differences for all the patients, conditional on their individual covariates.

7.2.2 Step II

The goal of Step II is to identify treatment effect heterogeneity based on the estimated ITE in Step I.GUIDE is used to identify treatment effect patterns because of its reliable and robust performance in pattern recognition (Loh 2002).

Figure 7.1 shows the procedures of GUIDE, in which ITE is the outcome, X_k is the k-th covariate, p is the total number of covariates. X_s is the covariate that has the smallest p-value from fitting the univariate models ITE ~ X_{kl} $k = 1, \ldots, p$. $ITE_s(a)$, $ITE_s(b)$, $X_s(a)$ and $X_s(b)$ are the outcomes and covariate X_s that were split by value x_s, respectively. SSE(a) and SSE(b) are the two SSEs obtained from two model fittings of $ITE_s(a)$, $ITE_s(b)$, $X_s(a)$ and $X_s(b)$ (Loh 2002). Once GUIDE has identified the subgroups, the mean ITE difference for each subgroup is calculated. According to the algorithm, we can interpret the mean ITE difference as the difference between the outcome if the patient received treatment and the outcome if the patient received placebo, while adjusting for all the patient's covariates. When the original outcome variable Y is binary (0 or 1), the above algorithm is used to predict the probability of $Y = 1$, which will be ITE. And thus the interpretation of ITE difference would be in terms of the difference of two probabilities : probability of $Y = 1$ if the patient received treatment and probability of $Y = 1$ if the patient received placebo.

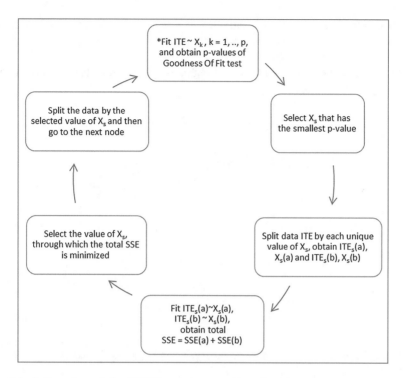

Fig. 7.1 A flow chart of the GUIDE pattern recognition algorithm in the setting of finding heterogeneous patterns of ITE. *The procedure starts from here

7.3   Simulations

7.3.1   Set-up

In order to evaluate the statistical properties of the VG method and to compare it to other methods, we performed a simulation study.

The targeted subgroup is the one with the corresponding signature(s) decided by the predictive variable interacting with the treatment variable. In other words, the treatment effect observed in the targeted subgroup is larger than that observed outside the targeted subgroup due to the interaction between the predictive variable and the treatment variable.

Thus, we first defined the true subgroup in our simulation. The signature was decided through predictive variable(s). To simplify, we considered cases where there was only one predictive variable, X_{pred}, in our simulation. Moreover, we defined one prognostic variable, X_{prog}, which had only main effect to the outcome but no interaction with the treatment.

We defined T as the treatment variable and Y as the outcome variable. In addition, we created $\mathbf{Z_{n,p}}$, a matrix containing p variables that are independent with the response and treatment variables.

In our simulation, we used the following models to generate the treatment vector and a covariate matrix:

$$T \ (0 \text{ or } 1) \sim BIN \ (n, 0.5)$$

$$\mathbf{X_{n,p+2}} = \begin{pmatrix} x_{\text{pred},1} & x_{\text{prog},1} \\ \vdots & \vdots & \mathbf{Z_{n,p}} \\ x_{\text{pred},n} & x_{\text{prog},n} \end{pmatrix} \ MVN \left(0, \sum_{p+2,p+2} \right)$$

if all $\mathbf{X_{n,p+2}}$ are continuous and

$$\sum_{p+2,p+2} = \begin{pmatrix} 1 & 0.5 & \cdots & 0.5 \\ 0.5 & 1 & \cdots & 0.5 \\ \vdots & \vdots & \ddots & \vdots \\ 0.5 & \cdots & 0.5 & 1 \end{pmatrix}$$

For the case where we simulated binary X for X_{pred}, X_{prog} and/or $\mathbf{Z_{n,p}}$, we first simulated

$$p_x \sim Beta \ (2, 3)$$

and we used

$$X \sim BIN \ (n, p_x.)$$

to generate the values for binary covariates. And for outcome variable Y, we used the following models:

- when Y is continuous:

$$Y = \beta_{\text{pred}} \times I \left(X_{\text{pred}} > x_0 \right) \times T + \beta_{\text{prog}} \times X_{\text{prog}} + \beta_{\text{trt}} \times T + e$$

$$e \sim N \ (0, 0.25)$$

- when Y is binary:

$$\mu = \beta_{\text{pred}} \times I \left(X_{\text{pred}} > x_0 \right) \times T + \beta_{\text{prog}} \times X_{\text{prog}} + \beta_{\text{trt}} \times T$$

$$p_y = \frac{\exp(\mu)}{1 + \exp(\mu)}$$

$$Y \sim BIN(n, p_y)$$

where x_0 is the cut $-$ off point, in our simulation, we defined x_0 as the mean of X_{pred} and e is the noise that follows a normal distribution with mean 0 and variance 0.25. We'd like to simulate a dataset close to that from an observed clinical trial. Since we utilize a clinical trial with approximately 15 covariates, and 400 subjects, the resulting simulated datasets additionally contained 13 noise variables ($p = 13$), and 400 subjects ($n = 400$) within each of the iterations.

Therefore, the dataset contains all the components below:

$$Y = \begin{pmatrix} y_1 \\ \vdots \\ y_{400} \end{pmatrix} \quad T = \begin{pmatrix} t_1 \\ \vdots \\ t_{400} \end{pmatrix} \quad X = \begin{pmatrix} x_{pred,1} & x_{prog,1} \\ \vdots & \vdots & Z_{n,p} \\ x_{pred,400} & x_{prog,400} \end{pmatrix}$$

To simulate different scenarios, we selected different values for β_{pred}, β_{prog} and β_{trt}. The different scenarios that we have simulated are summarized in Table 7.1.

Through the three scenarios defined above, we used the following metrics to compare VG, Gi and VT methods:

1. Type I error: probability of identifying a subgroup when there are no subgroups.
2. Power: probability of identifying a subgroup when there is a subgroup.
3. Conditional true discovery rate: conditional probability of correctly identifying the predictive variable when a subgroup is identified.

For the purpose of fair comparison, we compared the power and true discovery rate for the three methods under the same type I error rate. We have simulated 500 iterations for each of the scenarios.

7.3.2 Results

According to the simulation results (Fig. 7.2), all three methods behave similarly and demonstrate above 90% power and almost 100% conditional true discovery rate

Table 7.1 Simulation scenarios

Scenarios	β_{pred}	X_{pred}	β_{prog}	X_{prog}	β_{trt}	Noise variables ($p = 13$)
No prognostic	0.5	Continuous	0	None	0.2	Continuous
No prognostic mix	0.2	Binary	0	None	0.2	Binary/continuous[a]
Prognostic	0.5	Continuous	0.5	Continuous	0.2	Continuous

[a]Includes 1 binary variable and 12 continuous variables

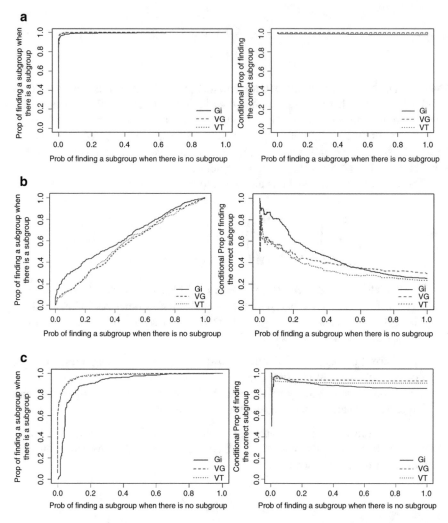

Fig. 7.2 Plot of power (Left) and conditional true discovery rate (Right) vs Type I error. (**a**) No prognostic scenario. (**b**) No prognostic Mix scenario. (**c**) Prognostic scenario

under most of Type I error rates when the predictive variable was continuous and there was no prognostic effect.

In the scenario where the predictive variable was binary with an effect size of 0.2 and one of the noise covariates was also binary, Gi demonstrated higher power and conditional true discovery rate than the other two methods, especially when the Type I error was controlled between 0 and 0.4.

When the prognostic effect was added to the simulation as a continuous variable, Gi had lower power and similar or lower conditional true discovery rate compared to the other two methods.

Since the VG and VT methods are very similar with respect to the background framework, these two methods behave very much alike. However, one can still notice about 5%improvement with VG compared to VT in the simulation results regarding the conditional true discovery rate, especially in the 'Prognostic' scenario.

7.4   Case Study

7.4.1   Type I Error Control

In this paper, we define Type I error as the probability of identifying a subgroup when there are no subgroups. When conducting an analysis using real data, it can be challenging to obtain Type I error control as if the analysis was conducted using simulated data. Hence, we implemented the permutation method for the analysis involving real data.

Specifically, we break the association between treatment and covariate as well as between treatment and response; while keeping the association between covariate and response. In other words, we eliminated the predictive effect while keeping the prognostic effect. Other ways to control Type I error include the Šidák-based multiplicity adjustment method (Hochberg and Tamhane 1987) as implemented in SIDES (Lipkovich et al. 2011) and some more complicated permutation methods described in Foster et al. (2016).

7.4.2   Bootstrap

In the analysis involving real data, one can easily obtain a naïve confidence interval for the point estimate of the treatment effect. However, such a confidence interval may not be valid because it does not take into consideration uncertainties from the selection of the predictive variable and the split value of the predictive variable. In the case study section of this paper, we implement a more complicated bootstrap method that is proposed by Loh et al. (2015) to obtain confidence intervals for the point estimate of the treatment effect.

The bootstrap sample was drawn (with replacement) from the original dataset with the same size, and the VG method was applied. However, we ignored the identified signature based on the bootstrap sample. Instead, we obtained newly predicted ITE, which are different from ITE predicted during the first step of VG method. The new ITE can be obtained directly from GUIDE. Then, by using the identified signature of the subgroup from the original dataset, the bootstrap sample can be separated into subgroups and the mean of new ITE for these subgroups can be obtained. After these procedures have been repeated B times (B = 500 in our case), the distribution and the confidence interval of the mean of new ITE for the identified subgroup can be obtained.

7.4.3 Application

We applied the VG method to a real world example from a clinical study evaluating an experimental treatment for patients with Alzheimer's disease. The endpoint was the change from baseline to week 12 in ADAS-Cog 11 subscale score (0–70), which measures the change in severity of the disease. Thus, at the end of week 12, negative changes indicate improvement from baseline. There were two treatment arms: experimental treatment and placebo. In this case, the ITE for patient i would be calculated as

$$\text{ITE}_i = y_i \,|\, Placebo, X_i - y'_i \,|\, Treatment, X_i$$

so that the larger the ITE, the better the treatment effect compared to placebo.

We have included 17 covariates after consulting with medical professionals, including but not limited to age, sex, race, baseline Mini-Mental State Examination (MMSE, a disease staging measure, range 0–30), the change of ADAS-Cog 11 subscale score from screening to baseline, and Apolipoprotein E4 (APOE4).

Since there were total of three datasets in this project, and we needed to control the Type I error while analyzing the data. Thus, we followed the steps below to conduct the subgroup identification analysis on the first two datasets.

1. Use permutation method on the first dataset to find the Type I error control;
2. Analyze the first dataset while controlling the Type I error, identify the signature(s); and
3. Find the subgroup in the second dataset according to the signature(s) identified in step 2, and evaluate the treatment effect in the subgroup to see if it differs from the other subgroup.

Unfortunately, when the Type I error was controlled at the 0.05 level, no subgroup was identified. Thus, we ignored Type I error control allowing for exploration of results that can be found. As shown in Fig. 7.3, the covariate 'Years Since Onset of the Symptom' (YearOnset) was found as the predictive variable with a cut-off value at 3.55 years.

Fig. 7.3 Subgroup identified from the first dataset

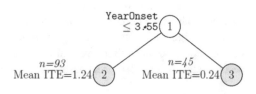

Table 7.2 Estimated
treatment effects

Patient group	Estimated treatment effect	
	First dataset	Second dataset
Overall	0.91	0.07
YearOnset $\leq$3.55	1.24	0.00
YearOnset >3.55	0.24	0.15

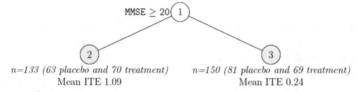

MMSE $\geq$ 20 ①

②   ③

n=133 (63 placebo and 70 treatment)   *n=150 (81 placebo and 69 treatment)*
Mean ITE 1.09   Mean ITE 0.24

Fig. 7.4 Subgroup identified from the third dataset

The cartoon in Fig. 7.3 can be interpreted as follows (GUIDE reference): group '1' contains the overall patient population from this dataset. The patients who satisfy the criteria 'Years Since Onset of the Symptom $\leq$ 3.55 years' are classified into subgroup '2' (go along with the left line to circle 2), otherwise, the patients are classified into subgroup '3' (go along with the right line to circle 3).

According to the result, subgroup '2' had mean ITE = 1.24 with sample size 93, which means, on average, a larger treatment effect was observed in this subgroup of patients compared to the treatment effect observed in the rest of the patients (mean ITE = 0.24, n = 45). However, since Type I error was not controlled, we attempted to validate this result by observing the treatment effect in the subgroup of patients who were selected according to the identified signature ('Years Since Onset of the Symptom $\leq$3.55 years) from the second dataset, and compared it with the treatment effect observed from the first dataset.

As shown in Table 7.2, the observed treatment effect in the subgroups from the first dataset cannot be replicated in the subgroups from the second dataset by using the same signature. Therefore, the result obtained from the VG method was not valid in this case given there was no Type I error control.

The third dataset in this project had a larger sample size than either of the first two datasets. Also, the experimental drug (Treatment) utilized in this dataset was different from the one utilized in the first two datasets. We applied the VG method on this dataset while controlling the Type I error at 0.05, and one covariate was identified as the predictive variable with a cut-off 20.

As shown in Fig. 7.4, subgroup '2' was identified with a larger ITE compared to subgroup '3', although subgroup '3' had larger sample size. The identified covariate was MMSE, and the ITE appears to represent a clinically meaningful difference (Perneczky et al. 2006). The cut-off value 20 is suggested to be used as the separation of disease staging between moderate ($\leq$20) and mild (>20). For this dataset, we have also calculated the 95% confidence interval in order to obtain an estimation of the validity of the results.

Table 7.3 Estimated treatment effects and 95% CI

Subgroup	Mean effect	95% CI
MMSE ≥20 (group '2')	1.09	$(-0.11, 2.29)$
MMSE <20 (group '3')	0.24	$(-1.19, 1.67)$
Difference between above two subgroups	0.85	$(-0.79, 2.50)$

As shown in Table 7.3, although the treatment effect was numerically different from 0, the 95% confidence intervals for both subgroups contain 0. However, the 95% confidence interval in subgroup '2' is suggestive of a positive trend, while that in subgroup '3' is not. Moreover, the 95% confidence interval for the difference of the treatment effects between the two subgroups is also suggestive of a positive effect favoring subgroup 2 over subgroup 3. Therefore, these results were felt to be clinically meaningful. Additional exploration (i.e., studies) may be necessary to demonstrate whether this is truly a clinically meaningful effect.

7.5   Discussion

Precision medicine attempts to improve the safety and/or efficacy of a drug by tailoring the treatment according to the patient's characteristics. Subgroup identification is a critical step to realize the potential of precision medicine. However, current realizations of subgroup analysis in clinical trials are often limited within pre-defined subgroups. The current state of conducting analyses according to pre-defined subgroups while ignoring Type I error control may result in true predictive variable(s) and/or true cut-off value(s) being missed. Some data-mining based subgroup identification methods also exist. Most of these methods are used to prospectively search for subgroups given a dataset. By using data mining techniques, one can avoid pitfalls of common one variable at a time subgroup analyses.

In this paper, we have proposed a novel method of prospective subgroup identification, the VG method, which combines the advantages of two existing methods (i.e., Virtual Twins and GUIDE). However, the VG method is not a simple combination of the two methods, it replace the CART part of the VT method by GUIDE. In other words, the VG method first calculates the Individual Treatment Effect (ITE) according to the counterfactual concept in causal inference; it then applies GUIDE to identify the subgroup(s) based on the ITE. Results from our simulation studies show that the VG method outperforms Virtual Twins when there are binary and continuous covariates in the data and also outperforms GI when prognostic effect is as strong as predictive effect. The key advantage of the VG method compared to Gi is that it targets directly on the treatment effect and can identify a predictive variable in the presence of a prognostic effect. Also, the VG method has less potential for selection bias when compared to Virtual Twins given the latter's reliance on CART. However, in our simulation, we have assumed there is only one predictive variable with one cut-off value due to the limitation of the tools we are using. In fact, there could be more than one predictive variable and there can be more than one cut-off value for a predictive variable in a given dataset.

Through the case study, without Type I error control, the identified subgroup is not valid and the results cannot be reproduced. When the Type I error is controlled, although the identified subgroup is not statistically significant based on the 95% confidence intervals calculated using the bootstrap method, it demonstrates a trend related to the treatment effect that might be clinically meaningful. In other words, with conservatively controlled Type I error, the results might not be statistically significant, but might provide some clinically helpful information. In this case, additional research (i.e., clinical trials) would be needed to confirm this result.

Our work has provided a clearly defined framework to compare three different subgroup identification methods according to type I error control, power and the conditional true discovery rate. It also provides two applications for controlling type I error and estimating 95% confidence intervals in the analysis of a real dataset, which are permutation and bootstrap methods, respectively. The performance of VG method relies on datasets, case by case simulation is suggested to be tailored to the study. Generalization the conclusion to other studies should be careful.

Our future work involves the improvement of the prediction accuracy when calculating the ITE, which is a critical factor that impacts the performance of the VG method. Moreover, we are trying to extend the VG method to both binary and time-to-event endpoints.

Acknowledgement This manuscript was sponsored by AbbVie. AbbVie contributed to the design, research, and interpretation of data, writing, reviewing, and approving the publication. Jia Jia, and Wangang Xie are employees of AbbVie, Inc. Qi Tang is an employee of Sanofi US, Inc. and a former employee of AbbVie, Inc. We thank Richard Rode, a former AbbVie employee, for reviewing and editing the manuscript.

References

Dusseldorp E, Mechelen IV (2014) Qualitative interaction trees: a tool to identify qualitative treatment-subgroup interaction. Stat Med 33:219–237

Foster JC et al (2016) Permutation testing for treatment–covariate interactions and subgroup identification. Stat Biosci 8(1):77–98

Foster JC, Taylor JM, Ruberg SJ (2011) Subgroup identification from randomized clinical trial data. Stat Med 30:2867–2880

Hochberg Y, Tamhane AC (1987) Multiple comparison procedures. Wiley, Hoboken

Lipkovich I, Dmitrienko A, Denne J, Enas G (2011) Subgroup identification based on differential effect search—a recursive partitioning method for establishing response to treatment in patient subpopulations. Stat Med 30:2601–2621

Loh W-Y (2002) Regression trees with unbiased variable selection and interaction detection. Stat Sin 12:361–386

Loh W-Y, He X, Man M (2015) A regression tree approach to identifying subgroups with differential treatment effects. Stat Med 34:1818–1833

Negassa A et al (2005) Tree-structured subgroup analysis for censored survival data: validation of computationally inexpensive model selection criteria. Stat Comput 15:231–239

Perneczky et al (2006) Mapping scores onto stages: mini-mental state examination and clinical dementia rating. Am J Geriatr Psychiatry 14(2):139–144

Su X et al (2009) Subgroup analysis via recursive partitioning. J Mach Learn Res 10:141–158

Su X et al (2008) Interaction trees with censored survival data. Int J Biostat 4(1):2

Chapter 8
Subgroup Identification for Tailored Therapies: Methods and Consistent Evaluation

Lei Shen, Hollins Showalter, Chakib Battioui, and Brian Denton

Abstract In contrast to the "one-size-fits-all" approach of traditional drug development, the paradigm of tailored therapeutics seeks to identify subjects with an enhanced treatment effect. In this chapter, we describe a statistical approach (TSDT) to subgroup identification that utilizes ensemble trees and resampling. For each potential subgroup identified, TSDT produces a multiplicity-adjusted strength of the subgroup finding as well as a bias-adjusted estimate of the treatment effect in the identified subgroup, both of which are important for decision-making in the development of tailored therapeutics. We describe a careful examination of simulation studies in a number of related publications, in order to determine the ideal framework to compare subgroup identification methods. A simulation study is performed to evaluate the performance of TSDT. The method has been implemented in a publicly available R package.

8.1 Background

In randomized clinical trials, individuals are assigned randomly to a treatment group and a control group. Efficacy and safety outcomes are measured and compared between the two groups. The main interest of the study investigators is to evaluate the treatment effect on the overall population. However, some subgroups of patients may have greater response to treatment than the overall population. It is well known that patients respond to drugs differently, with many factors affecting the response to any given drug, such as genetic makeup, phenotypic, pharmacokinetic, social, and disease severity, as well as demographic factors. Increasingly in pharmaceutical drug development, it is not sufficient to merely show that the mean effect of a new treatment is statistically significantly better than the control. Patients, physicians,

L. Shen (✉) · H. Showalter · C. Battioui · B. Denton
Eli Lilly and Company, Indianapolis, IN, USA
e-mail: shen_lei@lilly.com; hdhshowalter@lilly.com; battioui_chakib@lilly.com; denton_brian_david@lilly.com

© Springer Nature Switzerland AG 2020
N. Ting et al. (eds.), *Design and Analysis of Subgroups with Biopharmaceutical Applications*, Emerging Topics in Statistics and Biostatistics,
https://doi.org/10.1007/978-3-030-40105-4_8

and payers want and, in fact, are demanding, to know more about individual patient outcomes (Ruberg et al. 2010), so that the right drug can be selected to properly fit each patient. It has therefore become important to improve on the traditional "one size fits all" paradigm of drug development (Wong et al. 2019), and there are now examples of marketed compounds that make tailored therapeutics a reality, such as trastuzumab/Herceptin and imatinib/Gleevec (Ruberg et al. 2010).

This challenge of identifying subgroups of patients with more desirable clinical outcomes has also been a complex problem for statisticians. Traditional subgroup analyses are based on interaction tests where differential treatment effects among subgroups are analyzed by testing treatment-by-subgroup interactions in regression models. Such analyses have many drawbacks, such as the inability to consider more complex subgroups involving multiple markers. These limitations have led to many recommendations and generated much caution on the interpretation of results. Many researchers proposed that subgroup analysis should be (1) limited to a few clinically important questions proposed in advance; (2) based on formal tests of interaction; (3) adjusted for multiplicity; and (4) fully reported (including all analyses performed) and not over-interpreted (Brookes et al. 2001; Wang et al. 2007; Rothwell 2005). However, inappropriate analyses continue to appear in the literature, and there have been many examples of apparently important findings on treatment effect heterogeneity that are subsequently shown to be false (Rothwell 2005).

Recently, a number of approaches to subgroup identification have been proposed (Negassa et al. 2005; Su et al. 2008, 2009; Lipkovich et al. 2011; Foster et al. 2011; Loh et al. 2015) that utilize more advanced statistical methodologies. Shen et al. (2015) provides an overview of a number of subgroup identification methods.

Two techniques found in many of these approaches are recursive partitioning and resampling. In the next section, we propose a rigorous and sophisticated approach to apply these techniques in order to identify subgroups with enhanced treatment effects with controlled type I error rate and improved power. The method produces two consistency measures for each potential subgroup identified. We compare simple ways to combine these measures into an overall summary of strength. Using stratified permutations and out-of-bag samples, the approach also provides a multiplicity-adjusted p-value and bias-corrected estimate of treatment effect.

In order for development of tailored therapeutics to be successful, it is imperative to determine the best subgroup identification methods to be applied. Given the increasing number of subgroup identification methods that have been, and are being, developed, and the fact that no single method can be expected to be optimal across all scenarios, it is valuable to have a consistent framework to evaluate and compare the performance of various subgroup identification methods. In the third section, we carefully review the simulations studies in a number of publications on subgroup identification, and describe a framework for consistent evaluation.

In the fourth section, a simulation study is performed to evaluate the performance of the proposed TSDT method, which has been implemented in a publicly available R package. We then end the chapter with some concluding remarks.

8.2 A Resampling-Based Ensemble Tree Method to Identify Patient Subgroups with Enhanced Treatment Effect

In this section, we propose a subgroup identification method, which we named TSDT (Treatment-Specific Subgroup Detection Tool). It uses recursive partitioning, which has been shown to be extremely useful in modern data mining problems thanks to many attractive features, including minimal assumptions on distributions and models (Breiman and Stone 1984). Furthermore, the fact that it directly leads to patient subgroups—as opposed to regression models from some other types of analyses—closely matches the needs for drug development. For tailored therapeutics, a subgroup definition is required to enable the design of a subsequent trial, labeling of the drug by regulatory agencies, and medical decision making by prescribers.

A single analysis of recursive partitioning on a given dataset may not be very stable, small changes in the dataset can lead to quite different results. Significant improvements can be made in this regard by using an ensemble approach enabled by resampling techniques such as bootstrap, cross-validation, and subsampling (Breiman 1996; Breiman 2001; Friedman 2001). Although these are similar, subsampling (that is, sampling without replacement) is preferable since it provides the most flexibility in terms of the number and dimensions of the resampled datasets. The same recursive partitioning analysis is performed for each resampled dataset, and we further enrich the ensemble of candidate subgroups by harvesting multiple subgroups from a given tree and consider multiple competing trees for each subsample dataset. By aggregating similar subgroups identified in this ensemble approach, we can easily summarize the frequency by which a given subgroup is identified among resampled datasets, which provides a highly robust measure that we will refer to as "internal consistency".

An additional benefit of subsampling that we also take advantage of is the out-of-bag sample, consisting of observations not included in a given resampled dataset. Since these data are entirely distinct from the corresponding resampled dataset, they can be used to assess, in an unbiased manner, any subgroup findings from the subsample dataset. Similar to the internal consistency, the results of this assessment can be averaged across subsampled datasets pairs (of in-bag and out-of-bag samples) to yield an "external consistency". While more complex choices can be made, we have found it both simple and useful to assess a subgroup finding from an in-bag sample by determining whether it is directionally consistent in the corresponding out-of-bag sample, and calculating the percentage of consistent out-of-bag samples among all the times this potential subgroup was selected in in-bag samples.

Once we have obtained both the internal and external consistency measures, which contain distinct and complementary information, it is natural to ask how we can best combine the two in order to measure the strength of an identified subgroup. For the remainder of this paper, we will use M_i ("i" for "internal") and M_e ("e" for "external") to denote these two measures for a given subgroup finding. Specifically:

M_i = the frequency by which the subgroup is identified among resampled datasets.

M_e = the percentage of consistent out-of-bag samples among all the times this potential subgroup was selected in in-bag samples.

Perhaps the most intuitive and obvious choices for combining the two are: $\min(M_i, M_e)$ and $M_i \times M_e$. The rationale for the first is to require a subgroup to have a minimum level of both internal and external consistencies. It should be more beneficial than using M_i or M_e alone, provided that they manifest on comparable scales, so that one is not always greater than the other. The second combination is the product of the two measures, which would always utilize information contained in both measures. There are, of course many other reasonable ways to combine the two measures—for example, we can weigh the two unequally—but it will be more complex to investigate those, which is an interesting area of future research.

As an initial investigation of the performance of different measures, we performed a simulation study using virtual datasets with 240 subjects and 20 markers, including one marker that defines a subgroup with enhanced treatment effect. Recursive partitioning with subsampling, as described above, yielded a number of potential subgroups for each dataset, all with various M_i and M_e values. Four overall consistency summaries for these subgroups were considered: (1) M_i alone (2) M_e alone (3) $\min(M_i, M_e)$, and (4) $M_i \times M_e$. For each summary, the type I error rate and power were estimated for each possible "critical value" of the summary by calculating the respective numbers of correct and incorrect subgroups whose summary exceeded the critical value. The power curves (power vs. type I error rate) for the four summaries (Fig. 8.7) demonstrate the superiority of the last summary, $M_i \times M_e$, in this setting.

When it is desirable to produce a multiplicity-adjusted p-value for the strongest subgroup identified, our method utilizes permutation testing that is stratified by treatment groups. That is, a permuted dataset is obtained by shuffling the observed responses within each treatment arm. While all permutation methods require (often implicitly) assumptions to be valid, this specific permutation scheme preserves the overall treatment effect and is an ideal match to the tree construction method (that is, trees are constructed first from one of the two treatment arms), hence is expected to be quite robust. As is standard, after performing a large number of simulations, the summary of the best subgroup identified from each permuted dataset provides a reference distribution, with which the summary of the top subgroup from the actual data is compared to yield a multiplicity-adjusted p-value.

Besides the external consistency measure described above, an additional benefit provided by the out-of-bag samples is an unbiased estimate of the differential treatment effect associated with the subgroup. It is well known that the "naïve" estimate of the size of an effect from the same data that led to the identification of this effect is upwardly biased, sometimes severely so. In the data mining literature, the best option to replicate a finding, including the size of the effect, is to utilize an independent dataset. Put in the drug development context, this means either a new clinical study, or having sufficient amount of data from the current clinical study so that part of that data is set aside as "testing data" to be used, not in the identification of subgroups, but only in validation of an identified subgroup. However, given the fact that a clinical study is typically powered to detect a main treatment effect, coupled with the lower power of detecting a treatment-by-subgroup interaction, in practice, not surprisingly, it is rare to have the luxury of a sufficiently large study to enable the setting aside of a testing dataset. In such situations, the out-of-bag samples made possible by bootstrap or subsampling provide the next best solution to obtaining an unbiased estimate of the differential treatment effect.

The proposed method can therefore be described by the following algorithm:

1. Sample the original data B times (done separately for each treatment arm), each time creating a pair of mutually exclusive datasets (in-bag and out-of-bag samples) with size as specified percentages (such as 50–50% or 70–30%) of the original dataset.
2. Harvest potential subgroups of enhanced treatment effect for each in-bag dataset by first building a tree with a specified maximum depth using a specified treatment arm, and then combining with the other treatment arm and applying specified selection criteria (for example the observed treatment effect in the subgroup needs to be enhanced beyond a certain threshold as compared to the observed overall treatment effect). The number of potential subgroups identified from each resampled dataset also depends on the specified number of competing markers (i.e. other top candidate markers) to be considered; for example if one competing marker is considered, then after the strongest subgroup is identified the analysis is re-run without the corresponding marker. The purpose of considering competing markers is to avoid "masking" of markers and subgroups.
3. Each identified subgroup is assessed for consistency and differential treatment effect in the corresponding out-of-bag sample.
4. Combining results across subsampled dataset pairs, the internal and external consistency measures are calculated for each identified subgroup. The two can be combined (we use the product $M_i \times M_e$) to produce an overall summary.
5. Using permutation stratified by treatment arms, a large number of permuted datasets are obtained, each analyzed as described above. This provides a reference distribution of the summary measure, against which the observed result from the actual dataset is compared to yield a multiplicity-adjusted p-value.

The architecture of the method lends itself naturally to parallel computing that can dramatically improve the speed if a large number of computing nodes are utilized.

8.3 Consistent Assessment of Biomarker and Subgroup Identification Methods

Before the simulation study in the next section, we here execute a careful examination of simulation studies performed in a number of publications on subgroup

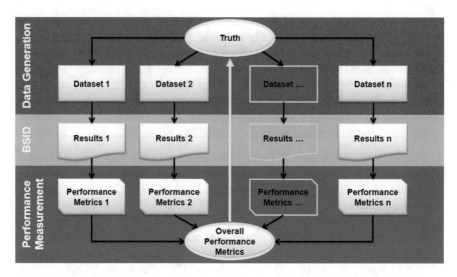

Fig. 8.1 Framework for evaluating subgroup identification methods

identification methods. These methods should ideally be examined within a common simulation framework that includes consistently generated virtual trial data and consistent measurement of performance using the same set of metrics (Fig. 8.1).

On a broad level, this approach enables comprehensive comparison of competing subgroup identification methods. The situations under which each performs well or poorly are more accurately identified, potentially leading to further improvements and/or synergies derived by combining ideas across multiple methods. Within the context of a single clinical trial or development program, this approach allows researchers to optimize the application of specific subgroup identification methods by calibrating various tuning parameters. After assessment via simulation, subgroup identification methods may be applied to real trial data with increased assurance and reliability.

8.3.1 Data Generation

All trial simulations require virtual trial data. To generate virtual data, assumptions are made in an attempt to emulate real data. These assumptions may be based upon results observed in past trials, knowledge of the disease state, and/or specifics of the therapy under examination. Because the "truth" underlying the virtual data is known, the methodology, assay, or tool applied during simulation may be assessed for its ability to provide accurate and useful inference. Because there is likely uncertainty associated with the assumptions, it is common to explore multiple scenarios of "truth" in order to examine trade-offs and consequences should the real trial data not reflect the most strongly held set of beliefs. This gives researchers confidence that the methodology, assay, or tool in question can provide reasonable results under a broad range of outcomes. These considerations make it challenging, if not impossible, to directly use real trial data for simulations, primarily because we do not know the true data-generating mechanism behind a real dataset.

In the case of subgroup identification methodology assessment, virtual trial data must incorporate embedded markers and subgroups. Key attributes of the virtual trial data include:

- Sample size
- Treatment assignment
- Number of predictors
- Response type (e.g., continuous, dichotomous, time-to-event)
- Predictor type (e.g., prognostic, predictive) and correlation
- Subgroup size
- Size of effects: placebo response, overall treatment effect, prognostic effect(s), predictive effect(s)
- Missing data, etc.

These attributes are then tied together via a data generation model. Ideally, code used to produce the virtual trial data should be flexible enough to accommodate any/all attributes and models. Furthermore, it is best practice to format the data consistently, and to ensure that all individual virtual datasets are reproducible (via documentation of settings, including seeds used for random number generation). Depending upon the number of simulation scenarios, the size of datasets generated, the computational intensity of the subgroup identification method(s) under investigation, and the scope of the intended performance measurement, utilization of a nested dataset structure (i.e. crossing the attributes in a factorial fashion) might be considered. Not only does this allow for a more economical use of resources, it removes a source(s) of variability, enabling cleaner comparisons. In order to facilitate subsequent performance measurement, it is paramount to clearly identify the attributes and models used to generate the virtual trial data. Moreover, capturing the underlying "truth" in a structured way makes automated performance measurement possible.

Figure 8.2 contains the attributes and models utilized for virtual trial data generation in a selection of papers describing various subgroup identification methods:

Methods:

1. SIDES (Lipkovich et al. 2011)
2. SIDES (Lipkovich and Dmitrienko 2014)
3. VT (Foster et al. 2011)
4. GUIDE (Loh et al. 2015)
5. QUINT (Dusseldorp and Van Mechelen 2014)
6. IT (Su et al. 2008)

It should be noted that not all attributes were clearly identified in the papers; some had to be inferred from context or calculated.

Attribute	SIDES (2011)[1]	SIDES (2014)[2]	VT[3]	GUIDE[4]	QUINT[5]	IT[6]
n	900	300, 900	400 - 2000	100	200 - 1000	300, 450
p	5 - 20	20 - 100	15 - 30	100	5 - 20	4
response type	continuous	continuous	binary	binary	continuous	TTE
predictor type	binary	binary	continuous	categorical	continuous	ordinal, categorical
predictor correlation	0, 0.3	0, 0.2	0, 0.7	0	0, 0.2	0
treatment assignment	1:1	1:1	?	~1:1	~1:1	?
# predictive markers	0 - 3	2	0, 2	0, 2	1 - 3	0, 2
predictive effect(s)	higher order	higher order	higher order	N/A, simple, higher order	simple, higher order	simple
predictive M+ group size (% of n)	15% - 20%	50%	N/A, ~25%, ~50%	N/A, ~36%	~16% - ~50%	N/A, ~25%, ?
# prognostic markers	0	0	3	0 - 4	1 - 3	0, 2
prognostic effect(s)	N/A	N/A	simple, higher order	N/A, simple, higher order	simple, higher order	simple
	"contribution model"		logit model (w/o and with subject-specific effects)	linear model (on probability scale)	"tree model"	exponential model

Fig. 8.2 A survey of simulation studies

Fig. 8.3 Three levels of performance measures

8.3.2 Performance Measurement

Performance measurement is the process of quantifying how well (or poorly) a methodology, assay, or tool recaptures the "truth" underlying generated virtual data. In the case of subgroup identification, researchers are eager to understand how accurately a method identifies: markers, information about subgroups (such as size and expected effect), and subgroup membership (i.e., the subjects that comprise subgroups). These measures can be thought of as representing the three types of problems dealt with in the field of Statistics: testing, estimation, and prediction, respectively (Fig. 8.3).

Along with the "truth" underlying the virtual data, most performance measurement metrics can be calculated given the following input:

- A list of the predictive biomarkers identified
- The proposed subgroup membership (i.e., an indication of whether each subject in the virtual data is in the proposed subgroup or not)
- The estimated treatment effect in the proposed subgroup

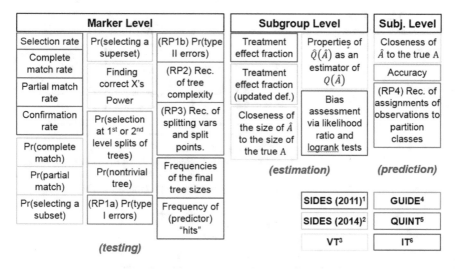

Fig. 8.4 A survey of performance measures in simulation studies

(Note that the input may reflect a "null scenario". This is a special case whereby the virtual data does not contain any markers/subgroups.)

Within the simulation framework, the subgroup identification method under investigation must produce this collection of input for each of the virtual datasets to which it is applied. Overall performance is then summarized by averaging metrics across all simulations.

Although there are many different metrics that may be utilized in measuring subgroup identification performance, in general each can be placed into one of the aforementioned buckets. Figure 8.4 describes how the metrics utilized by the selection of papers referenced in the *Data Generation* section may be classified.

Although each group of metrics attempts to answer a similar question, the lack of consistency makes it difficult to compare subgroup identification methods. Therefore, the following metrics are recommended:

Marker Level Sensitivity, specificity, positive predictive value (PPV), and negative predictive value (NPV) given the true markers versus those identified (Fig. 8.5).

Subgroup Level Size of the true subgroup versus size of the proposed subgroup; true treatment effect in the true subgroup versus true treatment effect in the proposed subgroup (using the potential outcomes framework) versus estimated treatment effect in the proposed subgroup. Note that, given operational input, these metrics may be translated into implications on sample size, time, and cost of future trials.

Subject Level Sensitivity, specificity, PPV, and NPV given the true subgroup membership versus the proposed subgroup membership (Fig. 8.6).

An advantage of the recommended metrics—particularly those at the marker and subject levels – is that they are used in many areas of scientific research. This

Fig. 8.5 Marker level
performance measures

		Predictive Biomarker	
		True	**False**
Identified as Predictive	**Yes**	True Positive	False Positive
	No	False Negative	True Negative

- *Sensitivity* = True Positive / True Predictive Biomarkers
- *Specificity* = True Negative / False Predictive Biomarkers
- *PPV* = True Positive / Identified as Predictive
- *NPV* = True Negative / Not Identified as Predictive

		Potential to Realize Enhanced Treatment Effect*	
		True	**False**
Membership Classification	**M+**	True Positive	False Positive
	M-	False Negative	True Negative

*at a meaningful or desired level

- *Sensitivity* = True Positive / True Enhanced Treatment Effect
- *Specificity* = True Negative / False Enhanced Treatment Effect
- *PPV* = True Positive / Classified as M+
- *NPV* = True Negative / Classified as M-

Fig. 8.6 Subject level performance measures

creates efficiencies by (a) harnessing well-accepted concepts and terminology, and (b) allowing subgroup identification developers to spend less time defining and explaining niche (perhaps even project-specific) metrics.

Markers/subgroups can be very difficult to find. As a result, the subgroup identification method under investigation may propose many null scenarios when applied to the virtual trial datasets. This can make it even harder to compare multiple methods, since the performance measures will be washed out by the null submissions. To combat this problem, the aforementioned metrics may be conditionally produced with null submissions removed. This informs researchers of how accurate a method is when it *does* manage to find markers/subgroups.

No single metric can comprehensively describe the accuracy or usefulness of a subgroup identification method. Therefore, the entire collection of metrics must be considered. However, the relative importance associated with each category of performance measure is dependent upon the objective of the researcher. Marker level measures are useful for determining in which marker(s) to further invest. Subgroup level measures can provide information about tailoring subsequent trials/designs.

Subject level measures help to assess impact in clinical practice. Measurement of performance with a consistent set of metrics not only enables comparison of subgroup identification methods, it clarifies the purpose for which methods are optimally employed.

8.4 Simulation Study

A simulation study was performed to assess the proposed method. Each generated dataset consists of a number (represented by p) of 3-level genetic markers (taking values 0, 1, and 2, representing the number of miner alleles a subject is carrying for a given SNP), a continuous outcome Y, and a binary treatment variable T (representing "Treated" and "Placebo" groups). The outcome Y was generated from a linear model, where the mean placebo response is -0.1, and the standard deviation conditional on all markers is 1.13. The number of markers (p) can be 5, 20, or 50 and the sample size (n) is either 240 or 480. In terms of marker effects, datasets were generated under both the "null" scenario (that is, no true predictive markers) for evaluation of type 1 error rate, and "alternative" scenarios with one or two true predictive biomarkers for assessment of statistical power. When true predictive biomarkers are present, the mean treatment effect in the weakest-responding subgroup is -0.1, and each predictive marker is associated with a differential treatment effect of -0.45 (applied additively). Therefore, when a single true predictive marker is present, the datasets contain subgroups with treatment effects of -0.1 and -0.55. When two true predictive markers are present, the datasets contain subgroups with treatment effects of -0.1, -0.55, and -1. A subgroup is considered to be identified if the multiplicity-adjusted p-value is less than 0.1.

The 12 simulation scenarios are summarized in Table 8.1. Results of the simulations are presented in Tables 8.2, 8.3, 8.4, 8.5, 8.6, 8.7, and 8.8 and Figs. 8.7, 8.8, and 8.9.

Table 8.2 compares the performance under the null scenario when the number of markers ranges from 5 to 50 (scenarios A, B, C), and the results show that type I errors are controlled at close to the nominal level.

Results for scenarios D, E, and F are presented in Table 8.3 and Fig. 8.7. Here datasets were generated with 1 true predictive marker, and the number of markers again ranges from 5 to 20 to 50. Figure 8.7 demonstrates that the simple combined consistency measure, $M_i \times M_e$, had the best performance among the four considered. The overall statistical power is low, in that no subgroup was identified for 55–74% of datasets across the scenarios. However, when looking at the instances when at least one subgroup is identified (i.e., conditionally), the performance of the method is good, especially when the number of markers is small. In other words, when a subgroup is identified it tends to be a correct subgroup. Table 8.4 provides additional performance summaries for these three scenarios at the subgroup level.

Table 8.1 Summary of simulation scenarios

Scenario	A	B	C	D	E	F	G	H	I	J	K	L
# subjects	240	240	240	240	240	240	240	240	480	480	480	480
# markers	5	20	50	5	20	50	20	50	20	50	20	50
Effect size in subgroup	NA	NA	NA	−0.55	−0.55	−0.55	−0.55 and −1	−0.55 and −1	−0.55	−0.55	−0.55 and −1	−0.55 and −1
# predictive markers	0	0	0	1	1	1	2	2	1	1	2	2

Table 8.2 Estimated type I error rate for identifying predictive markers (n = 240)

Scenario	Markers	Type I error rate
A	5	0.10
B	20	0.11
C	50	0.12

Table 8.3 Summaries of estimated power for identifying predictive markers (n = 240, 1 true marker)

Scenario	Markers	No subgroup identified	(Cond.) sensitivity	(Cond.) specificity	(Cond.) PPV	(Cond.) NPV
D	5	0.550	0.955	0.988	0.955	0.988
E	20	0.620	0.769	0.986	0.763	0.988
F	50	0.740	0.538	0.989	0.519	0.990

Table 8.4 Summaries of identified subgroups (n = 240, 1 true marker)

Scenario	Markers	No subgroup identified	(Cond.) proportion of subjects in subgroup	(Cond.) average treatment effect
D	5	0.550	0.500	−0.540
E	20	0.620	0.521	−0.495
F	50	0.740	0.535	−0.452

Table 8.5 Summaries of estimated power for identifying predictive markers (n = 240)

Scenario	Markers	True markers	No subgroup identified	(Cond.) sensitivity	(Cond.) specificity	(Cond.) PPV	(Cond.) NPV
E	20	1	0.620	0.769	0.986	0.763	0.988
G	20	2	0.520	0.395	0.988	0.791	0.936
F	50	1	0.740	0.538	0.989	0.519	0.990
H	50	2	0.600	0.462	0.996	0.850	0.978

Table 8.6 Summaries of identified subgroups (n = 240)

Scenario	Markers	True markers	No subgroup identified	(Cond.) proportion of subjects in subgroup	(Cond.) average treatment effect
E	20	1	0.620	0.521	−0.495
G	20	2	0.520	0.528	−0.737
F	50	1	0.740	0.535	−0.452
H	50	2	0.600	0.518	−0.752

To assess the performance under different combinations of p and number of predictive markers, in Table 8.5 we summarized results when the total number of markers is 20 or 50, and the number of true predictive markers is 1 or 2. Comparing scenarios E and G, we can see that the conditional power of identifying both predictive markers is about half of that of identifying the lone predictive marker, when there are 20 markers overall. The drop is smaller when there are 50 markers overall (Fig. 8.8). Table 8.6 provides additional subgroup-level performance summaries.

Table 8.7 Summaries of estimated power for identifying predictive markers

Scenario	Markers	True markers	Sample size	No subgroup identified	(Cond.) sensitivity	(Cond.) specificity	(Cond.) PPV	(Cond.) NPV
E	20	1	240	0.620	0.769	0.986	0.763	0.988
I	20	1	480	0.370	0.968	0.994	0.936	0.998
F	50	1	240	0.740	0.538	0.989	0.519	0.990
J	50	1	480	0.490	0.941	0.997	0.921	0.998
G	20	2	240	0.520	0.395	0.988	0.791	0.936
K	20	2	480	0.150	0.688	0.996	0.958	0.967
H	50	2	240	0.600	0.462	0.996	0.850	0.978
L	50	2	480	0.240	0.638	0.998	0.956	0.985

Table 8.8 Summaries of identified subgroups

Scenario	Markers	True markers	Sample size	No subgroup identified	(Cond.) proportion of subjects in subgroup	(Cond.) average treatment effect
E	20	1	240	0.620	0.521	−0.495
I	20	1	480	0.370	0.522	−0.539
F	50	1	240	0.740	0.535	−0.452
J	50	1	480	0.490	0.510	−0.536
G	20	2	240	0.520	0.528	−0.737
K	20	2	480	0.150	0.508	−0.772
H	50	2	240	0.600	0.518	−0.752
L	50	2	480	0.240	0.516	−0.779

The impact of sample size is illustrated in Tables 8.7 and 8.8 and Fig. 8.9, where for each p (20 or 50) and true predictive markers (1 or 2), sample sizes of 240 and 480 are compared. Large gain of statistical power is seen across the board.

8.5   Concluding Remarks

We have described a resampling-based ensemble tree approach to identify subgroups of patients with enhanced treatment effect in clinical trials. It has a number of advantages:

- The recursive partitioning approach determines subgroups, a good match with drug development and medical and regulatory decision making;
- By using an ensemble approach, the results are robust to outliers, which reduces spurious findings to which some other methods are prone;
- Criteria such as minimum subgroup size can be applied to eliminate subgroups that do not meet the need of a specific project, thus reducing the scope of the overall "search space" and lessening the severity of multiplicity, leading to increased statistical power;

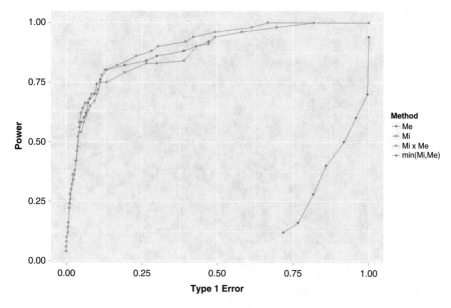

Fig. 8.7 Comparing performance of different "summaries of strength"

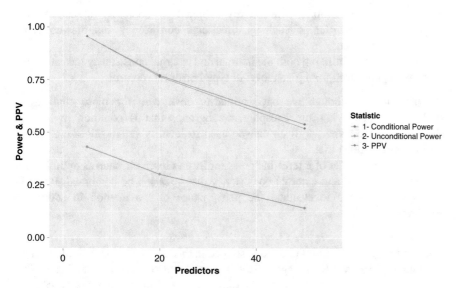

Fig. 8.8 Relationship between power and total number of markers

- By allowing a specified number of competing markers in the "harvesting" of trees, the issue of collinearity is easily handled, so that a potentially useful marker is not masked by others;

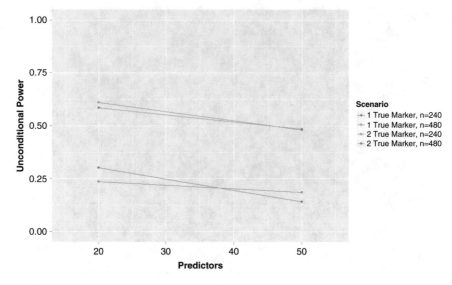

Fig. 8.9 Power comparison between various scenarios

- The out-of-bag samples conveniently supplied by bootstrap or subsampling provide key information such as directional consistency and bias-corrected estimate of effect;
- By intelligently utilizing both the internal and external consistency measures, the power is improved for a given level of type I error rate control.

Furthermore, although we have primarily dealt with the more challenging problem of identifying super-responder subgroups that is common in tailored therapeutics, the same approach can be used to identify prognostic factors from a single-arm clinical trial.

There are a number of interesting areas for further research, such as optimization of how the internal and external consistency measures can be combined. It would also be informative to evaluate the performance of the method in additional scenarios.

Resources

The subgroup identification method described in this chapter has been implemented in a publicly available R package, TSDT. The main function call is given below. The package and full documentation can be found at https://cran.r-project.org/package= TSDT

```
TSDT(response = NULL, response_type = NULL, survival_model
= "kaplan-meier", percentile = 0.5, tree_builder = "rpart",
tree_builder_parameters = list(), covariates, trt = NULL,
trt_control = 0, permute_method = NULL, permute_arm = NULL,
n_samples = 1, desirable_response = NULL, sampling_method
= "bootstrap", inbag_proportion = 0.5, scoring_function = NULL,
```

```
scoring_function_parameters = list(), inbag_score_margin = 0,
oob_score_margin = 0, eps = 1e-05, min_subgroup_n_control = NULL,
min_subgroup_n_trt = NULL, min_subgroup_n_oob_control = NULL,
min_subgroup_n_oob_trt = NULL, maxdepth = .Machine$integer.max,
rootcompete = 0, strength_cutpoints = c(0.1, 0.2, 0.3),
n_permutations = 0, n_cpu = 1, trace = FALSE)
```

Because of the need to perform nested resampling (for example, permutation and subsampling), this approach can be computationally intensive. However, the architecture of the approach lends itself naturally to parallel computing, which can be leveraged to dramatically improve the computing speed.

References

Breiman L (1996) Bagging markers. Mach Learn 24:123–140

Breiman L (2001) Random forests. Mach Learn 45:5–32

Breiman L, Stone CJ (1984) Classification and regression trees. Chapman & Hall, New York

Brookes ST, Whitley E, Peters TJ, Mulheran PA, Egger M, Davey Smith G (2001) Subgroup analyses in randomised controlled trials: quantifying the risks of false-positives and false-negatives. Health Technol Assess 5(33):1–56

Dusseldorp E, Van Mechelen I (2014) Qualitative interaction trees: a tool to identify qualitative treatment-subgroup interactions. Stat Med 33:219–237

Foster JC, Taylor JM, Ruberg SJ (2011) Subgroup identification from randomized clinical trial data. Stat Med 30(24):2867–2880

Friedman J (2001) Greedy function approximation: a gradient boosting machine. Ann Stat 29(5):1189–1232

Lipkovich I, Dmitrienko A (2014) Strategies for identifying predictive biomarkers and subgroups with enhanced treatment effect in clinical trials using SIDES. J Biopharm Stat 24:130–153

Lipkovich I, Dmitrienko A, Denne J, Enas G (2011) Subgroup identification based on differential effect search- a recursive partitioning method for establishing response to treatment in patient subpopulations. Stat Med 30(21):2601–2621

Loh W-Y, He X, Man M (2015) A regression tree approach to subgroup identification for censored data. Stat Med 34(11):1818–1833

Negassa A, Ciampi A, Abrahamowicz M, Shapiro S, Boivin J-F (2005) Tree-structured subgroup analysis for censored survival data: validation of computationally inexpensive model selection criteria. Stat Comput 15:231–239

Rothwell PM (2005) Subgroup analysis in randomized controlled trials: importance, indications, and interpretation. Lancet 365:176–186

Ruberg SJ, Chen L, Wang Y (2010) The mean does not mean as much anymore: finding subgroups for tailored therapeutics. Clin Trials 7:574–583

Shen L, Ding Y, Battioui C (2015) A framework of statistical methods for identification of subgroups with differential treatment effects in randomized trials. In: Applied statistics in biomedicine and clinical trials design: selected papers from 2013 ICSA/ISBS Joint Statistical Meetings, Chapter 25. Springer, Cham, pp 411–425

Su XG, Zhou T, Yan X, Fan J, Yang S (2008) Interaction trees with censored survival data. Int J Biostat 4(1):2

Su X, Tsai CL, Wang H, Nickerson DM, Li B (2009) Subgroup analysis via recursive partitioning. J Mach Learn Res 10:141–158

Wang R, Lagakos SW, Ware JH et al (2007) Statistics in medicine – reporting of subgroup analyses in clinical trials. New Eng J Med 357:2189–2194

Wong CH, Siah KW, Lo AW (2019) Estimating clinical trial success rates and related parameters in oncology. https://doi.org/10.2139/ssrn.3355022

Chapter 9
A New Paradigm for Subset Analysis in Randomized Clinical Trials

Richard Simon and Noah Simon

Abstract There are numerous methods for identifying subsets of patients in a randomized clinical trial who appear to benefit from the test treatment to a greater or lesser extent than average. Generally such claims are based multiple hypothesis testing and re-substitution estimates of treatment effect that are known to be highly optimistically biased. In this chapter we describe a new paradigm for subset analysis. Rather than being based on multiple hypothesis testing, it is based on training a single predictive classifier and provides an almost unbiased estimate of treatment effect for the selected subset.

Keywords Predictive classifier · Re-sampling · Pre-validation · Personalized treatment · Bootstrap sampling · Cross-validation · Subset analysis

9.1 Introduction

The main objective of most randomized clinical trials is to determine whether the test treatment is beneficial on average for the population of all eligible patients with regard to the primary endpoint. For biologically heterogeneous diseases like most forms of cancer, it has become increasingly apparent that for most treatments the treatment effect is not uniform across the eligible population. Consequently the average treatment effect is an imperfect guide for basing treatment strategies and there is often interest in identifying subsets of patients who have treatment effects greater than or less than the average. Statisticians often dismiss this objective as "exploratory" because they are not familiar with reliable methods which can perform discovery and inference on the same dataset. This problem of reliably

R. Simon (✉)
R Simon Consulting, Bethesda, MD, USA

N. Simon
Department of Biostatistics, University of Washington, Seattle, WA, USA

© Springer Nature Switzerland AG 2020

199

N. Ting et al. (eds.), *Design and Analysis of Subgroups with Biopharmaceutical Applications*, Emerging Topics in Statistics and Biostatistics, https://doi.org/10.1007/978-3-030-40105-4_9

characterizing treatment effect heterogeneity is not a hypothesis testing problem although it is often treated as if it were.

There is no lack of subset identification methods. Because of the high false positive rate for tests of treatment effect in subsets selected from the data, such analyses usually elicit skepticism and are viewed as hypothesis generation to be tested on independent data. Often however such independent data is not available. In this chapter we shall describe a new paradigm for subset analysis based on developing a "predictive classifier" (Freidlin and Simon 2005). We shall also describe how this predictive classifier can be internally validated using measures of performance appropriate for classifiers.

9.2 Methods

9.2.1 Predictive Classifiers

Let D denote the data from a randomized clinical trial comparing a test treatment to a control regimen. The data consists of covariate vectors (X) for the patients, treatment indicators (z) and outcomes (y). If our clinical trial is "negative" with regard to average treatment effect for all eligible patients, then our objective may be to identify and validate a subset of patients who benefit from the test treatment. If our clinical trial is "positive" overall, the objective may be to identify an "intended excluding patients" who do not seem to benefit from the test treatment. More generally, we may want to stratify the population with regard to the likelihood that they benefit from the test treatment.

A predictive classifier is not like the usual prognostic classifier relating baseline covariates to prognosis. When there are two treatments, a predictive classifier is a function which indicates whether the patient is likely to benefit from the test treatment or not. Here we will discuss tri-level classifiers with $C(X) = 2$ indicating that a patient with covariate vector X is very likely to benefit from the test treatment, $C(X) = 1$ meaning that the patient is moderately likely to benefit and $C(X) = 0$ meaning that the patient is unlikely to benefit or may have better outcome on the control treatment.

We may denote the classifier as $C(X; \mathcal{A}, D)$ meaning it is a function of the covariate vector X and that it was developed by applying a predictive classifier development algorithm $\mathcal{A}$ to the dataset D. Specifying $\mathcal{A}$ means that the user is required to specify in advance the types of analyses that will be performed to develop a fully specified classifier. This is essential for using re-sampling methods for evaluating classifiers because the same classifier development algorithm must be applied to several re-sampled training sets.

A predictive classifier is not a "risk classifier". Instead it classifies patients with regard to their likelihood of benefit from the test treatment relative to the control regimen. Predictive classifiers have been called "regimes" by some investigators (Bai et al. 2017).

There are many types of predictive classifiers. For example one could develop separate prognostic models for the test treatment T and for the standard treatment S. Denoting these models as f(X;T) and f(X;S), they provide expected outcome or a function of expected outcome for a patient with covariate vector X. These models might be based on penalized logistic regression, random forest, support vector machines, etc. Our predictive classifier C might be defined based on these models as

$$
\begin{aligned}
&C\,(X;\mathcal{A}, D) = 2 \ \ \text{if} \ \ f\,(X; T) - f\,(X; S) > k_2 \\
&C\,(X;\mathcal{A}, D) = 1 \ \ \text{if} \ \ k_2 > f\,(X; T) - f\,(X; S) > k_1 \\
&C\,(X;\mathcal{A}, D) = 0 \ \ \text{otherwise.}
\end{aligned} \tag{9.1}
$$

The set of covariate vectors

$$
\mathcal{S}_2 = \{X : C\,(X; \mathcal{A}, D) = 2\}
$$

might be taken as the intended use population for the new treatment. $\mathcal{S}_1$ and $\mathcal{S}_0$. can be analogously defined. The characterization of the covariate vectors in these subsets can be used for product labeling if T is a new treatment. Otherwise the subsets can be used for patient management; i.e. patients with covariate vectors in $\mathcal{S}_2$ would generally receive the test treatment and those with covariate vectors in $\mathcal{S}_0$ generally would receive the control. For patients with covariate vectors in $\mathcal{S}_1$, treatment selection would be influenced by secondary endpoints and patient preference. The constants k_1 and k_2 can be specified based on clinical significance, cost or adverse effects of the test treatment. For example, with survival outcome k_2 might be defined as the natural logarithm of 0.90 taking a 10% decrease in hazard as minimally clinically significance. Defining k_1 as zero would identify $\mathcal{S}_2$ as the class in which expected outcome on the control is better than on the test treatment.

With survival modeling, one might fit a proportional hazards model

$$
\log \frac{h\,(t; X, z)}{h_0(t)} = \alpha z + z\beta' X + (1 - z)\,\gamma' X
$$

where z is a (0,1) treatment indicator. The treatment effect on the log hazard ratio scale is the value of the log hazard ratio for $z = 1$ minus the value for $z = 0$; that is $\alpha + (\beta - \gamma)' X$. The three class classifier described above is $C = 2$ if $\alpha + (\beta - \gamma)' X \leq k_2$ and the other classes defined similarly (sign reversed because lower hazard is better). $C = 1$ if $k_1 \leq \alpha + (\beta - \gamma)' X < k_2$ and $C = 0$ otherwise. In classifying cases we use the maximum likelihood estimates of the parameters. The sets S_0 S_1 and S_2 thus are a partition of the cases. If there are a large number of candidate covariates, then penalized regression methods can be utilized in training the classifier.

Our objective here is not to provide advice about what types of predictive classifiers are best nor to develop a new type of predictive classifier, but to show how to internally validate a predictive classifier once it has been defined.

9.2.2 De-biasing the Re-substitution Estimates

The usual approach to subset analysis involves some type of analysis of the full dataset D to identify a subset S_2 for which the treatment effect seems large. The empirical estimate of treatment effect for S_2 in these circumstances is called a "re-substitution estimate". S_2 was used as part of D for subset identification and then as the basis for computing treatment effect and this often results in a large bias in the estimate of treatment effect.

Although the re-substitution estimates of treatment effect based on the sets S_2, S_1, and S_0 are biased estimates, they can be de-biased in the following manner as suggested by Zhang et al. (2017).

Let D_b denote a non-parametric bootstrap sample of cases and let $C_b = C(X; A, D_b)$ denote the predictive classifier developed on D_b using the classifier development algorithm A. Define

$$\Delta (C_b, D_b)$$

to be the empirical average treatment effect for patients in D_b for whom $C_b = 2$. Since D_b was the data on which classifier C_b was trained, this is a re-substitution estimate of treatment effect.

We can also use the classifier C_b to classify the withheld cases $\overline{D}_b = D - D_b$ i.e. those not used to develop the classifier. That classification determines $\Delta (C_b, \overline{D}_b)$ the empirical estimate of treatment effect for the subset of the hold-out subset for which $C_b = 2$. Since the hold-out set was not included with the bootstrap data used to train C_b, $\Delta (C_b, \overline{D}_b)$ is an unbiased estimate of the treatment effect to be expected in the future for cases with $C_b = 2$. Also, the differences

$$\eta_b = \Delta (C_b, D_b) - \Delta (C_b, \overline{D}_b)$$

are estimates of the re-substitution bias in estimating treatment effect in S_2 using our algorithm A for classifier development. These estimates can be averaged over bootstrap samples and then used to debias the re-substitution estimates. We have described it here for S_2 but it can be done similarly for S_1 and S_0.

9.2.3 Pre-validated Estimates of Treatment Effect

An alternative approach for estimating the treatment effects is to classify each patient i using a classifier trained on a dataset not including case i. This approach was first developed for use in the Cross-Validated Adaptive Signature Design (Freidlin et al. 2010). Suppose we perform a leave-one-out cross validation. When case i is omitted we train a classifier and use it to classify the omitted case i. Let $C(X_i; \mathcal{A}, D^{(-i)})$ denote the classification of this omitted observation. This is called "pre-validated" classifications because each observation i is classified using a classifier trained on a dataset not containing case i.

After all the folds of the cross-validation are completed, we have pre-validated classifications for all the cases. We can thus collect together the cases classified $C(X_i; \mathcal{A}, D^{(-i)}) = 2$. These cases define S_2 and we can compute the empirical treatment effect within this subset. Pre-validated subsets S_1 and S_0 can be analyzed analogously.

We simulated clinical trials to illustrate the bias of the re-substitution estimate of treatment effect on the S_2 subset and the effectiveness of defining S_2 based on pre-validated classifications. The simulations involved 300 patients with exponentially distributed survival and 40 binary covariates each with equal prevalence. The intended use subset S_2 was determined by fitting a full proportional hazards model (2). For each patient the predictive index was computed for the patient receiving the test treatment and for receiving the control. If the difference was less than -0.2 then the patient was classified in S_2. Table 9.1 shows the results of 10 simulated clinical trials with no treatment effect. For the first three columns the classifier was trained on the full dataset and then applied to the same full data to obtain S_2. Consequently it provides biased re-substitution estimates. Column 2 shows the hazard ratio estimates of treatment effect in these S_2 subsets and column 3 shows the computed log-rank test statistics of treatment effect which should have a chi-square distribution on one degree of freedom for the usual setting of no treatment effect and an independent test set. It is seen that the hazard ratios are not close to 1.0 as they should be and the log-rank distribution looks shifted to larger values.

Columns 4 and 5 of Table 9.1 show results of cross-validation for the same ten simulated clinical trials. The classifiers were fit to the training sets of each fold of a tenfold cross validation. Those ten classifiers were used to classify the patients in the ten respective hold-out sets. That is, for purposes of cross-validated evaluation, the classifier used to classify a case was trained on a subset of the full dataset with that target case omitted. These cross-validation based classifiers are not used for classifying future patients, but they provide a way of evaluating the classifier developed on the full dataset that avoids the bias of the re-substitution estimator. The patients classified in S_2 in this way were taken as constituting the pre-validated S_2 set. The empirical treatment effect was computed on these pre-validated sets and the hazard ratios and log-rank statistics are shown. The hazard ratios are all expressed as less than 1. The estimated hazard ratios are closer to 1.0 and the log-rank statistics are smaller.

Table 9.1 Simulation of 10 null clinical trials

Trial	Re-substitution		Cross-validated	
	HR	LR-chisq	HR	LR-chisq
1	0.59	4.5	0.82	1.0
2	0.60	4.0	0.73	2.7
3	0.64	1.7	0.83	0.67
4	0.62	2.4	0.94	0.07
5	0.68	2.0	0.84	0.66
6	0.72	1.3	0.83	0.67
7	0.49	9.8	0.77	2.0
8	0.48	12.2	0.69	4.9
9	0.54	6.7	0.69	3.7
10	0.56	6.7	0.77	2.1

Estimated HR and log-rank chi-squared in adaptively determined intended use subset

Table 9.2 Simulation of 10 clinical trials with treatment effect for subset with marker 1 equal to 1

Trial	Re-substitution		Cross-validated	
	HR	LR-chisq	HR	LR-chisq
1	0.49	11.8	0.62	7.2
2	0.28	46.9	0.38	34.0
3	0.54	4.8	0.72	2.0
4	0.55	7.3	0.60	8.2
5	0.41	20.3	0.62	7.8
6	0.43	21.2	0.61	9.4
7	0.54	7.6	0.79	1.5
8	0.38	24.0	0.56	11.0
9	0.38	20.9	0.53	12.8
10	0.63	4.7	0.72	3.1

True HR = 0.6 in subset
Estimated HR and log-rank chi-squared in adaptively determined intended use subset

Table 9.2 shows analogous results for 10 clinical trials simulated with a treatment effect of hazard ratio 0.6 for the half of patients with covariate 1 equal to 1. The same type of proportional hazards predictive classifier was fit as before. The cross-validated chi-square values for treatment effect within the adaptively determined intended use subset is not as inflated as the re-substitution values and the hazard ratio estimates within the intended use subset are closer to the true 0.6 values used for simulating the data. The R software used to compute Tables 9.1 and 9.2 are available from the first author.

9.2.4 Testing Treatment Effects in Subsets S_2, S_1 and S_0

We can estimate the expected treatment effects in these subsets as described in the previous section but we would also like to test the null hypothesis that these treatment effects are zero. We can test the null hypothesis that the expected treatment effect is zero in S_2 by permuting the treatment assignments, re-computing the adaptively determined S_2 and using the empirical treatment effect in the new S_2 as a test statistic for the permutation test.

9.2.5 PPV and NPV of the Predictive Classifier

If we take classification into subset S_2 as indicating that the patient is more likely to benefit from the new treatment, then what is the PPV and NPV of the classifier? If outcomes are survival times and the treatments have proportional hazards within each subset, then the probability that a patient classified in S_2 benefits from the test treatment is approximately

$$PPV = \frac{1}{1 + e^{\delta_2}}$$

where δ_2 is the hazard ratio of the test treatment to control in S_2. This is shown by Simon (2015) under the assumption of independence of treatment effects for a patient. Similarly, the NPV for a case classified in S_2 is approximately

$$NPV_0 = \frac{e^{\delta_0}}{1 + e^{\delta_0}}$$

where δ_0 denotes the hazard ratio for cases in S_0. For a case classified in S_1 the NPV is approximately

$$NPV_1 = \frac{e^{\delta_1}}{1 + e^{\delta_1}}.$$

9.2.6 Calibration of Pre-Validated Treatment Effects

The development above enables the classification of future patients into the three subsets, S_2 representing very likely to benefit from the test treatment, S_0, very unlikely to benefit from the test treatment and an intermediate group S_1. The cases in S_0 may have better outcomes on the control. This is individualized prediction because it is based on the covariate vector X. These estimates are discretized into three sets, however, and are based on the parametric prognostic models f(X,T) and f(X, S). An alternative approach is to focus on the pre-validated treatment effect difference $f(X,T) - f(X,S)$ for each case. Then, if our outcome is survival, these difference scores can be smoothed by fitting a proportional hazards model containing a main effect of treatment. By using a spline we can estimate the relationship of difference score to treatment effect. This is similar to the approach as suggested by (Matsui et al. 2012).

Instead of fitting the proportional hazards model with the splines of the pre-validated $d_i^{(p)}$ values, a simple window smoother can possibly be used. For every small window on the d axis we compute an estimate of the hazard ratio of the two treatments. The empirical hazard ratio is $(e_1/m_1)/(e_0/m_0)$ where e_1 and e_0 denote the number of events in the window for the treatment and control groups respectively and m_1 and m_0 are the numbers of patients at risk at the start of the window for those groups. This is only used for windows for which m_1 and m_0 are both positive. This is related to the approach suggested by Cai (2011).

9.3   Discussion

In the new paradigm of subset analysis that we have described multiple hypothesis testing is replaced with the development of a single predictive classifier. We have shown how to obtain approximately unbiased estimates of the treatment effect for the set of future patients selected based on this predictive classifier and testing the significance of this treatment effect. Simulation studies have shown that the residual bias is very small (Simon and Simon 2019). We have also shown how to estimate the PPV, NPV for the predictive classifier.

The bootstrap de-biasing approach described provides a method of estimating and correcting the bias of the re-substitution estimate of treatment effect in an adaptively defined subset like S_2. The re-substitution estimate is the empirical treatment effect in S_2. It is biased because S_2 was included in the application of the algorithm $\mathcal{A}$. The estimate of bias is based on comparing the re-substitution estimate for each bootstrap sample to the treatment effect in the subset of the "out of box" cases which have covariate vectors characteristic of S_2. These bias estimates are averaged over the bootstrap samples. The method will fail, however, if the sample size is too small because there will be insufficient "out of box" cases to estimate the treatment effect in the S_2 subset.

The method based on pre-validated classification of the cases remains effective with smaller sample sizes. This method evaluates treatment effect in the set of cases which were classified in S_2 during the fold of the cross-validation in which they were left out. Under the null, those expected treatment effects should all be zero.

In a prospective randomized clinical trial, we recommend that this approach be part of the primary analysis. The other part is the usual test of average treatment effect for the entire eligible population. The threshold significance levels for the overall test and the test of treatment effect in the adaptively defined intended use subset can be chosen to ensure that the overall type I error of the trial is limited to the desired 0.05. If the null hypothesis of no average treatment effect for the overall eligible population is rejected, one can still use the approach described above for identifying the subset of patients most likely to benefit from the test treatment. This can be clinically useful if the proportion with benefit is quite limited as it is in many clinical trials. The re-sampling procedure can also provide a de-biased estimate of the treatment effect the complement of the intended use subset.

Rather than use a binary classifier, one can use a three level classifier to identify patients most likely to benefit from the test treatment, those least likely and those intermediate. The pre-validated scores can be divided into three sets either based on the 33rd and 66th percentiles of the difference scores or on pre-specified constants representing clinical significance as shown here.

We have emphasized here valid evaluation of the predictive classifier, not advocating using one type of classifier or another as is more usual. Although predictive classifiers have not been nearly as extensively studied as prognostic classifiers, many approaches to predictive classification are possible. The prognostic methods literature can be utilized by training prognostic classifiers for the treatment and control groups and then combining them into a predictive classifier or predictive score. The prognostic models can be based on logistic regression, random forest, support vector machines, proportional hazards regression etc.

Although there are many subset identification methods in the literature, there are very few subset validation methods. Dixon and Simon (1991) described an empirical Bayesian method that can be used with proportional hazards or logistic modeling with a large number of binary covariates. Hierarchical priors are placed on the interaction effects. The posterior distributions of treatment effect for any subset defined by one or more covariates are easily computed. These distributions are shrunken towards zero thereby providing a type of internally validated subset analysis. The methods presented here, however, avoid the assumption of hierarchical prior distributions.

Two final points deserve emphasis. First, all aspects of the development should be described prospectively in the statistical analysis plan. Secondly, fully external validation of a "subset effect" is always valuable. Generally there is no valid internal evaluation of the treatment effect in adaptively defined subsets and the claims are based solely on the biased re-substitution estimates. With the paradigm proposed here, there will be much stronger evidence of the value of a predictive classifier based on the internal evaluation. This can guide investigators about whether a confirmatory study is warranted.

References

Bai X, Tsiatis AA, Lu W, Song R (2017) Optimal treatment regimes for survival endpoints using a locally-efficient doubly-robust estimator from a classification perspective. Lifetime Data Anal 23:585–604

Cai T, Tian L, Wong PH, Wei LJ (2011) Analysis of randomized comparative clinical trial data for personalized treatment selections. Biostatistics (Oxford, England) 12:270–282

Dixon DO, Simon R (1991) Bayesian subset analysis. Biometrics 47(3):871–881

Freidlin B, Jiang W, Simon R (2010) The cross-validated adaptive signature design. Clin Cancer Res 16:691–698

Freidlin B, Simon R (2005) Adaptive signature design: an adaptive clinical trial design for generating and prospectively testing a gene expression signature for sensitive patients. Clin Cancer Res 11:7872–7878

Matsui S, Simon R, Qu P, Shaughnessy JD, Barlogie B, Crowley J (2012) Developing and validating continuous genomic signatures in randomized clinical trials for predictive medicine. Clin Cancer Res 18:6065–6073

Simon R (2015) Sensitivity, specificity, PPV, and NPV for predictive biomarkers. J Natl Cancer Inst 107:153–156

Simon R, Simon N (2019) Finding the intended use population for a new treatment. J Biopharm Stat 29(4):675–684

Zhang Z, Li M, Lin M, Soon G, Greene T, Shen C (2017) Subgroup selection in adaptive signature designs of confirmatory clinical trials. J R Stat Soc C 66:345–361

Chapter 10
Logical Inference on Treatment Efficacy When Subgroups Exist

Ying Ding, Yue Wei, and Xinjun Wang

Abstract With rapid advances in understanding of human diseases, the paradigm of medicine shifts from "one-fits-all" to targeted therapies. In targeted therapy development, the patient population is thought of as a mixture of two or more subgroups that may derive differential treatment efficacy. To identify the right patient population for the therapy to target, inference on treatment efficacy in subgroups as well as in the overall mixture population are all of interest. Depending on the type of clinical endpoints, inference on a mixture population can be non-trivial and it depends on the efficacy measure as well as the estimation procedure. In this chapter, we start with introducing the fundamental statistical considerations in this inference procedure, followed by proposing suitable efficacy measures for different clinical outcomes and establishing a general logical estimation principle. Finally, as a step forward in patient targeting, we present a simultaneous inference procedure based on confidence intervals to demonstrate how treatment efficacy in subgroups and mixture of subgroups can be logically inferred. Examples from oncology studies are used to illustrate the application of the proposed inference procedure.

Keywords All-comers · Binary outcome · Clinical outcome · Delta method · Differential efficacy · Hazard ratio · Logical relationship · Least squares means · Logical inference · Marginal means · Mean survival · Median survival · Mixture population · Multiplicity adjustment · Non-proportional hazards · Odds ratio · Personalized medicine · Predictive marker · Prognostic marker · Randomized clinical trial · Relative response · Responder · Simultaneous confidence interval ·

Electronic Supplementary Material The online version of this chapter (https://doi.org/10.1007/978-3-030-40105-4_10) contains supplementary material, which is available to authorized users.

Y. Ding (✉) · Y. Wei · X. Wang
Department of Biostatistics, University of Pittsburgh, Pittsburgh, PA, USA
e-mail: yingding@pitt.edu

© Springer Nature Switzerland AG 2020
N. Ting et al. (eds.), *Design and Analysis of Subgroups with Biopharmaceutical Applications*, Emerging Topics in Statistics and Biostatistics,
https://doi.org/10.1007/978-3-030-40105-4_10

Simultaneous inference · Subgroup analysis · Subgroup identification · Subgroup mixable estimation · Suitable efficacy measure · Targeted therapy · Targeted treatment · Time-to-event outcome · Treatment efficacy

10.1 Introduction

The uptake of targeted therapies has significantly changed the field of medicine. Traditionally, a treatment with successful clinical trial results is recommended to the entire population to which the trial is targeted, but such treatment implementation is oversimplified and has many drawbacks due to treatment effect heterogeneity among patients. For example, a subpopulation may benefit more from standard of care than the new treatment, although the trial still meets clinical and statistical significance in the overall population. Realizing this, instead of the "one-fits-all" approach, people begin to put tremendous effort in precision medicine or targeted therapy. One aspect of targeted therapy research is to tailor existing therapies to individual patients so that the mean treatment response is optimized under such tailoring. This process identifies the so called "optimal treatment regime (OTR)" or "individualized treatment rule (ITR)" for each patient among existing treatments based on their individual characteristics such as patient's demographics, clinical measurements, and genetic/genomic markers. It is worthwhile to note that although such a process gives individualized treatment recommendation, it is still based on optimizing a reward function that is an expectation (average). Many statistical methods have been developed to find OTR, such as a two-step procedure with the use of l_1-penalized least squares (l_1-PLS) by Qian and Murphy (2011), outcome weighted learning (OWL) by Zhao et al. (2012), a general framework that shows OTR can be transformed into a classification problem by Zhang et al. (2012), a novel penalized minimization method based on the difference of convex functions algorithm (DCA) by Huang and Fong (2014), and tree-based treatment rules by Laber and Zhao (2015).

Another aspect of targeted therapy research is to develop new treatments that target a subgroup of patients. This process aims at identifying subgroup of patients with enhanced treatment benefit (compared to its complementary group) from the new therapy. In this chapter, we focus on statistical considerations and issues in the process of targeted treatment development, specifically in terms of the inference on treatment efficacy when subgroups exist. Traditional statistical methods for "subgroup analysis" rely on testing the interactions between treatment and subgroups defined by biomarkers in a regression model. However, this approach, although straightforward, has many deficiencies such as the need of identifying candidate covariates associated with subgroups in priory, difficulty of handling continuous markers, insufficient or inappropriate multiplicity adjustment and low power. Given this fact, numerous alternative methods have been developed, most of which take advantage of some existing tools or concepts from other statistical research fields such as machine learning and multiple testing. For example, virtual twins (Foster et al. 2011) uses potential outcome framework with random forest; interaction trees (Su et al. 2009), SIDES (Lipkovich et al. 2011; Lipkovich and Dmitrienko 2014) and GUIDE (Loh et al. 2015) utilize recursive partitioning, and Berger et al. (2014) adopts a Bayesian framework for subgroup identification. More methods are summarized and introduced in a tutorial article in subgroup analysis by Lipkovich et al. (2017). Shen et al. (2015) summarized three major challenges of subgroup identification, including how to (1) handle treatment-by-subgroup interaction; (2) search for candidate subgroups; and (3) adjust for multiplicity. They proposed a framework for subgroup identification by handling these three major challenges using different combinations of existing or emerging approaches. For the multiplicity challenge, a specific statistical consideration they raised is regarding how to find an "honest" estimate of subgroup treatment effect. Without a proper adjustment, the estimated treatment effect directly from the subgroup is usually biased.

From the regulatory perspective, subgroup analysis with proper use is necessary in clinical trials to meet the trial objectives, but interpreting the results from subgroup analysis is usually challenging mainly due to the chance of false discovery. Alosh et al. (2017) classified subgroup analysis into three categories: (1) Exploratory analyses aim to identify subgroups with differential treatment effect at early stage of trials or when trials fail to demonstrate treatment efficacy in overall targeted population; (2) Supportive analyses aim at assessing the consistency of treatment effect across subgroups after trials successfully demonstrate their treatment efficacy in overall population; and (3) Inferential analyses aim at demonstrating treatment efficacy in pre-specified subgroup and/or overall population. They provided guidance for the latter two categories including statistical considerations of using tests for the treatment-by-subgroup interaction as the investigational tool, and concluded that such analyses should limit the number of subgroups to be assessed which are pre-defined and of reasonable sample size. For the category (1), Dmitrienko et al. (2016) mentioned that adjustments of the target population at later stages would require more convincing data and arguments such as biological arguments, strict adherence to key scientific and statistical principles. Therefore, the exploratory subgroup analysis at early stage is usually beneficial and may help construct adaptive trial designs. In general, the advances in precision medicine have already led to FDA-approved treatments which are tailored to patient's characteristics such as genetic markers. For example, Larotrectinib (VITRAKVI) (Scott 2019), a treatment for patients with tumors containing a neurotrophic receptor tyrosine kinase (NTRK) gene fusion, was approved by FDA in November 2018.

In Sect. 10.2, we illustrate two fundamental statistical issues and considerations when subgroups exist. Then in Sect. 10.3, we first introduce an efficacy estimation principle for the situation when the population is a mixture (of subgroups with differential treatment efficacy), followed by a simultaneous inference procedure through confidence intervals. We demonstrate the application of the method using two real data examples. Finally, we discuss and conclude in Sect. 10.4.

10.2 Fundamental Statistical Considerations When Subgroups Exist

In this section, we discuss several key statistical issues that are fundamental in targeted treatment development. These issues should be well considered in both the stage of discovering subgroups (that exhibit differential treatment effects) and the stage of assessing the treatment efficacy once subgroups being identified.

10.2.1 Treatment Efficacy Measures

In a randomized clinical trial (RCT), the new treatment (denoted by Rx) is typically compared with a control (denoted by C) such as a standard of care (SoC). The "relative" treatment effect between Rx and C measures the treatment efficacy. Targeted treatment development process involves the measurement of treatment efficacy in subgroups and in *combination* of subgroups. In this chapter, we use "marker positive" group (g^+) to denote the targeted subgroup and use "marker negative" group (g^-) to denote the non-targeted subgroup.

10.2.1.1 Which Group(s) Need to Be Assessed?

Information such as patients in g^+ derive extra efficacy from the therapy, patients in g^- derive little or no benefit from the therapy, or do even worse under the new therapy than under the SoC, all are useful to the patients and prescribers. In Lin et al. (2019), it has been made clear that, the efficacy in g^+, g^- and the overall population are all needed for a decision making. Figure 10.1 illustrates three hypothetic situations. In scenarios (a) and (b), the efficacy for the overall and g^+ populations are both significant, and the treatment in g^- is still efficacious but not significant. In (a), g^+ has enhanced efficacy compared to the overall, indicated by the non-overlapping confidence intervals. Thus it might be reasonable to target the

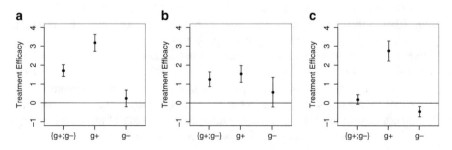

Fig. 10.1 Treatment efficacy for all-comers, g^+ and g^- in three hypothetic situations (**a–c**)

overall population in both scenarios, but have a statement in the label indicating the treatment has better efficacy in the g^+ group for scenario (a). In scenario (c), the treatment is efficacious in g^+ but not efficacious in g^-. Thus, targeting g^+ is a reasonable decision for this case. These hypothetic examples demonstrate the efficacy in all three groups are important in the targeted treatment development.

The above considers the situation when subgroups are already identified. In fact, in the development process of targeted treatment, in order to find the right patient population for the new treatment, it is still necessary to infer treatment efficacy in subgroups and *combination* of subgroups too. For example, for a SNP that separates patients into three subgroups (denoted by AA, Aa and aa), one may have to decide whether the new treatment should target a single subgroup (e.g., aa) or a combination of subgroups (e.g., $\{Aa, aa\}$). In this case, the treatment efficacy in both single genetic subgroups and combinations of subgroups need to be assessed.

10.2.1.2 Logic-Respecting Efficacy Measures

When the population is a mixture of g^+ and g^- with (potential) differential efficacy, then the efficacy measure has to be carefully chosen so that it respects the logical relationships among the subgroups and their combinations, that is, if the treatment efficacy for g^+ is a, and for g^- is b, then the efficacy for the combined group should be within $[a, b]$ (assuming $a \leq b$) (Lin et al. 2019). This is intuitive, implying if a treatment is truly efficacious in g^+ and in g^-, then it should be truly efficacious in their mixture $\{g^+, g^-\}$. However, this issue has not been fully recognized and some commonly used efficacy measures are not logic-respecting for mixture populations. The efficacy measures depend on outcome types. Frequently used clinical outcomes can be binary, continuous and time-to-event. We will demonstrate this issue with examples in different types of outcomes.

Efficacy Measures for Continuous Outcomes In the RCT setting that we consider, the probability of being g^+ is independent of the random assignment of patients to Rx or C. Denote this probability by γ^+. For a continuous outcome, if efficacy is measured by the *difference* of mean treatment and control effects, then

$$\mu_{g^+} = \mu_{g^+}^{Rx} - \mu_{g^+}^{C} \text{ and } \mu_{g^-} = \mu_{g^-}^{Rx} - \mu_{g^-}^{C}$$

represent the efficacy in g^+ and g^-, respectively. Then since

$$\mu^{Rx} = \gamma^+ \times \mu_{g^+}^{Rx} + (1 - \gamma^+) \times \mu_{g^-}^{Rx},$$
$$\mu^{C} = \gamma^+ \times \mu_{g^+}^{C} + (1 - \gamma^+) \times \mu_{g^-}^{C},$$

the efficacy in the combined population is

$$\mu_{\{g^+, g^-\}} = \mu^{Rx} - \mu^{C} = \gamma^+ \times \mu_{g^+} + (1 - \gamma^+) \times \mu_{g^-},$$

Table 10.1 An example of responders (R) and non-responders (NR) probabilities given Rx and C, in g^- and g^+ subgroups and the all-comers $\{g^+, g^-\}$ population

	g^- subpopulation				g^+ subpopulation				population		
	R	NR			R	NR			R	NR	
Rx	1/16	3/16		+	3/16	1/16		=	1/4	1/4	
C	1/40	9/40		+	1/8	1/8		=	3/20	7/20	
			1/2				1/2				1

which is guaranteed to be within $[\mu_{g^-}, \mu_{g^+}]$. Thus, *the difference of mean* effects is a logic-respecting efficacy measure for continuous outcomes.

Efficacy Measures for Binary Outcomes Binary outcomes are often used in clinical trials to capture the response in terms of "Yes" or "No" to a treatment. Binary data are typically analyzed using either a logistic model or a log-linear model, where the Odds Ratio (OR) or Relative Response (RR) (between treatment and control) can be directly obtained through model parameters. Thus, they are often selected as the efficacy measures for binary case. However, OR is not logic-respecting, as we show below.

Table 10.1 gives a hypothetic example for the probabilities of responding or non-responding in each g^+ and g^- subgroup and the overall $\{g^+, g^-\}$. We can calculate the OR for each group and the all-comers as follows,

$$OR_{g^+} = \frac{\frac{3}{16 \times 8}}{\frac{1}{(16 \times 8)}} = 3, \ OR_{g^-} = \frac{\frac{9}{(16 \times 40)}}{\frac{3}{(16 \times 40)}} = 3, \ OR_{\{g^+, g^-\}} = \frac{\frac{7}{(4 \times 20)}}{\frac{3}{(4 \times 20)}} = 2\frac{1}{3}.$$

Therefore, OR is not logic-respecting and cannot be used to measure efficacy for a mixture population as it can yield paradoxical conclusions. For example, suppose a clinically meaningful efficacy is defined as $OR > 2.5$, then Rx is efficacious in both g^+ and in g^- relative to C, but is lack of efficacy in $\{g^+, g^-\}$.

The other commonly used efficacy measure, RR, is the ratio of response probability between Rx and C. Replacing the numbers in Table 10.1 by the corresponding mathematical notation, for example, $p_{g^+}^{Rx}(R)/p_{g^+}^{Rx} = P(R|Rx, g^+)$ is the response probability given the patients are g^+ and treated by Rx, where $p_{g^+}^{Rx}$ ($p_{g^+}^{Rx}(R)$) is the joint probability of being marker positive and being assigned to treatment Rx (and responding), then we can derive the RR for g^+, g^- and the all-comers $\{g^+, g^-\}$ respectively:

$$RR_{g^+} = \frac{p_{g^+}^{Rx}(R)p_{g^+}^C}{p_{g^+}^C(R)p_{g^+}^{Rx}}, \ RR_{g^-} = \frac{p_{g^-}^{Rx}(R)p_{g^-}^C}{p_{g^-}^C(R)p_{g^-}^{Rx}}, \ RR_{\{g^+, g^-\}} = \frac{p^{Rx}(R)p^C}{p^C(R)p^{Rx}}.$$

As shown in Lin et al. (2019), the RR for the overall population can be represented as

$$RR_{\{g^+,g^-\}} = \frac{p_{g^+}^C(R)}{p^C(R)} \times RR_{g^+} + \frac{p_{g^-}^C(R)}{p^C(R)} \times RR_{g^-}. \tag{10.2.1}$$

So $RR_{\{g^+,g^-\}}$ is a mixture of RR_{g^+} and RR_{g^-} and thus guaranteed to be logic-respecting.

Efficacy Measures for Time-to-Event Outcomes   Time-to-event outcomes are frequently used in oncology studies. Since Cox regression is the most popular model to analyze time-to-event data, where the Hazard Ratio (HR) (between Rx and C) can be directly obtained from the model parameter, it has been taken as a "natural" statistic to measure efficacy. However, as shown in Ding et al. (2016), HR is not suitable for a mixture population, since even if both g^+ and g^- have constant HRs, the overall population typically does not have a constant HR. In fact, the HR of the mixture population is usually a complex function of time, with values at some time points outside of $[HR_{g^+}, HR_{g^-}]$. Figure 10.2 gives such an example. The data are generated from a Weibull distribution where $HR_{g^+} = \exp(-0.5) = 0.61$ and $HR_{g^-} = \exp(0.5) = 1.65$. The true HR for the combined group is a smooth function of t and goes below 0.61 for larger t.

In Ding et al. (2016), it has been shown that the ratio of median (or mean) survival and difference of median (or mean) survival (between Rx and C) are suitable efficacy measures for mixture population. Not only these measures are logic-respecting, they have more direct clinical interpretations compared to HR. For example, if the median survival time for patients randomized to Rx is 36 months and is 24 months for patients randomized to C, then the interpretation is that the Rx extends the median survival time for 12 (= 36 − 24) months or for 1.5 (36/24) times as compared with C. Ding et al. (2016) provides a graphical tool, namely, the

Fig. 10.2 The plot of HR for g^+, g^-, and $\{g^+, g^-\}$, illustrating the illogical issue

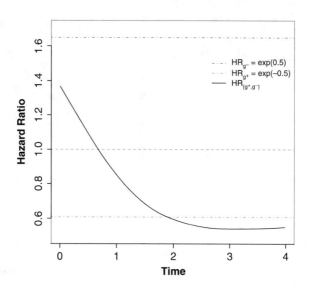

M&M plot (M stands for mean or median), to visualize the efficacy in g^+, g^- and the overall $\{g^+, g^-\}$ population.

10.2.1.3 Prognostic or Predictive?

In targeted treatment development process, often times researchers aim to discover "predictive" biomarkers, rather than "prognostic" biomarkers. The definition of prognostic or predictive biomarkers can be easily found in the literature. The prognostic biomarker is a disease-related biomarker, and it provides information on how such a disease may develop or progress in a patient population regardless of the type of treatment. While the predictive biomarker is a drug-related biomarker, it helps assess whether a particular treatment will be more effective in a specific patient population. However, the definitions usually do not specify how the effectiveness or equivalently, the efficacy of a treatment is defined. In fact, whether a biomarker is prognostic or predictive depends on the efficacy measure. We use the following example to illustrate.

Assume a biomarker divides the patient population into two groups g^- and g^+ and each of them receives Rx or C randomly. The median overall survival (OS) for the g^- group is 45 weeks if receiving C and 90 weeks if receiving Rx. While the median OS for the g^+ group is 25 weeks if receiving C and 70 weeks if receiving Rx. Is the marker prognostic or predictive? The answer will be different depending on how the efficacy is defined. We plot the data in Fig. 10.3. If the efficacy is measured by the difference in median OS, both marker groups demonstrate the

Fig. 10.3 The M&M plot illustrating whether a biomarker is prognostic or predictive depends on the efficacy measure

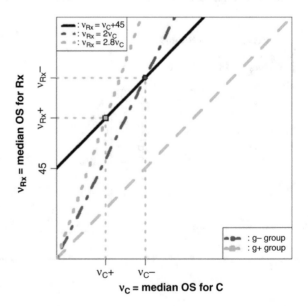

same treatment efficacy ($Rx - C = 45$ weeks), indicated by the solid 45° line in the figure. Therefore, the marker is not predictive for this particular treatment Rx. It is prognostic instead since the g^+ patients have a worse median survival outcome as compared to the g^- patients, regardless of whether they receive C or Rx. However, if the efficacy is measured by the ratio in median OS, the g^+ group demonstrates better efficacy than the g^- group ($Rx/C = 2.8$ in g^+ vs $Rx/C = 2$ in g^-), indicated by the dotted and dashed-dotted lines. Therefore, the marker is predictive for Rx in this case.

10.2.2 Inference on Mixture Populations

After suitable efficacy measures being chosen, another separate issue is the estimation procedure for obtaining the efficacy estimates for the mixture population. One common approach is to ignore the subgroup labels and use the marginal means. However, this is incorrect and can cause a Simpson's Paradox to occur. Another popular approach is to apply the Least Squares means (LSmeans) technique indiscriminately for any type of outcome. For example, applying LSmeans on the log RR from a log linear model to estimate the combination group's RR. Unfortunately none of these approaches is correct. To yield valid estimates of the treatment efficacy in subgroups and their combinations, the estimation procedure has to respect the logical relationship among efficacy parameters.

10.2.2.1 Marginal Means

First, we use the M&M plot to illustrate how marginal means can cause paradoxical result. Denote by μ^{Rx} and μ^C the true mean responses in the overall population if the overall population had received Rx or C, respectively. Denote by $\mu_{g^+}^{Rx}$, $\mu_{g^-}^{Rx}$, $\mu_{g^+}^C$, $\mu_{g^-}^C$ the corresponding mean responses in the g^+ and g^- subgroups. Figure 10.4 draws an example of such mean responses. Assume the efficacy is defined as the difference in mean response between Rx and C, then efficacy in the g^+ and g^- subgroups are perpendicular distances from those two points to the 45° line, after scaled by $\sqrt{2}$. If g^+ and g^- patients are equally prevalent ($\gamma^+ = 50\%$), then true efficacy for the combined g^+ and g^- population should be the perpendicular distance from the mid-point (of the two dots) to the 45° line (denoted by the purple line and arrow in the middle). However, in an extreme case in finite samples, suppose Rx patients are mostly g^+, while C patients are mostly g^-, then the marginal means estimate for efficacy in the combined g^+ and g^- population will be close to the perpendicular distance from the upper left corner (denoted by 'x' in the graph) of the shaded rectangle to the 45° line. Or similarly, if Rx patients are mostly g^- and C patients are mostly g^+, then the marginal means estimate for g^+ and g^- combined will be close to the perpendicular distance from the lower right

Fig. 10.4 M&M plot
illustrating the Simpson's
paradox

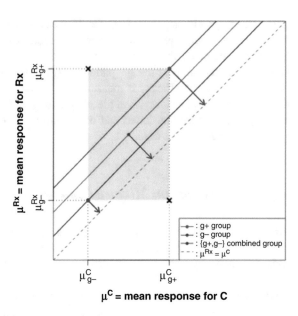

corner 'x' to the 45-degree line. Either reflects a paradox that the efficacy in the combined group does not lie in between the efficacy of the two subgroups, which is illogical.

10.2.2.2 LSmeans

Continuous outcomes are usually modeled by a linear model with i.i.d. normal errors. By the Gauss-Markov theorem, the estimation that respects the logical relationship coincides with the LSmeans estimation. However, the LSmeans estimation cannot be simply extended to other types of outcomes when efficacy is defined differently. For example, in the case of a binary outcome modeled by a log-linear model, the LSmeans estimation (that is analogous to the continuous case) estimates the logRR for the mixture by the weighted average of the logRR for the subgroups, which turns to be incorrect as we show below.

Using the same hypothetic example given in Table 10.1, we have $RR_{g+} = \frac{3/16}{1/8} = \frac{3}{2}$ and $RR_{g-} = \frac{1/16}{1/40} = \frac{5}{2}$. Assume $\gamma^+ = \frac{1}{2}$, then

$$\log(RR_{\{g+,g-\}}) = \frac{1}{2} \times \log\left(\frac{3}{2}\right) + \frac{1}{2} \times \log\left(\frac{5}{2}\right),$$

which is not equal to $\log(\frac{1/4}{3/20}) = \log(\frac{5}{3})$, the true logarithm of RR for $\{g^+, g^-\}$, as given directly from Table 10.1. Taking logarithm or not is not the key issue, as combining the RRs by γ^+ without taking logarithms results in

$$RR_{\{g^+,g^-\}} = \frac{1}{2} \times \left(\frac{3}{2}\right) + \frac{1}{2} \times \left(\frac{5}{2}\right) = 2,$$

which still is not equal to the true RR $\frac{5}{3}$ for $\{g^+, g^-\}$. The issue is that the true RR in $\{g^+, g^-\}$ cannot be determined by RRs in the g^+ and g^- alone. Applying the LSmeans technique indiscriminately for any type of outcome is not a right procedure to estimate efficacy for a mixture population.

10.3   Subgroup Mixable Inference Procedure

Given those fundamental issues presented in Sect. 10.2, in this section, we will introduce a general estimation principle, namely, the Subgroup Mixable Estimation (SME) principle, which is originally proposed in Ding et al. (2016). The key of the SME principle is: if an efficacy measure is logic-respecting, then its *estimation* should be logic-respecting as well. As shown in Sect. 10.2, the logical relationships depend on the model for the outcome variable and how efficacy is defined. They usually need to be established on a case by case basis. However, the SME principle is general, independent of the model and efficacy measure, so long as the efficacy measure is suitable and logic-respecting for mixture populations.

10.3.1   The General SME Principle

The general SME principle contains three key steps, we will use the same notation as in the description of the M&M plot in Sect. 10.2 to describe.

1. Fit the model for the clinical outcome Y, obtain the estimate for the model parameter $\boldsymbol{\theta}$ and its associated variance covariance estimate.
2. Within each treatment Rx and C, estimate the clinical response in each subgroup through a function of the model parameter: $\mu_{g^+}^{Rx} = r(\boldsymbol{\theta}; g^+, Rx)$, $\mu_{g^-}^{Rx} = r(\boldsymbol{\theta}; g^-, Rx)$ and $\mu_{g^+}^{C} = r(\boldsymbol{\theta}; g^+, C)$, $\mu_{g^-}^{C} = r(\boldsymbol{\theta}; g^-, C)$. Moreover, estimate the responses in $\{g^+, g^-\}$ combined, $\mu_{\{g^+,g^-\}}^{Rx}$ and $\mu_{\{g^+,g^-\}}^{C}$, where the estimation needs to respect the logical relationships between $\mu_{\{g^+,g^-\}}^{Trt}$ and $\mu_{g^+}^{Trt}$, $\mu_{g^-}^{Trt}$, with $Trt = Rx$ or C.
3. Calculate the efficacy in each subgroup: $\mu_{g^+} = h(\mu_{g^+}^{Rx}, \mu_{g^+}^{C})$ and $\mu_{g^-} = h(\mu_{g^-}^{Rx}, \mu_{g^-}^{C})$, and in their mixture: $\mu_{\{g^+,g^-\}} = h(\mu_{\{g^+,g^-\}}^{Rx}, \mu_{\{g^+,g^-\}}^{C})$, based on a pre-defined efficacy measure through the function h.

In step 1, adjusting for imbalance in sample sizes and other covariates (such as baseline measurements) can be done under a model for which the LSmeans technique suitably applies. In step 2, instead of "mixing" the subgroup's efficacy

μ_{g+} and μ_{g-} to directly compute the mixture's efficacy $\mu_{\{g+,g-\}}$, one has to estimate the clinical response within each treatment (Rx and C) first, for each subgroup and their combinations. The "mixing" involved in step 2 (when estimating the response for the combination group) should always be on the probability scale, rather than directly on the efficacy measure scale. The mixture (conditional) density functions for the outcome Y, given by (10.3.1) and (10.3.2), are the basis for obtaining correct estimation for treatment efficacy in the mixture population that respects the logical relationships among all efficacy measures. Here, γ^+ is the population proportion of g^+ subjects, and it is assumed to be independent of the treatment assignment, which is typically satisfied in the setting of RCTs.

$$f(y|Rx; \boldsymbol{\theta}) = (1 - \gamma^+)f(y|g^-, Rx; \boldsymbol{\theta}) + \gamma^+ f(y|g^+, Rx; \boldsymbol{\theta}), \quad (10.3.1)$$

$$f(y|C; \boldsymbol{\theta}) = (1 - \gamma^+)f(y|g^-, C; \boldsymbol{\theta}) + \gamma^+ f(y|g^+, C; \boldsymbol{\theta}). \quad (10.3.2)$$

10.3.2 Simultaneous Confidence Intervals

With the SME principle, the logic relationship among the true efficacy in g^+, g^- and $\{g^+, g^-\}$ is guaranteed to be preserved among their *point estimates*. For inference, we need additional information besides point estimates. In targeted treatment development, clinical effect size matters and confidence intervals (CIs) are a lot more informative than mere p-values. For example, a reduction in glycosylated hemoglobin (HbA1c, a typical primary endpoint for diabetes trials) between 0.8 and 1.2 is much more clinically meaningful than a reduction between 0.4 and 0.6. Yet the confidence intervals (0.8, 1.2) and (0.4, 0.6) can have identical *p*-values.

We propose to use the simultaneous CIs to infer efficacy in g^+, g^- and $\{g^+, g^-\}$. Besides the reason in terms of "more informative", another main reason is, the form of two-sided level $1 - \alpha$ simultaneous CIs I^+, I^-, $I^\pm$:

$$\inf P\{\mu_{g+} \in I^+ \text{ and } \mu_{g-} \in I^- \text{ and } \mu_{\{g+,g-\}} \in I^\pm\} \geq 1 - \alpha$$

will also reflect the logic relationships among the true efficacy in g^+, g^- and $\{g^+, g^-\}$. On the contrary, it is difficult to convey this logic relationship in terms of *p*-values, because a *p*-value has already reduced the data to a point estimate divided by its estimated standard error. Detailed explanation with a counter example has been given in Lin et al. (2019).

Providing confidence intervals also allows direct and flexible decision-making in targeting patients. Suppose a larger μ is better, and a population can be targeted if $\mu > \delta$. An appropriate δ value may depend on whether the patients are in g^+, g^-, or $\{g^+, g-\}$. For example, δ can be two separate superiority margins δ^+ and $\delta^\pm$ for patients in g^+ and $\{g^+, g^-\}$, and a non-inferiority margin δ^- for patients in g^-. One possible decision could be to target the g^+ subpopulation if I^+ is entirely larger than δ^+, and to indicate the drug (in the label) for all-comers $\{g^+, g^-\}$ as well if $I^{\pm}$ is larger than $\delta^{\pm}$ and I^- is entirely larger than δ^-.

In the following, we demonstrate the application of the SME principle, together with the simultaneous CIs inference procedure, on binary and time-to-event outcomes.

10.3.3 Application of SME on Binary Outcomes

We use RR (relative response) to be the efficacy measure for binary outcomes, as it is shown to be logic-respecting.

10.3.3.1 Theoretical Derivations

Denote by $Y = 1$ for "Responder" and $Y = 0$ for "Non-Responder". Denote conditional responder probability for each treatment $\times$ subgroup combination by:

$$p_1 = P(Y = 1|C, g^+), \quad p_2 = P(Y = 1|C, g^-)$$
$$p_3 = P(Y = 1|Rx, g^+), \quad p_4 = P(Y = 1|Rx, g^-). \tag{10.3.3}$$

With the assumption that the probability of being g^+ is the same in the treatment group and the control group, $P(g^+|C) = P(g^+|Rx) = \gamma^+$, the response rates in $\{g^+, g^-\}$ under Rx and C are then:

$$p_5 = P(Y = 1|C) = \gamma^+ p_1 + (1 - \gamma^+) p_2$$
$$p_6 = P(Y = 1|Rx) = \gamma^+ p_3 + (1 - \gamma^+) p_4. \tag{10.3.4}$$

Following the SME principle, after obtaining the sample estimates of $p_1, \ldots, p_4$ by $\widehat{p}_1, \ldots, \widehat{p}_4$ from either a logistic or a log-linear model for binary data (step 1 of SME), we can obtain the estimates of p_5 and p_6 by

$$\widehat{p}_5 = \gamma^+ \widehat{p}_1 + (1 - \gamma^+) \widehat{p}_2, \quad \widehat{p}_6 = \gamma^+ \widehat{p}_3 + (1 - \gamma^+) \widehat{p}_4. \tag{10.3.5}$$

The variance-covariance matrix of $\widehat{p}_1, \ldots, \widehat{p}_6$ can be also obtained by using the Delta method on original model parameters.

Then followed by the step 3 of SME, the RR for each subgroup and the combined group, as functions of $p_1, \ldots, p_6$,

$$RR_{g^+} = \frac{p_3}{p_1}, \quad RR_{g^-} = \frac{p_4}{p_2}, \quad RR_{\{g^+, g^-\}} = \frac{p_6}{p_5} = \frac{\gamma^+ p_3 + (1 - \gamma^+) p_4}{\gamma^+ p_1 + (1 - \gamma^+) p_2},$$

can be estimated by

$$\widehat{RR}_{g^+} = \frac{\widehat{p}_3}{\widehat{p}_1}, \quad \widehat{RR}_{g^-} = \frac{\widehat{p}_4}{\widehat{p}_2}, \quad \widehat{RR}_{\{g^+,g^-\}} = \frac{\widehat{p}_6}{\widehat{p}_5}.$$

The variance-covariance matrix of these three RR estimates can be also obtained by the Delta method (applying on $p_1, \ldots, p_6$).

Note that this principled SME computations produce the correct asymptotic distribution for efficacy estimates without going through the formula (10.2.1), which gives the true logic relationship between RR_{g^+}, RR_{g^-}, and $RR_{\{g^+,g^-\}}$. This is in fact the key feature of SME, which requires to "mix" within each treatment arm before deriving the efficacy. Therefore, the true logic relationship among efficacy parameters is not needed. In Lin et al. (2019), they also provided a different estimation procedure by directly using the formula (10.2.1). However, as pointed out by the authors, this alternative procedure is not simpler and is specific to RR being the efficacy measure only. It is important to point that the mixing coefficient in (10.2.1), i.e., the proportion of C responders in the g^+ subgroup among all C responders ($\frac{p_{g^+}^C(R)}{p^C(R)}$) cannot be simply treated as a constant, and its estimate is a random variable that in fact estimates the treatment effect of C. Therefore, the principled SME procedure is preferred.

In practice, we recommend to derive the asymptotic joint distribution of three efficacy estimates on the logarithm scale since the normality approximation works better on the log scale than on the original scale. Then we transform them back to the original scale for the simultaneous CIs. Specifically, we obtain the critical value q based on the multivariate normal distribution such that

$$P\{|\log(\widehat{RR}_{g^+}) - \log(RR_{g^+})|/se(\log(\widehat{RR}_{g^+})) < q,$$

$$|\log(\widehat{RR}_{g^-}) - \log(RR_{g^-})|/se(\log(\widehat{RR}_{g^-})) < q,$$

$$|\log(\widehat{RR}_{\{g^+,g^-\}}) - \log(RR_{\{g^+,g^-\}})|/se(\log(\widehat{RR}_{\{g^+,g^-\}})) < q\} = 1 - \alpha,$$

for a desired α level. This can be done using the R function {qmvnorm}. The input for the {qmvnorm} function contains the simultaneous coverage probability $1 - \alpha$ and the correlation matrix from the joint multivariate normal distribution of the three efficacy estimates (on log scale). Finally, the simultaneous confidence intervals for RR_{g^+}, RR_{g^-} and $\overline{RR}$ can be obtained as $I^+ = \exp\{\log(\widehat{RR}_{g^+}) \pm q \times se(\log(\widehat{RR}_{g^+}))\}$, $I^- = \exp\{\log(\widehat{RR}_{g^-}) \pm q \times se(\log(\widehat{RR}_{g^-}))\}$, and $I^\pm = \exp\{(\log(\widehat{RR}_{\{g^+,g^-\}})) \pm q \times se(\log(\widehat{RR}_{\{g^+,g^-\}}))\}$.

10.3.3.2 A Real Example

We use a Phase 3 oncology study to illustrate the SME inference. This study compares an immunotherapy OPDIVO (Nivolumab) with a chemotherapy Docetaxel for Non-Small-Cell Lung Cancer (NSCLC) (Borghaei et al. 2015). Patients were retrospectively stratified into two groups according to their tumor PD-L1 protein

Table 10.2 The objective response rates in g^+, g^- and $\{g^+, g^-\}$, separated by treatment arms.

	g^+ ($\geq 1\%$)	g^- ($<1\%$)	$\{g^+, g^-\}$ (overall)
Nivolumab (Rx)	38/123	10/108	48/231
Docetaxel (C)	15/123	15/101	30/224

Table 10.3 Relative response estimates in g^+, g^- and $\{g^+, g^-\}$

Group	$\widehat{RR}$	$\log(\widehat{RR})$	95% simultaneous CI for $\widehat{RR}$
g^+	2.53	0.93	[1.33, 5.14]
g^-	0.62	-0.47	[0.26, 1.52]
$\{g^+, g^-\}$	1.55	0.44	[0.96, 2.54]

expression level, measured with the use of a validated automated immunohisto-chemistry (IHC) assay. While the primary endpoint was overall survival, a key secondary endpoint was objective response rate (ORR). Table 10.2 presents the ORRs given OPDIVO (Novimumab) and Docetaxel, separately for patients with tumor PD-L1 levels $< 1\%$ and $\geq 1\%$.

We analyzed the data using a logistic regression and applied the SME principle to derive the estimates of RR (in terms of the ORRs). We also obtained the simultaneous CIs for g^+, g^- and $\{g^+, g^-\}$ using the formulas in Sect. 10.3.3.1. Table 10.3 presents the result. As we can see, both the g^+ (PD-L1 expression $\geq 1\%$) group and the overall population group have estimated RRs greater than 1. However, only the g^+ group has an CI not covering 1. The g^- (PD-L1 expression $< 1\%$) group has the estimated RR less than 1, although its CI covering 1. This suggests that, the efficacy of OPDIVO (Nivolumab) (relative to Docetaxel), based on the secondary endpoint OOR, is (only) established in patients with PD-L1 levels $\geq 1\%$.

10.3.4 Application of SME on Time-to-Event Outcomes

For time-to-event outcomes, with Weibull model, we will use the ratio of median survival as the efficacy measure (which has been proved to be logic-respecting in Ding et al. (2016)) to apply the SME principle.

10.3.4.1 Theoretical Derivations

Assume no other covariates, the Weibull model is then given by

$$h(t|Trt, M) = (kt^{k-1}/\lambda^k) \exp\{\beta_1 Trt + \beta_2 M + \beta_3 Trt \times M\},$$

where $Trt = 0$ (C) or $= 1$ (Rx), $M = 0$ (g^-) or $= 1$ (g^+), and $h_0(t) = kt^{k-1}/\lambda^k$ is the hazard function for the g^- subgroup receiving C, which is from a Weibull distribution with scale λ and shape k.

Let $\theta_1 = e^{\beta_1}$, $\theta_2 = e^{\beta_2}$ and $\theta_3 = e^{\beta_3}$. Then the survival function for each of the subgroups has the following form:

$$S_{g-}^{C}(t) = e^{-(t/\lambda)^k}, \quad S_{g-}^{Rx}(t) = e^{-\theta_1(t/\lambda)^k},$$

$$S_{g+}^{C}(t) = e^{-\theta_2(t/\lambda)^k}, \quad S_{g+}^{Rx}(t) = e^{-\theta_1\theta_2\theta_3(t/\lambda)^k}.$$

Then the median survival for each subgroup can be directly obtained as

$$v_{g+}^{Rx} = \lambda(\frac{\log 2}{\theta_1\theta_2\theta_3})^{\frac{1}{k}}, \quad v_{g+}^{C} = \lambda(\frac{\log 2}{\theta_2})^{\frac{1}{k}} \text{ and } v_{g-}^{Rx} = \lambda(\frac{\log 2}{\theta_1})^{\frac{1}{k}}, \quad v_{g-}^{C} = \lambda(\log 2)^{\frac{1}{k}}.$$

$$(10.3.6)$$

For $\{g^+, g^-\}$, the median survival for Rx and C are the solutions for the following two equations, respectively,

$$t = v^{Rx} : (1 - \gamma^+)e^{-\theta_1(t/\lambda)^k} + \gamma^+ e^{-\theta_1\theta_2\theta_3(t/\lambda)^k} = 0.5, \quad (10.3.7)$$

$$t = v^{C} : (1 - \gamma^+)e^{-(t/\lambda)^k} + \gamma^+ e^{-\theta_2(t/\lambda)^k} = 0.5. \quad (10.3.8)$$

Therefore, the ratio of median for $\{g^+, g^-\}$ is an *implicit* function of the model parameters $(\lambda, k, \theta_1, \theta_2, \theta_3)$.

With the preparation above, we illustrate the three key steps of SME below.

1. First, fit the Weibull model and obtain the point estimates and their estimated variance-covariance for all model parameters.
2. Then, within each treatment Rx and C, compute the median survival estimates for g^+ and g^- and their mixture based on Eqs. (10.3.6), (10.3.7) and (10.3.8), and compute their estimated variance covariance matrix by the Delta method (for the implicitly defined random variables).
3. Finally, calculate the ratio of median for g^+, g^- and $\{g^+, g^-\}$, and compute their estimated variance covariance matrix based on the Delta method.

As indicated in step 2, the Delta method for implicitly defined random variables (Benichou and Gail 1989) needs to be applied since the median survival for the combination group in Rx and C are implicitly defined by Eqs. (10.3.7) and (10.3.8). In step 3, similar to the binary data case, the asymptotic normal approximation in the (standard) Delta method can be applied on the logarithm of ratios (instead of the original ratios). Then we go back to the original scale when computing the simultaneous CIs for the ratios.

10.3.4.2   A Real Example

We use a phase 2 oncology study for patients with advanced non-small-cell lung cancer to illustrate the use of SME on time-to-event data (Spigel et al. 2013). The study compared a dual treatment to a single treatment to test whether the

Table 10.4 The SME result on the PFS data of the MET$^+$ patients in Spigel et al. (2013)

Population	Median in C	Median in Rx	Ratio and 95% CI
2+	2.22 (n=25)	4.97 (n=26)	2.24 (1.12, 4.48)
3+	2.00 (n=6)	3.19 (n=9)	1.59 (0.51, 5.05)
{2+, 3+}	2.17 (n=31)	4.47 (n=35)	2.06 (1.14, 3.74)

Estimates of median survival and their ratios are provided

dual treatment was more efficacious. The progression-free survival (PFS) is a (co-)primary endpoint. In this study, the patients were first separated into four groups by their MET expression level measured by the IHC test (0, 1+, 2+, and 3+) and then combined into two groups, namely, MET$^-$ (0, 1+) and MET$^+$ (2+, 3+).

We applied the SME procedure on the MET$^+$ patients using the "reversed-engineered" PFS data by Ding et al. (2016). More details regarding how the data were reverse-engineered can be found in Ding et al. (2016). The results are shown in Table 10.4. All three groups show positive efficacy (indicated by ratio of median > 1) for the dual treatment. However, only the 2+ and the combined group are statistically significant (with CIs not covering 1). Note that, with our SME procedure, the estimated efficacy for the combined group stays between the estimated efficacy of 2+ and 3+, in terms of both point estimates and simultaneous CIs.

10.4   Discussion

In Sect. 10.2, we presented two key fundamental statistical issues in the targeted treatment development process, especially during the inference on treatment efficacy when subgroups exist. The Rcode and example data sets for performing the SME approach for both time-to-event and binary outcomes can be found in Github: https://github.com/yingding99/SME as well as from the online supplementary materials.

Besides these two issues, there are certainly additional issues or challenges in the analysis or identification of subgroups. We briefly discuss two more in this section.

10.4.1   Additional Issues or Challenges

Multiplicity Adjustment Multiplicity is perhaps the most frequently mentioned challenge in statistical methods dealing with subgroups, especially in subgroup identification. It can come from several sources. First, the number of predictors/markers to be tested can range from tens to thousands (or even more). Second, if a marker is measured on a continuous scale, evaluating different cut points will introduce multiplicity. Third, in the analysis exploring subgroups defined by more than a single marker, the number of potential subgroups defined by the same set of markers increases exponentially with higher complexity of the subgroups being considered. In the recently released guidance papers by U.S. Food and Drug Administration and European Medicines Agency (US Food and Drug Administration 2012; Committee for Medicinal Products for Human Use and others 2010), it has been made clear that all the statistical approaches for subgroup analysis need to adjust for multiplicity.

In general, there are (at least) two levels of multiplicity when searching for subgroups. One is "across" markers and the other is "within" a marker. Appropriate types of error rates (to be controlled) should be considered for each level. For example, in Ding et al. (2018) for finding SNPs that are predictive of treatment efficacy, they proposed to control the *familywise* error rate (FWER) within a SNP and the *per family* error rate across SNPs. This sounds reasonable, since within a SNP, the consequence of an incorrect inference may directly lead to targeting a wrong patient population, which is very serious. Thus, controlling a stringent error rate is appropriate. For inference across a large number of SNPs, controlling a less stringent error rate is acceptable. How the two error rates are controlled usually depends on each other. Note that in our proposed simultaneous CI inference under the SME principle, the error rate being (strongly) controlled is the FWER. A proof can be found in Lin et al. (2019).

Re-sampling approach (such as permutation or bootstrap) is an option for multiplicity adjustment, provided it is done carefully to satisfy the conditions needed for its validity. For example, if permutation methods are used to produce the null distribution, then one has to check the stringent conditions in Huang et al. (2006) and Kaizar et al. (2011).

Formulation of Null Hypothesis In many subgroup analysis or identification proce-dures, the null hypothesis is often formulated as "all the subgroups have identical treatment efficacy". For example, if the marker is a SNP that separates the patients into three groups AA, Aa, aa, the null hypothesis is often stated as $AA = Aa = aa$, which is tested against specific alternatives such as the SNP has a dominant, recessive, or additive effect. Such a null hypothesis is called a "zero-null". In fact, controlling the Type I error rate in testing zero-nulls sometimes offers little protection against false discoveries. For example, in the SNP testing case, if the truth is a allele is dominant, then the rejection infers a recessive is counted positively towards power (when the null is formulated as $\mu_{AA} = \mu_{Aa} = \mu_{aa}$). In drug development, correct inference matters, with an incorrect inference possibly worse than no inference at all. For example, when the truth is a dominant, then inferring a recessive leads to targeting only the aa subgroup, missing out on targeting the combined $\{Aa, aa\}$ subgroup.

In Ding et al. (2018), based on the foundations of multiple testing principle, they proposed to formulate the null hypothesis as the *complement* of each desired inference as a separate null hypothesis. In the SNP testing case, a null hypothesis could be formulated as *a* is NOT dominant. In this way, a rejection of the null hypothesis can lead to the desired inference unambiguously.

10.4.2   Moving Forward

Targeted therapies are becoming more and more common. Such medicine can be large molecules such as antibodies targeting specific antigens on cell surfaces or extracellular growth factors. They can also be small molecules penetrating cell membrane to interact with enzymatic activities. The development of targeted treatment is a very complex process, of which the key is to identify the right patients for the drug to target. Any patient subgroup with significantly better efficacy could be identified for tailoring with appropriate labeling language and reimbursement considerations in the market. Correct and useful statistical inference is critical in decision-making for such a process. When developing new statistical methods motivated by the needs and challenges in this process, some fundamental issues mentioned in this chapter shall be well considered.

References

Alosh M, Huque MF, Bretz F, D'Agostino Sr RB (2017) Tutorial on statistical considerations on subgroup analysis in confirmatory clinical trials. Stat Med 36(8):1334–1360

Benichou J, Gail MH (1989) A delta method for implicitly defined random variables. Amer Statist Assoc 43:41–44

Berger JO, Wang X, Shen L (2014) A bayesian approach to subgroup identification. J Biopharm Stat 24(1):110–129

Borghaei H, Paz-Ares L, Horn L, Spigel DR, Steins M, Ready NE, Chow LQ, Vokes EE, Felip E, Holgado E, Barlesi F, Kohlhäufl M, Arrieta O, Burgio MA, Fayette J, Lena H, Poddubskaya E, Gerber DE, Gettinger SN, Rudin CM, Rizvi N, Crinò L, Blumenschein GRJ, Antonia SJ, Dorange C, Harbison CT, Graf Finckenstein F, Brahmer JR (2015) Nivolumab versus docetaxel in advanced nonsquamous non–small-cell lung cancer. N Engl J Med 373:1627–1639

Committee for Medicinal Products for Human Use and others (2010) Guideline on the investigation of bioequivalence. European Medicines Agency website, London

Ding Y, Lin HM, Hsu JC (2016) Subgroup mixable inference on treatment efficacy in mixture populations, with an application to time-to-event outcomes. Stat Med 35(10):1580–1594

Ding Y, Li YG, Liu Y, Ruberg SJ, Hsu JC et al (2018) Confident inference for SNP effects on treatment efficacy. Ann Appl Stat12(3):1727–1748

Dmitrienko A, Muysers C, Fritsch A, Lipkovich I (2016) General guidance on exploratory and confirmatory subgroup analysis in late-stage clinical trials. J Biopharm Stat 26(1):71–98

Foster JC, Taylor JM, Ruberg SJ (2011) Subgroup identification from randomized clinical trial data. Stat Med 30(24):2867–2880

Huang Y, Fong Y (2014) Identifying optimal biomarker combinations for treatment selection via a robust kernel method. Biometrics70(4):891–901

Huang Y, Xu H, Calian V, Hsu JC (2006) To permute or not to permute. Bioinformatics 22(18):2244–2248

Kaizar EE, Li Y, Hsu JC (2011) Permutation multiple tests of binary features do not uniformly control error rates. J Am Stat Assoc 106(495):1067–1074

Laber E, Zhao Y (2015) Tree-based methods for individualized treatment regimes. Biometrika 102(3):501–514

Lin HM, Xu H, Ding Y, Hsu JC (2019) Correct and logical inference on efficacy in subgroups and their mixture for binary outcomes. Biom J 61(1):8–26

Lipkovich I, Dmitrienko A (2014) Strategies for identifying predictive biomarkers and subgroups with enhanced treatment effect in clinical trials using sides. J Biopharm Stat 24(1):130–153

Lipkovich I, Dmitrienko A, Denne J, Enas G (2011) Subgroup identification based on differential effect search—a recursive partitioning method for establishing response to treatment in patient subpopulations. Stat Med 30:2601–2621

Lipkovich I, Dmeitrienko A, D'Agostino RB (2017) Tutorial in biostatistics: data-driven subgroup identification and analysis in clinical trials. Stat Med 36:136–196

Loh WY, He X, Man M (2015) A regression tree approach to identifying subgroups with differential treatment effects. Stat Med 34:1818–1833

Qian M, Murphy SA (2011) Performance guarantees for individualized treatment rules. Ann Statist 39(2):1180

Scott LJ (2019) Larotrectinib: first global approval. Drugs 79:1–6

Shen L, Ding Y, Battioui C (2015) A framework of statistical methods for identification of subgroups with differential treatment effects in randomized trials. In: Applied statistics in biomedicine and clinical trials design. Springer, Berlin, pp. 411–425

Spigel DR, Ervin TJ, Ramlau RR, Daniel DB, Goldschmidt Jr JH, Blumenschein Jr GR, Krzakowski MJ, Robinet G, Godbert B, Barlesi F, Govindan R, Patel T, Orlov SV, Wertheim MS, Yu W, Zha J, Yauch RL, Patel PH, Phan S, Peterson AC (2013) Randomized phase ii trial of onartuzumab in combination with erlotinib in patients with advanced non-small-cell lung cancer. J Clin Oncol 31(32):4105–4114

Su X, Tsai CL, Wang H, Nickerson DM, Li B (2009) Subgroup analysis via recursive partitioning. J Mach Learn Res 10(Feb):141–158

US Food and Drug Administration (2012) Guidance for industry: Enrichment strategies for clinical trials to support approval of human drugs and biological products. FDA, Silver Spring

Zhang B, Tsiatis AA, Davidian M, Zhang M, Laber E (2012) Estimating optimal treatment regimes from a classification perspective. Stat 1(1):103–114

Zhao Y, Zeng D, Rush AJ, Kosorok MR (2012) Estimating individualized treatment rules using outcome weighted learning. J Am Stat Assoc 107(499):1106–1118

Chapter 11
Subgroup Analysis with Partial Linear Regression Model

Yizhao Zhou, Ao Yuan, and Ming T. Tan

Abstract In clinical trials it is common that the treatment has different effects on different subjects. This motivates the precision medicine, the goal is to identify the treatment favorable or unfavorable subgroups, if they exist, and classify the subjects into one of the subgroups based on their covariate values. In practice, some covariate(s) is known to affect the response non-linearly, in this case the existing linear model is not adequate. To address this issue, we use a partial linear model, in which the effect of some specific covariates is a non-linear monotone function, along with a linear part for the rest of the covariates. This approach not only makes the model more flexible than the parametric linear model, and more interpretable and efficient than the full nonparametric model. The Wald statistics is used to test the existence of subgroups, and the Neyman-Pearson rule is used to classify the subjects. Simulation studies are conducted to evaluate the performance of the method, and then the method is used to analyze a real clinical trial data.

11.1 Introduction

In clinical studies, often treatment effect is not uniform over all the patients, some subgroup of patients may benefit significantly from the treatment and others may not so. Thus one of goals of precision medicine is to find out if such subgroups exist or not, and if existence is justified, identify the subgroups of patients according to their covariate values. For example, in IBCSG (2002), patients with ER-negative tumors were likely to benefit from chemotherapy, while those with ER-positive tumors did not.

Y. Zhou (✉) · A. Yuan · M. T. Tan
Department of Biostatistics, Bioinformatics and Biomathematics, Georgetown University, Washington, DC, USA
e-mail: yz459@georgetown.edu

© Springer Nature Switzerland AG 2020   229
N. Ting et al. (eds.), *Design and Analysis of Subgroups with Biopharmaceutical Applications*, Emerging Topics in Statistics and Biostatistics,
https://doi.org/10.1007/978-3-030-40105-4_11

Subgroup analysis is recently a very active research area see, e.g., Sabine (2005), Song and Chi (2007), Ruberg et al. (2010), Foster et al. (2011), Lipkovich et al. (2011), Friede et al. (2012), Shen and He (2015), Fan et al. (2017), and Ma and Huang (2017). Rothmann et al. (2012) discussed issues for subgroups testing and analysis. Fokkema (2018) used generalized linear mixed-effect model tree (GLMM tree) algorithm detecting treatment-subgroup interactions in clustered datasets. Yuan et al. (2018, 2020) proposed semiparametric methods for this problem.

Existing methods for this problem often use linear model. In practice, sometimes it is known that some covariate has non-linear effect on the response, incorporating such information can improve the quality of the analysis. Here we consider such case and apply a more featured partial linear model to identify the existence of subgroups and to classify the subjects into different subgroups if the existence of subgroup is confirmed. This model assumes a monotone non-linear effect of some covariate, and linear effects from the rest covariates. First, a partial model with individual subgroup membership as latent variable and with a covariate whose effect are known as non-linear are formulated and the model regression parameters is estimated with expectation-maximization algorithm (E-M algorithm), and isotonic regression method is used for the maximum likelihood of the nonparametric non-linear part. Then null hypothesis of non-existence of subgroups are tested with Wald Statistics. If the existence of subgroup is confirmed, we use the Neyman-Pearson rule to classify each subject so that the misclassification error for the treatment favored group is under control while the misclassification error for the other subgroup is minimized.

The rest of the chapter is organized as follows. In Sect. 11.2 we describe the model and parameter estimation, Sect. 11.3 elaborates the testing and classification method, and Sect. 11.4 illustrates the simulation study and real data analysis.

11.2   The Method

The observed data is denoted as $D_n = \{(y_i, x_i, z_i), i = 1, \ldots, n\}$, where $y_i \in R$ is the response variable of i-th subject, $x_i = (x_{i1}, \ldots, x_{id})' \in R^d$ and z_i is another covariate, which is known to have a non-linear monotone effect on the response. Each subject i receives the same treatment, and we assume that bigger value of the response corresponds to better treatment effects. We want to test if there are treatment favorable and non-favorable subgroups in the patients. If subgroup does exist, we need to classify each subject into corresponding subgroup based on his/her covariate profile. In this paper, we assume that there are only two potential subgroups: treatment-favorable and treatment-nonfavorable subgroups. We need first to specify the model, estimate the model parameters, and then perform the hypothesis test and classification of subjects.

11.2.1 The Semiparametric Model Specification

We specify the semiparametric partial linear model as

$$y_i = \boldsymbol{\beta}' \boldsymbol{x}_i + g(z_i) + \delta_i \eta + \epsilon_i, \qquad \epsilon \sim N(0, 1), \qquad g \in \mathcal{G},$$

where δ_i is a latent indicator for whether subject i belongs to the treatment favorable subgroup ($\delta_i = 1$) or not ($\delta_i = 0$). $\boldsymbol{\beta}$ is a d-vector of unknown parameters, η is the effect of treatment favorable subgroup, and the constraint $\eta \geq 0$ is used for the identifiability with the intercept vector term in $\boldsymbol{\beta}$. It is assumed that the covariate z_i has a non-linear effect $g(\cdot)$ to the response y_i, we only know that $g(\cdot) \in \mathcal{G}$, the collection of all monotone increasing functions on R.

Denote the i.i.d. copy of the $(y_i, \boldsymbol{x}_i, z_i, \delta_i, \varepsilon_i)$'s as $(\boldsymbol{y}, \boldsymbol{x}, \boldsymbol{z}, \boldsymbol{\delta}, \boldsymbol{\varepsilon})$. Let $\lambda = P(\delta = 1)$ and $\boldsymbol{\theta} = (\boldsymbol{\beta}', \eta, \lambda)'$ be the vector of all the Euclidean parameters. Conditioning on $(\boldsymbol{x}, z)$, the density of y is the mixture

$$h(y|\boldsymbol{x}, z, \boldsymbol{\theta}) = \lambda \phi \Big(y - \boldsymbol{\beta}' \boldsymbol{x} - g(z) - \eta \Big) + (1 - \lambda) \phi \Big(y - \boldsymbol{\beta}' \boldsymbol{x} - g(z) \Big).$$

where $\phi(\cdot)$ is the density function of the standard normal distribution. The log-likelihood of the observed data is

$$\ell(\boldsymbol{\theta}, g|D_n) = \sum_{i=1}^{n} \log \Big(\lambda \phi(y_i - \boldsymbol{\beta}' \boldsymbol{x}_i - g(z_i) - \eta) + (1 - \lambda) \phi(y_i - \boldsymbol{\beta}' \boldsymbol{x}_i - g(z_i)) \Big),$$

$$\boldsymbol{\theta} \in \boldsymbol{\Theta}, \quad g \in; \mathcal{G}. \tag{11.1}$$

Direct computation of the maximum likelihood estimate (MLE) from a mixture model (11.1) is not convenient, especially in the presence of the nonparametric component $g(\cdot)$, and it is known that E-M algorithm (Dempster et al. 1977) is typically easy to use. For this, we treat the latent variable δ_i's as missing data, with $\delta_i = 1$ if the i-th subject belongs to the treatment-favorable subgroup, otherwise $\delta_i = 0$. The likelihood based on the 'complete data' $D_n^c = \{(y_i, \boldsymbol{x}_i, z_i, \delta_i) : i = 1, \ldots, n\}$ is

$$L(\boldsymbol{\theta}, g|D_n^c) = \prod_{i=1}^{n} \Big(\lambda \phi(y_i - \boldsymbol{\beta}' \boldsymbol{x}_i - g(z_i) - \eta) \Big)^{\delta_i} \Big((1 - \lambda) \phi(y_i - \boldsymbol{\beta}' \boldsymbol{x}_i - g(z_i)) \Big)^{1 - \delta_i},$$

the corresponding log-likelihood is

$$\ell(\boldsymbol{\theta}, g|D_n^c) = \sum_{i=1}^{n} \Big(\delta_i \log \phi(y_i - \boldsymbol{\beta}' \boldsymbol{x}_i - g(z_i) - \eta)$$

$$+ (1 - \delta_i) \log \phi(y_i - \boldsymbol{\beta}' \boldsymbol{x}_i - g(z_i)) + \delta_i \log \lambda + (1 - \delta_i) \log(1 - \lambda) \Big). \tag{11.2}$$

The semiparametric MLE $(\hat{\boldsymbol{\theta}}_n, \hat{f}_n)$ of the true parameter $(\boldsymbol{\theta}_0, f_0)$ is given by

$$(\hat{\boldsymbol{\theta}}_n, \hat{g}_n) = \arg \max_{(\theta, g) \in (\Theta, \mathcal{G})} \ell(\boldsymbol{\theta}, g | D_n^c). \tag{11.3}$$

11.2.2   Estimation of Model Parameters

As the δ_i's are missing, $(\hat{\boldsymbol{\theta}}_n, \hat{g}_n)$ in (11.3) cannot be computed directly, the EM algorithm is used instead. For this a starting value $\boldsymbol{\theta}^{(0)}$ of $\boldsymbol{\theta}$ is needed, then find $g^{(1)}(\cdot) \in \mathcal{G}$ as the maxima of $\ell(\boldsymbol{\theta}^{(0)}, g | D_n^c)$, then fix $g^{(1)}$, find $\boldsymbol{\theta}^{(1)} \in \Theta$ as the maxima of $\ell_n(\boldsymbol{\theta}, g^{(1)})$, and so on.... until convergence of the sequence $\{(\boldsymbol{\theta}^{(r)}, g^{(r)})\}$, which is increasing the likelihood at each iteration, and will converge to at least some local maxima of $\ell_n(\boldsymbol{\theta}, g)$. In fact, the increasing likelihood property is obvious, as for all integer r,

$$\ell(\boldsymbol{\theta}^{(r+1)}, g^{(r+1)} | D_n^c) \geq \ell(\boldsymbol{\theta}^{(r)}, g^{(r+1)} | D_n^c) \geq \ell(\boldsymbol{\theta}^{(r)}, g^{(r)} | D_n^c).$$

A formal justification of the convergence of the above iterative algorithm is a case of the block coordinate descent methods in Bertsekas (2016).

Our algorithm is a semiparametric version of EM algorithm, see also Tan et al. (2009, chap. 2) for bio-medical applications of this algorithm. The semiparametric and nonparametric EM algorithm was used in a large number of literatures, such as in Muñoz (1980), Campbell (1981), Hanley and Parnes (1983), Groeneboom and Wellner (1992, Section 3.1), and see the argument there for the convergence of such algorithm (p. 67–68). Chen et al. (2002) applied the EM algorithm to a semiparametric random effects model, Bordes et al. (2007) applied the EM algorithm to a semiparametric mixture model, using simulation studies to justify the convergence of the algorithm. Balan and Putter (2019) developed an R-package of EM algorithm for semiparametric shared frailty models.

Now we give the detail of the algorithm. At each iteration r, do the following:

Step 0. For fixed $(g^{(0)}, \boldsymbol{\theta}^{(0)})$, compute $\{\delta_i^{(0)}\}$ with E-step of E-M algorithm.
Step 1. For fixed $(g^{(r)}, \boldsymbol{\theta}^{(r)})$, compute

$$H_n(\boldsymbol{\theta}, g | \boldsymbol{\theta}^{(r)}, g^{(r)}) = E_\delta[\ell(\boldsymbol{\theta}, g | D_n^c) | D_n, \boldsymbol{\theta}^{(r)}, g^{(r)}]$$

$$= \sum_{i=1}^n \Big(\delta_i^{(r)} \log \phi(y_i - \boldsymbol{\beta}' \boldsymbol{x}_i - g(z_i) - \eta)$$

$$+ \delta_i^{(r)} \log \lambda + (1 - \delta_i^{(r)}) \log \phi(y_i - \boldsymbol{\beta}' \boldsymbol{x}_i - g(z_i))$$

$$+ (1 - \delta_i^{(r)}) \log(1 - \lambda) \Big), \tag{11.4}$$

where the expectation is taken with respect to the missing δ, and as if the true data is generated from parameters $(\theta^{(r)}, g^{(r)})$. In particular, the r-th step estimates of the δ_i's (for $i = 1, \ldots, n; r = 0, 1, 2, \ldots$), are

$$
\begin{aligned}
\delta_i^{(r)} &= E(\delta_i | y_i, x_i, z_i, g^{(r)}, \theta^{(r)}) = P(\delta_i = 1 | y_i, x_i, z_i, g^{(r)}, \theta^{(r)}) \\
&= \frac{P(y_i | \delta_i = 1, x_i, z_i, g^{(r)}, \theta^{(r)}) P(\delta_i = 1 | x_i, z_i, g^{(r)}, \theta^{(r)})}{P(y_i | x_i, z_i, g^{(r)}, \theta^{(r)})} \\
&= \frac{\lambda^{(r)} \phi\left(y_i - \boldsymbol{\beta}'^{(r)} x_i - g^{(r)}(z_i) - \eta^{(r)}\right)}{\lambda^{(r)} \phi\left(y_i - \boldsymbol{\beta}'^{(r)} x_i - g^{(r)}(z_i) - \eta^{(r)}\right) + (1 - \lambda^{(r)}) \phi\left(y_i - \boldsymbol{\beta}'^{(r)} x_i - g^{(r)}(z_i)\right)}.
\end{aligned}
$$

Step 2. In the M-step for θ, compute

$$
\theta^{(r+1)} = \arg \sup_{\theta \in \Theta} H_n(\theta, g^{(r)} | \theta^{(r)}, g^{(r)}).
$$

This step can be computed by standard optimization packages. Especially,

$$
\lambda^{(r+1)} = \frac{1}{n} \sum_{i=1}^{n} \delta_i^{(r)}.
$$

Step 3. For fixed $(\theta^{(r+1)}, \delta_i^{(r+1)})$ compute

$$
g^{(r+1)}(\cdot) = \arg \max_{g \in \mathcal{G}} H_n(\theta^{(r+1)}, g | \theta^{(r)}, g^{(r)}).
$$

This step computes the nonparametric maximum likelihood estimate of $\hat{g}$ under shape restriction, which is non-trivial, we describe it below.

11.2.2.1 Computation of $g^{(r+1)}$

The pool adjacent violators algorithm (PAVA, see for example, Best and Chakravarti (1990)) is a convenient computational tool to perform such order restricted maximization or minimization, and is available in R. Patrick et al. (2009) gives a review of the algorithm history and computational aspects. In particular, the computation of $\hat{g}(z_i) = \hat{g}_i$ is as follows.

$$g^{(r+1)}(\cdot) = \underset{g \in \mathcal{G}}{\arg\max}\, H_n(\boldsymbol{\theta}^{(r+1)}, g | \boldsymbol{\theta}^{(r)}, g^{(r)})$$

$$= \underset{g \in \mathcal{G}}{\arg\min} \sum_{i=1}^{n} \left(\delta_i^{(r)} \left(y_i - \beta^{(r)} x_i - \eta^{(r)} - g_i \right)^2 \right.$$

$$\left. + (1 - \delta_i^{(r)}) \left(y_i - \boldsymbol{\beta}^{(r)} \boldsymbol{x}_i - g_i \right)^2 \right)$$

$$= \underset{g \in \mathcal{G}}{\arg\min} \sum_{i=1}^{n} \left(y_i - \beta^{(r)} x_i - \eta^{(r)} \delta_i^{(r)} - g_i \right)^2$$

Generally, let $v_i = y_i - \boldsymbol{\beta}' \boldsymbol{x}_i - \delta_i \eta$, $w_i = 1$, then

$$\hat{g} = \underset{g \in \mathcal{G}}{\arg\min} \sum_{i=1}^{n} w_i (v_i - g_i)^2$$

The above is the standard form of isotonic regression procedure, and $\hat{g}$ can be computed using the R-function $isoreg(\cdot)$.

11.2.3 Asymptotic Results of the Estimates

Zhou et al. (2019) derived asymptotic results for $\hat{\boldsymbol{\theta}}$ and $\hat{g}(\cdot)$, as presented below. Detailed regularity conditions and proofs can be found there.

Theorem 11.1 *Under regularity conditions, as $n \to \infty$*

$$\|\hat{\boldsymbol{\theta}} - \boldsymbol{\theta}_0\| \overset{a.s.}{\to} 0, \quad \int |\hat{g}(z) - g_0(z)| dz \overset{a.s.}{\to} 0.$$

Denote $\overset{D}{\to}$ for convergence in distribution.

Theorem 11.2 *Under regularity conditions, as $n \to \infty$,*

$$\sqrt{n}(\hat{\boldsymbol{\theta}} - \boldsymbol{\theta}_0) \overset{D}{\to} N(\boldsymbol{0}, I^{*-1}(\boldsymbol{\theta}_0 | g_0)),$$

where $I^*(\boldsymbol{\theta}_0 | g_0) = E[\ell^*(X, Z | \boldsymbol{\theta}_0, g_0) \ell^{*'}(X, Z | \boldsymbol{\theta}_0, g_0)]$ is the efficient Fisher information matrix of $\boldsymbol{\theta}$ for fixed g_0, and $\ell^*(X, Z | \boldsymbol{\theta}_0, g_0)$ is the efficient score for $\boldsymbol{\theta}$.

Let $\mathbb{B}(\cdot)$ be the two-sided Brownian motion originating from zero: a mean zero Gaussian process on R with $\mathbb{B}(0) = 0$, and $E(\mathbb{B}(s) - \mathbb{B}(h))^2 = |s - h|$ for all s, $h \in R$.

Theorem 11.3 *Denote $\dot{g}_0(z) = dg_0(z)/dz$ and density of z as $q(z)$. Assume $q(z) > 0$.*
Under regularity conditions, as $n \to \infty$,

$$n^{1/3}(\hat{g}_n(z) - g_0(z)) \overset{D}{\to} \left(\frac{4\dot{g}_0(z)}{q(z)}\right)^{1/3} \arg\max_{h \in R}\{\mathbb{B}(h) - h^2\}.$$

11.3   Testing the Null Hypothesis and the Classification Rules

11.3.1   Test the Null Hypothesis

After the model parameters are estimated, we need to test the existence of
subgroups, which is formulated as testing the null hypothesis $H_0 : \eta = 0$ vs the
alternative $H_1 : \eta \neq 0$. For parametric model, commonly used test statistic including
the likelihood ratio statistic, score statistic and the Wald statistic, and the three
statistics are asymptotically chi-squared distributed and equivalent. However, in our
case when $\eta = 0$, λ is non-identifiable in the model, although the other parameters
are still identifiable and estimable. In this case, the likelihood ratio statistic cannot
be applied. So we use the Wald statistic.

Denote $\boldsymbol{\theta} = (\boldsymbol{\theta}_1, \boldsymbol{\theta}_2)$ with $dim(\boldsymbol{\theta}) = d$ and $dim(\boldsymbol{\theta}_1) = d_1$, and $\hat{\boldsymbol{\theta}} = (\hat{\boldsymbol{\theta}}_1, \hat{\boldsymbol{\theta}}_2)$ is the
MLE of $\boldsymbol{\theta}$ under the full model. Consider the null hypothesis $H_0 : \boldsymbol{\theta}_1 = \boldsymbol{\theta}_{1,0}$. The
Wald test statistic is

$$W_n = (\hat{\boldsymbol{\theta}}_1 - \boldsymbol{\theta}_{1,0})' Var^{-1}(\hat{\boldsymbol{\theta}}_1)(\hat{\boldsymbol{\theta}}_1 - \boldsymbol{\theta}_{1,0}).$$

If $Cov(\hat{\boldsymbol{\theta}}_1)$ is known, then asymptotically $W_n \sim \chi^2_{d_1}$. If $Cov(\hat{\boldsymbol{\theta}}_1)$ is estimated,
asymptotically $W_n/d_1 \sim F_{d_1, n-d}$. For our problem, $\boldsymbol{\theta}_1 = \eta$, $\boldsymbol{\theta}_{1,0} = 0$, we treat
$Cov(\hat{\eta})$ to be known, so $W_n = \hat{\eta}_n Var^{-1}(\hat{\eta}_n)\hat{\eta}_n \sim \chi^2_1$ asymptotically, and if
$W_n > \chi^2_1(1 - \alpha)$, which is the upper $(1 - \alpha)$-th quantile of the χ^2_1 distribution,
then H_0 is rejected.

11.3.2   The Classification Rule

After the existence of subgroup is justified, or the null hypothesis above is
rejected, we need to classify the subjects. There are different classification rules.
In subgroup analysis, the correct classification of the treatment favorable subgroup
is of significant clinical meaning, so we use the Neyman-Pearson rule in Yuan et al.
(2018, 2020) as it can control the miss-classification error for the treatment favorable
subgroup.

To be specific, for each subject i, denote the i-th likelihood ratio

$$LR(y_i, \boldsymbol{x}_i) = \frac{f(y_i, \boldsymbol{x}_i, z_i | \hat{\boldsymbol{\theta}}, \delta = 1)}{f(y_i, \boldsymbol{x}_i, z_i | \hat{\boldsymbol{\theta}}, \delta = 0)} \approx \frac{\phi(y_i - \hat{\boldsymbol{\beta}}' \boldsymbol{x}_i - \hat{g}(z_i) - \hat{\eta})}{\phi(y_i - \hat{\boldsymbol{\beta}}' \boldsymbol{x}_i - \hat{g}(z_i))}.$$

Parallel to the NP uniformly most powerful test procedure for testing the simple hypothesis $H_0 : \eta = 0$ vs. $H_1 : \eta \neq 0$. For given significance level α, the optimal classification rule is: classify the i-th subject to subgroup S_1 if

$$LR(y_i, \boldsymbol{x}_i, z_i) \geq K(\alpha), \text{ with } K(\alpha) \text{ determined by } P_{H_0}\big(LR(Y, X, Z) \geq K(\alpha)\big) = \alpha,$$

or, with $\epsilon = y - \hat{\boldsymbol{\beta}}' \boldsymbol{x} - \hat{g}(z_i)$ generated under H_0,

$$P_{H_0}\left(\frac{\phi(y_i - \hat{\boldsymbol{\beta}}' \boldsymbol{x}_i - \hat{g}(z_i) - \hat{\eta})}{\phi(y_i - \hat{\boldsymbol{\beta}}' \boldsymbol{x}_i - \hat{g}(z_i))} \geq K(\alpha)\right) = \alpha.$$

We can find approximate solution for $K(\alpha)$. For simulated data, let $\{LR_j : j = 1, \ldots, n_0\}$ be the LR_j's of patients from the treatment unfavorable subgroup (for simulated data, the subgroup memberships are known), then set $K(\alpha)$ is estimated by the $(1 - \alpha)$-th upper quantile of $LR_1, \ldots, LR_{n_0}$, it is the cut-off beyond which patients will be classified to the treatment favorable subgroup, even though they are from the treatment unfavorable subgroup.

However, for real data $\{(y_i, \boldsymbol{x}_i, z_i) : i = 1, \ldots, n\}$, the subgroup memberships are unknown, we cannot use the above method to decide $K(\alpha)$, instead we obtain it by the following way. Set $LR_i = \phi(\epsilon_i - \hat{\eta})/\phi(\epsilon_i)$, let

$$Q_n(t) = \sum_{i=1}^{n} w_{ni} I(LR_i \leq t), \quad w_{ni} = (1 - \hat{\delta}_i)/\sum_{j=1}^{n}(1 - \hat{\delta}_j)$$

be a weighted empirical distribution of the LR_i's under the null hypothesis. Note that $1 - \hat{\delta}_i$ is the estimated membership of subject i belonging to group 0, corresponding to the null hypothesis, and $1 - \hat{\delta}_i$ scaled by $\sum_{j=1}^{n}(1 - \hat{\delta}_j)$ makes the w_{ni}'s a set of actual weights. So intuitively, $Q_n(\cdot)$ is a reasonable estimate of the distribution of the LR_i's under the null hypothesis. We set $K(\alpha) = Q_n^{-1}(1 - \alpha)$ to be the $(1 - \alpha)$-th upper quantile of Q_n.

For coming patient with covariate $\boldsymbol{x}$ but without response y, we define

$$LR(\boldsymbol{x}, z) = E_{H_0}\left(\frac{\phi(y - \hat{\boldsymbol{\beta}}' \boldsymbol{x} - \hat{\eta})}{\phi(y - \hat{\boldsymbol{\beta}}' \boldsymbol{x})} \Big| \boldsymbol{x}, z\right) \approx \frac{1}{n_0} \sum_{i=1}^{n_0} \frac{\phi(y_i - \hat{\boldsymbol{\beta}}' \boldsymbol{x} - \hat{g}(z_i) - \hat{\eta})}{\phi(y_i - \hat{\boldsymbol{\beta}}' \boldsymbol{x} - \hat{g}(z_i))},$$

where y_i ($i = 1, \ldots, n_0$) are the responses of the subjects already in the trail, and being classified to group 0, and classify this patient to group 1 if $LR(\boldsymbol{x}, z) > K(\alpha)$, with $K(\alpha)$ given above.

11.4 Simulation Study and Application

11.4.1 Simulation Study

We simulate four examples with non-linear effect of z_i to y_i. We simulate $n = 1000$ i.i.d. data with 1-dimensional response y_i's and with covariates $\boldsymbol{x}_i = (x_{i1}, x_{i2}, x_{i3})$. We first generate the covariates, sample the $\boldsymbol{x}_i$'s from the 3-dimensional normal distribution with mean vector $\boldsymbol{\mu} = (3.1, 1.8, -0.5)'$ and a given covariance matrix Γ. sample the z_i's from the normal distribution with mean $\mu = 0$ and $\sigma^2 = 1$. The ε_i are also sampled from normal distribution with mean $\mu = 0$ and $\sigma^2 = 1$. We will display estimation results with four different choices of $\boldsymbol{\theta}_0 = (\boldsymbol{\beta}_0, \eta_0, \lambda_0)$ and four choices of $g_0(\cdot)$ below. What is more, we fixed a point $(0, 0)$ for the non-linear effect.

Example 1 $g_0(z) = 6 \times Exponential(z + 2) - 6 \times Expnential(0 + 2)$;

Example 2 $g_0(z) = 5 \times Beta((z + 2)/4, 5, 1) - 5 \times Beta((0 + 2)/4, 5, 1)$;

Example 3 $g_0(z) = 6 \times I(z < 0) \times ((N(z, 0, 0.5)) - N(0, 0, 0.5)) + 6 \times I(z \geq 0) \times (N(z, 0, 0.2) - N(0, 0, 0.2)))$;

Example 4 $g_0(z) = 3 \times I(z < 0) \times (Beta((z + 2)/4, 0.2, 0.2) - Beta((0 + 2)/4, 0.2, 0.2)) + 7 \times I(z \geq 0) \times (Beta((z + 2)/4, 0.7, 0.7) - Beta((0 + 2)/4, 0.7, 0.7))$.

The estimated $\hat{g}$ and g_0 are shown in Fig. 11.1.

The parameter estimates from the proposed model are displayed in Tables 11.1, 11.2, 11.3 and 11.4, along with the estimates from commonly used linear model as comparison. The estimated standard errors are displayed as [se].

The hypothesis testing results from both partial linear and linear model are given in Table 11.5, and the classification results using the partial linear model are in Table 11.6.

From Table 11.5 we see that the partial linear model gives reasonable estimates, while the estimates from the linear model is not reasonable, may due to the fact that it seriously over-estimate the effect η for small value of it.

From Table 11.6, it is seen that the mis-classification error for the treatment favorable subgroup is well controlled around the specified level $\alpha = 0.05$, and the overall classification error depends on the effect size η. It is small when η is large and vice versa. Note that for $\eta = 0.95$ and 1.70, the N-P error is larger than 0.05 this is because the estimate of η is not that accurate when the true value of η is small.

Interpretation of the Results From Tables 11.1, 11.2, 11.3 and 11.4, we see that when the effect η of treatment favorable subgroup is tiny, the biases of the estimates

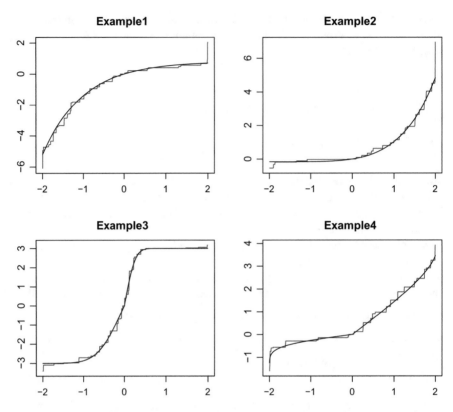

Fig. 11.1 Solid line: true $g_0(\cdot)$; Step line: estimate $\hat{g}(\cdot)$

Table 11.1 Parameter estimates under two models (example 1)

θ	β	η	λ
θ_0	$(1.300, 1.200, -1.600)$	1.650	0.700
Partial linear	$(1.291, 1.202, -1.598)$	1.619	0.723
[se]	[0.094, 0.087, 0.039]	[0.134]	[0.060]
95% CP	$(0.944, 0.945, 0.949)$	0.942	0.941
Linear model	$(0.995, 1.275, -1.520)$	2.053	0.653
[se]	[0.084, 0.100, 0.043]	[0.101]	[0.048]
95% CP	$(0.042, 0.885, 0.552)$	0.017	0.840
θ_0	$(1.200, -1.400, 3.200)$	7.740	0.300
Partial linear	$(1.198, -1.401, 3.202)$	7.740	0.300
[se]	[0.061, 0.068, 0.030]	[0.070]	[0.014]
95% CP	$(0.946, 0.953, 0.949)$	0.953	0.955
Linear model	$(0.966, -1.344, 3.263)$	7.742	0.300
[se]	[0.054, 0.081, 0.034]	[0.085]	[0.014]
95% CP	$(0.009, 0.896, 0.541)$	0.948	0.952

Table 11.2 Parameter estimates under two models (example 2)

θ	β	η	λ
θ_0	$(-1.300, 2.200, 1.400)$	1.700	0.700
Partial linear	$(-1.300, 2.202, 1.398)$	1.678	0.718
[se]	[0.085, 0.085, 0.039]	[0.126]	[0.055]
95% CP	(0.952, 0.947, 0.955)	0.951	0.946
Linear model	$(-1.033, 2.137, 1.328)$	2.029	0.549
[se]	[0.084, 0.103, 0.045]	[0.084]	[0.059]
95% CP	(0.105, 0.909, 0.643)	0.018	0.272
θ_0	$(-2.300, -1.400, 2.100)$	8.390	0.800
Partial linear	$(-2.295, -1.400, 2.098)$	8.392	0.801
[se]	[0.059, 0.070, 0.030]	[0.079]	[0.012]
95% CP	(0.942, 0.947, 0.942)	0.951	0.940
Linear model	$(-2.061, -1.458, 2.037)$	8.396	0.801
[se]	[0.061, 0.084, 0.035]	[0.099]	[0.012]
95% CP	(0.031, 0.897, 0.557)	0.949	0.939

Table 11.3 Parameter estimates under two models (example 3)

θ	β	η	λ
θ_0	$(-1.300, 1.400, 2.000)$	1.500	0.650
Partial linear	$(-1.282, 1.396, 1.995)$	1.454	0.666
[se]	[0.135, 0.090, 0.047]	[0.122]	[0.080]
95% CP	(0.955, 0.952, 0.951)	0.938	0.942
Linear model	$(-1.270, 1.389, 1.995)$	2.263	0.476
[se]	[0.083, 0.113, 0.048]	[0.079]	[0.046]
95% CP	(0.936, 0.945, 0.945)	0	0.023
θ_0	$(1.600, -1.200, 1.300)$	5.800	0.400
Partial linear	$(1.622, -1.204, 1.294)$	5.801	0.400
[se]	[0.108, 0.076, 0.039]	[0.067]	[0.015]
95% CP	(0.944, 0.941, 0.945)	0.948	0.949
Linear model	$(1.648, -1.209, 1.286)$	5.811	0.406
[se]	[0.076, 0.115, 0.046]	[0.090]	[0.015]
95% CP	(0.909, 0.956, 0.940)	0.948	0.918

from the linear model are much larger than those with the proposed partial linear model. That also can be used to explain the results of hypothesis testing with linear model. When the effect of treatment favorable subgroup is small, linear model tend to give an estimate with positive bias. So, type I error here is large and type II error is small. If the effect of treatment favorable subgroup is large, partial linear model and linear model tend to give similiar estimates of parameters.

Table 11.4 Parameter estimates under two models (example 4)

θ	β	η	λ
θ_0	(2.000,2.400,−2.500)	1.300	0.850
Partial linear	(1.977,2.403,−2.493)	1.402	0.853
[se]	[0.183,0.088,0.054]	[0.394]	[0.150]
95% CP	(0.948,0.946,0.952)	0.938	0.946
Linear model	(2.257,2.336,−2.566)	1.358	0.717
[se]	[0.116,0.083,0.041]	[0.190]	[0.122]
95% CP	(0.383,0.873,0.618)	0.931	0.825
θ_0	(−2.500,1.200, 1.700)	8.850	0.300
Partial linear	(−2.481,1.198,1.693)	8.851	0.299
[se]	[0.059,0.069,0.030]	[0.070]	[0.014]
95% CP	(0.940,0.946,0.947)	0.950	0.951
Linear model	(−2.287,1.151,1.643)	8.853	0.299
[se]	[0.049,0.072,0.030]	[0.074]	[0.014]
95% CP	(0.012,0.897,0.536)	0.945	0.951

Table 11.5 Hypothesis test using the partial linear and linear models (example 4)

η_0	Partial linear			Linear model		
	$\hat{\eta}$	Type I error	Power	$\hat{\eta}$	Type I error	Power
0	0.006	0.018		1.485	0.237	
0.02	0.003	0.016		1.496	0.234	
0.5	0.761		0.151	1.545		1
0.75	1.047		0.223	1.608		1
0.9	1.087		0.491	1.659		1
1.0	1.103		0.934	1.700		1
1.1	1.142		0.996	1.738		1
1.3	1.286		1	1.842		1

Table 11.6 Classification results using partial linear model (simulated data)

η_0	$\hat{\eta}$	Decision	Overall error	N-P Error	$K(0.05)$
0.95	0.680	H_1	0.165	0.390	1.093
1.70	1.531	H_1	0.198	0.182	1.000
3.50	3.592	H_1	0.042	0.032	1.000
5.00	5.036	H_1	0.003	0.004	1.000
7.74	7.785	H_1	0	0	1.000

11.4.2 Application to Real Data Problem

Now we analyze the real data ACTG175 with the proposed method. The trial was conducted by the AIDS Clinical Trials Group (ACTG), which was supported by the National Institute of Allergy and Infectious Diseases (NIAID). Participants were enrolled into the study between December 1991 and October 1992, and received treatment through December 1994. Follow-up and final evaluations of participants took place between December 1994 and February 1995.

The purpose of this data was to investigate whether treatment of HIV infection with one drug (monotherapy) was the same, better than, or worse than treatment with two drugs (combination therapy) in patients under some conditions. Three different drugs were used to conduct this study: (1) zidovudine (AZT), (2) didanosine (ddI), and (3) zalcitabine (ddC). The three drugs are nucleotide analogues that act as reverse transcriptase inhibitors (RT-inhibitors). The original study noted no clear differences between the ddI and AZT + ddI treatments—both appeared to be approximately equal effective in preventing HIV progressing. Treatment with AZT + ddC provided no additional benefit to continued treatment with AZT. However, the results of ACTG 175 together with the results from earlier studies demonstrate that antiretroviral therapy is beneficial to HIV-infected people who have less than 500 CD4+ T cells/mm3. This study also shows, for the first time, that an improvement in survival can be achieved in a sub-population.

We analyze this data using the proposed method on the combined therapy (ZDV+ddI). The number of patients is 522. The response variable is the CD4 counts after 20 weeks of the corresponding treatment, and the covariates are age, baseline CD4 counts, karnofsky score and number of days of previously received antiretroviral therapy. We assume the effect of baseline CD4 counts on the response variable is non-linear.

The analysis results are presented in Tables 11.7 and 11.8. We see that the null hypothesis of no subgroup is rejected, and there is a treatment favorable subgroup which is about 5% of the total patients. This is consistent with the result in Yuan et al. (2020). This case is of particular interest for hypothesis generating for developmental therapeutics. We can examine the small group of patients who are not benefiting from the treatment and identify underlying reasons and study them.

Table 11.7 Parameter estimates under two models (scaled real data)

θ	β	η	λ
Partial model	(0.073,0.0421,−0.105)	2.986	0.009
[se]	[0.042,0.041,0.044]	[0.610]	[0.031]
Linear model	(0.083,0.053,−0.123)	3.106	0.010
[se]	[0.042,0.041,0.043]	[0.585]	[0.004]

Table 11.8 Classification results (under scaled real data)

$\hat{\eta}$	Decision	$K(0.05)$	Group1-percent
2.986	H_1	0.349	0.052

11.5 Conclusion

A partial linear model is proposed for the analysis of subgroups in clinical trial, for the case one of the covariate has monotone non-linear effect on the response. The non-linear part is modeled by a monotone function along with the linear part of other covariates. The semiparametric maximum likelihood is used to estimate model parameters. Simulation study is conducted to evaluate the performance of the proposed method, and results show that the proposed model perform much better than linear models especially when treatment effect is relatively small. Then the model is applied to analyze a real data.

References

Balan TA, Putter H (2019) frailtyEM: an R package for estimating Semipaarametric shared frailty models. J Stat Softw 90(7)

Bertsekas DP (2016) Nonlinear programming, 3rd edn. Athena Scientific, Nashua

Best MJ, Chakravarti N (1990) Active set algorithms for isotonic regression; a unifying framework. Math Program 47:425–439

Bordes L, Chauveau D, Vandekerknove P (2007) A stochastic EM algorithm for a semiparametric mixture model. Comput Stat Data Anal 51:5429–5443

Campbell G (1981) Nonparametric bivariate estimation with ranodmly censored data. Biometrica 68:417–422

Chen J, Zhang D, Davidian M (2002) A Monte Carlo EM algorithm for generalized linear mixed models with ïňĆexible random effects distribution. Biometrics 3(3):347–360

Dempster AP, Laird NM, Rubin DB (1977) Maximum likelihood from incomplete data via the EM algorithm, J R Stat Soc Ser B 39:1–38

Fan A, Song R, Lu W (2017) Change-plane analysis for subgroup detection and sample size calculation. J Am Stat Assoc 112:769–778

Fokkema M, Smits N, Zeileis A et al (2018) Detecting treatment-subgroup interactions in clustered data with generalized linear mixed- effects model trees. Behav Res Methods 50:2016–2034

Foster JC, Taylor JMC, Ruberg SJ (2011) Subgroup identification from randomized clinical trial data. Stat Med 30:2867–2880

Friede T, Parsons N, Stallard N (2012) A conditional error function approach for subgroup selection in adaptive clinical trials. Stat Med 31:4309–4320

Groeneboom P, Wellner J (1992) Information bounds and nonparametric maximum likelihood estimation. Birkháuser Verlag, Basel

Hanley JA, Parnes MN (1983) Nonparametric estimation of a multivariate distribution in the presence of censoring. Biometrics 39:129–139

International Breast Cancer Study Group (IBCSG) (2002) Endocrine responsiveness and tailoring adjuvant therapy for postmenopausal lymph node-negative breast cancer: a randomized trial. J Natl Cancer Inst 94:1054–1065

Lipkovich I, Dmitrienko A, Denne J, Enas G (2011) Subgroup identification based on differential effect search (SIDES)—A recursive partitioning method for establishing response to treatment in patient sub-populations. Stat Med 30:2601–2621

Ma S, Huang J (2017) A concave pairwise fusion approach to subgroup analysis. J Am Stat Assoc 112:410–423

Muñoz A (1980) Nonparametric estimation from censored bivariate observations. Technical Report, Department of Statistics, Stanford University

Patrick M, Kurt H, Jan DL (2009) Isotonic optimization in R: pool-adjacent-violators algorithm (PAVA) and active set methods. J Stat Softw 32(5):1–24

Rothmann MD, Zhang J, Lu L, Fleming TR (2012) Testing in a pre-specified subgroup and the intent-to-treat population. Drug Inf J 46(2):175–179

Ruberg SJ, Chen L, Wang Y (2010) The mean doesn't mean as much any more: finding sub-groups for tailored therapeutics. Clin Trials 7:574–583

Sabine C (2005) AIDS events among individuals initiating HAART: do some patients experience a greater benefit from HAART than others? AIDS 19:1995–2000

Shen J, He X (2015) Inference for subgroup analysis with a structured logistic-normal mixture model. J Am Stat Assoc 110:303–312

Song Y, Chi GY (2007) A method for testing a pre-specified subgroup in clinical trials. Stat Med 26:3535–3549

Tan M, Tian G-L, Ng KW (2009) Bayesian missing data problems: EM, data augmentation and non-iterative computation. Chapman and Hall/CRC, London/Boca Raton

Yuan A, Chen X, Zhou Y, Tan MT (2018) Subgroup analysis with semiparametric models toward precision medicine. Stat Med 37(2):1830–1845

Yuan A, Zhou Y, Tan MT (2020) Subgroup analysis with a nonparametric unimodal symmetric error distribution. Comm Statist Theory Methods. Published online

Zhou Y, Yuan A, Tan MT (2019) Subgroup analysis with semiparametric partial linear regression model. Submitted to Statistical Methods in Medical Research

Chapter 12
Exploratory Subgroup Identification for Biopharmaceutical Development

Xin Huang, Yihua Gu, Yan Sun, and Ivan S. F. Chan

Abstract A major challenge in developing precision medicines is the identification and confirmation of patient subgroups where an investigational regimen has a positive benefit–risk balance. In biopharmaceutical development, exploring these patient subgroups of potential interest is usually achieved by constructing decision rules (a signature) using single or multiple biomarkers in a data-driven fashion, accompanied by rigorous statistical performance evaluation to account for potential overfitting issues inherent in subgroup searching. This chapter provides a comprehensive review of general considerations in exploratory subgroup analysis, investigates popular statistical learning algorithms for biomarker signature development, and proposes statistical principles for subgroup performance assessment. An example of subgroup identification for an immunology disease treatment leading to regulatory label inclusion will be provided.

12.1 Introduction

Patients may have different prognoses when experiencing the same disease and respond differently to the same treatment regimen, due to the heterogeneity of the biological system and its interaction with the environment. The use of biomarkers to understand the cause of this heterogeneity and to identify subgroups of patients with similar disease prognosis and treatment response is the key to the success of modern biopharmaceutical development. A biomarker, as defined by the FDA-NIH Biomarker Working Group, is "a characteristic that is objectively measured and evaluated as an indicator of normal biologic processes, pathologic processes, or biological responses to a therapeutic intervention"(Group 2016). The term *biomarker*, in clinical use for defining patient subgroups, refers to a broad range of markers which can have demographic, physiologic, molecular, histologic or

X. Huang · Y. Gu · Y. Sun · I. S. F. Chan (✉)
Data and Statistical Sciences, AbbVie, North Chicago, IL, USA
e-mail: ivan.chan@abbvie.com

© Springer Nature Switzerland AG 2020
N. Ting et al. (eds.), *Design and Analysis of Subgroups with Biopharmaceutical Applications*, Emerging Topics in Statistics and Biostatistics,
https://doi.org/10.1007/978-3-030-40105-4_12

radiographic characteristics or measurements that are thought to be related to some aspect of normal or abnormal biological functions or processes. A biomarker signature is a combination of one or more biomarkers that are measured at baseline or at an early disease progression/treatment time point and can predict an outcome of clinical interest via an empirical model or rule. From the perspective of subgroup identification in clinical usage, biomarkers signatures can be classified into two categories: prognostic biomarker signatures and predictive biomarker signatures, with the recognition that some biomarker signatures may fall into both categories. A prognostic biomarker signature focuses on patient risk classification, which is used to identify the likelihood of a clinical event, disease recurrence or progression in patients who have the disease or medical condition of interest, and thus usually aids in the decision of which patient subgroup needs an intensive treatment as opposed to no treatment or standard therapy. Examples of prognostic biomarker signatures are Breast Cancer genes 1 and 2 (BRCA1/2) mutations for assessing the likelihood of a second breast cancer (Basu et al. 2015); Oncotype Dx Breast Cancer Assay, measuring 21 genes to predict breast cancer recurrence in women with node negative or node positive, ER-positive, HER2-nagative invasive breast cancer (Mamounas et al. 2010; Paik et al. 2004); and C-reactive protein (CRP) level as a prognostic biomarker to identify patients with unstable angina (Ferreiros et al. 1999). A predictive biomarker signature focuses on treatment selection, which is used to identify a subgroup of patients who are more likely than similar individuals without the biomarker signature to experience a favorable or unfavorable effect from exposure to a medical product. Examples of predictive biomarker signatures are patients with advanced NSCLC with high PD-L1 IHC expressions having better penbrolizumab efficacy (Garon et al. 2015); BRCA1/2 mutations to identify patients likely to respond to PARP inhibitors (Ledermann et al. 2012); and NSCLC patients with high tumor mutation burden (≥ 10 mutations per megabase) having promising efficacy after being treated with nivolumab plus ipilimumab versus chemotherapy (Hellmann et al. 2018). Due to the complexity of biological systems, the causal mechanism of relationship between the clinical outcomes and the biomarkers are usually unknown and must be deduced empirically from experimental data. We focus our discussion in this chapter on the retrospective statistical development of biomarker signatures that provide a clear binary stratification of patients (signature positive versus signature negative) for exploratory subgroup identification. In Sect. 12.2, we discuss how to find the optimal and stable cutoff for both prognostic and predictive cases when a candidate biomarker is available but the cutoff to distinguish the signature positive versus signature negative subpopulation is unknown. In Sect. 12.3, we consider situations where no single candidate biomarker to identify patient subgroups is available and a complex biomarker signature needs to be derived. For these settings, we provide an overview of scoring-based and rule-based methods for the signature development for both prognostic and predictive cases. In Sect. 12.4, we propose a framework of internal validation for signature performance assessment, and emphasize that the goal of exploratory subgroup identification is to develop a stable signature with accurate performance assessment (which needs to be validated in future studies), instead of performing hypothesis testing. In Sect. 12.5, we present a successful example of subgroup identification for an immunology disease treatment leading to regulatory label inclusion.

12.2 Single Biomarker Signature

It is common and often desirable during pharmaceutical development that a single biomarker emerges as a promising predictor to select a subgroup of patients for an outcome of interest. It simplifies the biological interpretation as well as the downstream assay development. Many existing successful examples are based on a single biomarker, such as 17p deletion for venetoclax in patients with relapsed or refractory chronic lymphocytic leukemia (Stilgenbauer et al. 2016) and PD-L1 IHC expressions for penbrolizumab in patients with advanced NSCLC (Garon et al. 2015). The clinical outcome of interest can be both efficacy or safety endpoints, and the corresponding biomarker can be prognostic or predictive, depending on the objective of the exploratory analysis and the experimental design. We will begin this section with the prognostic case, introducing various tools for evaluating and determining the optimal cutoff. We will then extend the discussion to the predictive case, illustrating the difference in concepts and methods. Last, we will propose a framework of searching for robust and stable cutoff for a single biomarker. Some of the discussions in this section, especially the general concepts, also apply to the later section of complex biomarker signature development.

12.2.1 Prognostic Biomarker Signature Analysis

In this section, we consider the scenario where a single prognostic biomarker is used to identify a subgroup of patients that are more likely to experience an outcome under a given disease or treatment condition. Intuitively, this type of signature is useful for identifying a subgroup of patients who are more likely to progress in disease or develop an adverse event after treatment. In early phases when there is only a single arm, however, a signature can also be developed under the prognostic setting to generate a hypothesis for a certain patient subgroup who may respond better to the investigational new drug (usually combined with the support from biological interpretation and external/public data validation); the signature can be subsequently used for predictive purposes in later confirmative trials.

Many outcomes of interest during drug development are binary (e.g., 0 for event absence and 1 for event presence); and a cutoff on a biomarker is often needed to select a subgroup with significantly different event rate. This is often done through comparing the performance of different candidate cutoffs. Some common performance evaluation metrics for binary outcome can be shown in Table 12.1.

Table 12.1 Decision matrix for binary classifier

	Event present	Event absent		
Biomarker positive	a (true positive)	b (false positive, type I error)	Prevalence $= (a + c)/(a + b + c + d)$	Accuracy (ACC) $= (a + d)/(a + b + c + d)$
Biomarker negative	c (false negative, type II error)	d (true negative)	Positive predictive value (PPV), precision $= a/(a + b)$	False discovery rate (FDR) $= b/(a + b)$
	True positive rate (TPR), recall, sensitivity $= a/(a + c)$	False positive rate (FPR), fall-out $= b/(b + d)$	False omission rate (FOR) $= c/(c + d)$	Negative predictive value (NPV) $= d/(c + d)$
	False negative rate (FNR) $= c/(a + c)$	True negative rate (TNR), specificity $= d/(b + d)$	Positive likelihood ratio (LHR+) $=$ TPR/FPR	Negative likelihood ratio (LHR−) $=$ FNR/TNR
			Diagnostic odds ratio (DOR) $=$ LHR+/LHR−	F1 score $=$ $2 * ((\text{precision} * \text{recall})/(\text{precision} + \text{recall}))$

These metrics measure the cutoff performance from different perspectives, although some are more frequently used than others. A cutoff can then be selected through optimizing some of these metrics. Meanwhile, there are many graphical tools for evaluating the discriminatory accuracy of a continuous biomarker. The receiver operating characteristic (ROC) curve is the most commonly used method, which utilizes sensitivity and specificity to visualize and facilitate the cutoff selection. An ROC curve is created by plotting the sensitivity against the (1 − specificity) using empirical or fitted values at a series of ordered cutoffs as shown in Fig. 12.1.

The ROC curve describes the discriminatory accuracy of a biomarker, with the 45-degree diagonal line equivalent to random guessing. ROC-based cutoff determination is an important technique, and it has been widely used for not only subgroup identification but also assay and diagnostic test development. It is simple, intuitive, and easy to interpret. Common methods to determine optimal cutoff through ROC curves are listed below:

- Maximize the Youden index (sensitivity + specificity − 1) (Youden 1950), which corresponds to the vertical distance between a point on the curve and the 45 degree line.
- Minimize the Euclidean distance between a point on the curve and the ideal point (sensitivity = specificity = 1).
- Maximize the product of sensitivity and specificity (Liu 2012), which corresponds to the rectangular area under the ROC curve for a given point.

There are also methods that take into account costs (Cantor et al. 1999; McNeil et al. 1975; Metz 1978; Zweig and Campbell 1993). However, they are rarely used in the exploratory analysis during pharmaceutical development because it is often difficult to estimate the respective costs and prevalence.

The decision matrix in Table 12.1 essentially focuses on evaluating the concordance between the dichotomized biomarker and the binary outcome. Alternatively, one can use a model-based approach to select the best cutoff through optimization

Fig. 12.1 Receiver operating characteristic curve

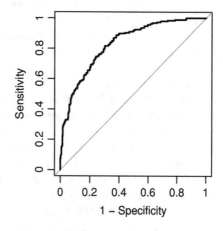

procedures. Consider a supervised learning problem with data (x_i, y_i), $i = 1, 2, \ldots,$ n, where x_i is the single biomarker variable and y_i is the binary response/outcome variable for the ith patient. We assume that the observed data are independently and identically distributed copies of (X, Y), and an appropriate model (e.g., logistic regression) incorporating a candidate cutoff can be built based on the data. Denote the observed log likelihood by $\sum_{i=1}^{n} \ell \{\eta(x_i), y_i\}$, where $\eta(x_i)$ is a function of the single biomarker that incorporates the cutoff. The following working model can be used for the development of a cutoff-based prognostic signature,

$$\eta(x) = \alpha + \beta \cdot \omega(x), \tag{12.1}$$

where $\omega(x)$ is the subgroup indicator, with 1 and 0 representing signature positive and negative subgroups respectively, so we have

$$\omega(x) = I(s \cdot x \geq s \cdot c), \tag{12.2}$$

where c is a candidate cutoff on the single biomarker x and $s = \pm 1$ indicates the direction of the cutoff. The best $\omega(x)$ along with the cutoff and the direction can then be decided by searching for the optimal cutoff via testing $\beta = 0$ based on score test statistics.

One of the major advantages of the model-based cutoff determination method is its flexibility of handling different types of endpoints and its potential to adjust for other covariates. Continuous and time-to-event outcomes are quite common in drug development. For example, in oncology clinical trials, we often have best tumor size change from baseline and survival time in addition to binary response. The aforementioned model-based method can be easily extended to derive cutoffs for the continuous and time-to-event endpoints by adopting a different model. For instance, we can use a linear regression model for the continuous outcome and a Cox regression model for the time-to-event outcome. The adjustment for potential covariates is also straightforward. We can simply add the covariates to the model in addition to $\alpha + \beta \cdot \omega(x)$, and search for the optimal cutoff by testing the statistical significance of $\beta = 0$ for $\omega(x)$ based on score test statistics.

It is important to have a comprehensive performance evaluation once a cutoff has been selected. While sensitivity, specificity and p-values for comparing subgroups are often used to derive the optimal cutoff, other metrics may also be important to gain a comprehensive assessment of performance of the final signature. For example, PPV and NPV are relevant for binary outcomes if we want to estimate the event rate within the selected and unselected subgroup. For continuous endpoints, we may want to conduct some robust non-parametric testing (e.g., rank or sign test) when the distribution of the biomarker within the subgroups is not normally distributed. For time-to-event endpoints, we may want to calculate hazard ratio or median/restricted mean survival time (Tian et al. 2014) for each subgroup. In addition to various tests and statistics for performance evaluation, it is also beneficial to utilize visualization tools to form a big picture of how the subgroup performance changes with a moving cutoff. The ROC curve discussed above is one such example,

Fig. 12.2 GAM plot on outcome versus biomarker cutoff

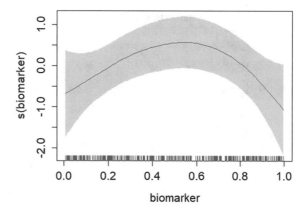

but it is not presented in the scale of the cutoff. An alternative option is to generate a plot of fitted curve from a generalized additive model (GAM) using the outcome and the biomarker data, as illustrated in Fig. 12.2.

The GAM plot shows the impact of a single biomarker on the outcome. Depending on the type of the outcome, the y-axis could represent the log odds ratio, log hazard ratio, or the difference compared to the population mean. Taking the continuous case as an example, the curve represents the difference between the outcome of a subject having a particular biomarker value on the x-axis and the mean outcome of the population. Such a GAM plot could be generated to provide additional insight for cutoff selection.

12.2.2 Predictive Biomarker Signature Analysis

A single predictive biomarker is used to identify a subgroup of patients who are more likely to experience a treatment difference given two different treatments. For example, we may want to identify a subpopulation that has a larger improvement in response rate over the standard care compared to the rest of the population when taking an investigational new drug. It is worth noting that, in the prognostic case, the focus on the outcome itself; while in the predictive case, the focus is on the difference of the outcome between the two treatment arms as illustrated in Fig. 12.3 (sig+ group corresponds to the selected subgroup using the single biomarker).

The prognostic and predictive cases are very different in nature, but the afore-mentioned model-based method can be easily modified for predictive biomarker cutoff derivation as in Eq. (12.3).

$$\eta(x) = \alpha + \beta \cdot [\omega(x) \times t] + \gamma \cdot t, \tag{12.3}$$

where t is the treatment indicator (e.g., 1 for treated and 0 for untreated subjects) and $\omega(x)$ follows the same definition as in Eq. (12.2). Similarly, the best $\omega(x)$ along

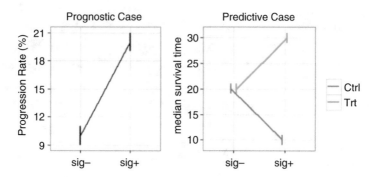

Fig. 12.3 Interaction plots for prognostic and predictive signature

with the cutoff and the direction can then be decided by searching for the optimal cutoff via testing $\beta = 0$ in (12.3) based on score test statistics. The above working model can be further modified to include the prognostic effect of the biomarker and other covariates by adding the corresponding terms into the model.

Subgroup performance needs to be further evaluated after a cutoff has been selected for the predictive biomarker. Similar to the prognostic case, various tests and statistics can be calculated. For example, it may be of interest to calculate the p-value of the treatment effect within each of the two subgroups, the p-value of the subgroup effect within each of the two treatment arms, or some other summary statistics within each subgroup. A GAM plot can also be generated to demonstrate the relationship between the outcome difference of the two arms and the single biomarker. Depending on the type of outcome, the treatment difference could be the log odds ratio, log hazard ratio, or the difference between the two treatments for subjects having a particular biomarker value on the x-axis. Alternatively, a so-called subpopulation treatment effect pattern plot (STEPP) (Bonetti and Gelber 2004) can be generated using a sliding window over the ordered values of the biomarker. Briefly speaking, to generate the STEPP plot, one needs to first choose a window size (sample size) and calculate a summary statistic within the window. Then a sliding window will be used along the ordered values of the biomarker at a fixed pace to generate the same summary statistic at each stop. The STEPP plot is simply the connected dots, with x values equivalent to the median of the biomarker values in the window, and y values equivalent to the corresponding summary statistic. The advantage of STEPP over GAM is that it does not rely on a specific model, and it can provide "model free" summary statistics within each window. The disadvantages of STEPP are also obvious: a sufficiently large window size is needed for a reliable estimate within each window; the total sample size is also needed to be large enough so that the window can actually slide at a meaningful pace for a meaningful distance; the visualization is affected by varying the window size or the sliding pace.

12.2.3   A Framework for Robust Cutoff Derivation

Due to the data variability and quality, a small perturbation in the dataset may result in a different cutoff selection, especially when the sample size is small. In this section, we introduce a general framework for the cutoff selection process, called Bootstrapping and Aggregating of Thresholds from Trees (BATTing) (Huang et al. 2017). The motivation of BATTing is that a single cutoff built on the original dataset may be unstable and not robust enough against small perturbations in the data, and prone to be over-fitting, thus resulting in lower prediction (stratification) power. We note that the idea of BATTing is closely related to Breiman's bagging method (Breiman 1996) for generating multiple versions of a predictor via bootstrapping and using these to get an aggregated predictor. We summarize the BATTing algorithm below:

BATTing procedure:

Step 1. Draw B bootstrap datasets from the original dataset.

Step 2. Build a single cutoff on the biomarker for each of these B datasets using appropriate cutoff derivation methods.

Step 3. Examine the distribution/spread of the B cutoffs, and use a robust estimate (e.g., median) of this distribution as the selected cutoff (BATTing cutoff estimate).

A simple simulation was performed with different sample sizes and effect sizes to demonstrate the advantage of BATTing in terms of the robustness of cutoff estimation. Specifically, in the simulation setting, data were generated from the predictive case following Eq. (12.3) x follows a normal distribution with mean $= 0$ and variance $= 2$; β is set according to the designated effect size; $\gamma = -\beta/2$; and subjects are randomized in a 1:1 ratio to receive treatment or placebo.

The benefit of BATTing on the threshold estimation under different scenarios (with different sample sizes and effect sizes) was investigated. Figure 12.4 shows the distribution of BATTing threshold estimates from 500 simulation runs across different numbers of bootstrapping for sample size $= 100$ and effect size $= 0.2$, with true optimal cutoff being 0 (red dashed vertical line). It shows that BATTing helps reduce the influence of data perturbations in the dataset and thus stabilizes the threshold estimate. Figure 12.5 shows the inter-quartile range of threshold estimates from 500 simulations runs across different effect sizes when $n = 100$ (top panel) and across different sample sizes when effect size is 0.2 (bottom panel). These plots also demonstrate that the BATTing procedure helps reduce the variation and stabilizes the threshold estimate. To further evaluate the advantage of using the more "stable" threshold in terms of the accuracy in identifying subgroups of patients of interest, a simulation was performed to compare the accuracy of identified subgroup labels via the BATTing procedure with 50 bootstraps versus a stub without bootstrap. The simulation result shows that the median accuracy is 85% versus 76% with

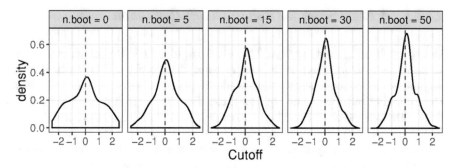

Fig. 12.4 Simulation comparison on BATTing threshold distribution (sample size = 100, effect size = 0.2; "n.boot" refers to the number of bootstraps)

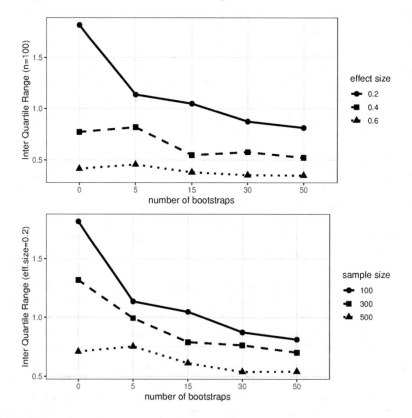

Fig. 12.5 Interval quartile ranges on the distribution of threshold estimates

and without BATTing, which represents a 12% improvement. As a rule of thumb, the number of bootstrap samples ≥50 is adequate and recommended in practice; and fewer bootstraps may be adequate as the sample size increases and for larger effect size.

12.3 Complex Biomarker Signature

With multiple candidate biomarkers available, one can develop statistical learning algorithms to combine multiple biomarkers into a single signature (decision rule). There are generally two types of signatures: (1) scoring-based methods first project the biomarkers from multiple dimensional spaces into a single composite score for each patient; then algorithms described in Sect. 12.2 can be applied to derive an optimal cutoff of this composite score in order to define the signature positive group; (2) rule-based algorithms directly define a (AND/OR) logic combination of multiple biomarkers and their cutoffs to define the signature positive subgroup.

The model-based approach introduced in Sect. 12.2 can be generalized to complex biomarker signature development. Consider a supervised learning problem with data (X_i, y_i), $i = 1, 2, \ldots, n$, where X_i is a p-dimensional vector of predictors and y_i is the response/outcome variable for the ith patient. We assume that the observed data are independently and identically distributed copies of (X, y). We consider three major applications: linear regression for continuous response, logistic regression for binary response, and Cox regression for time-to-event response, where $y_i = (T_i, \delta_i)$, T_i is a right censored survival time and δ_i is the censoring indicator. We denote the observed log likelihood or log partial likelihood by $\sum_{i=1}^{n} \ell \{\eta(X_i), y_i\}$, where $\eta(X_i)$ is a function of predictors. For example, $\eta(\cdot)$ may represent the mean response in simple linear regression, the log odds in logistic regression, or the log hazard ratio in the Cox proportional hazards model without intercept. We consider the following working model for the development of prognostic signatures (i.e., for identifying patient subgroups with favorable response, independent of the therapeutic),

$$\eta(X) = \alpha + \beta \cdot \omega(X) \tag{12.4}$$

Similarly, we consider the following working model for predictive signatures (i.e., for identifying patient subgroups with favorable response to a specific therapeutic),

$$\eta(X) = \alpha + \beta \cdot [\omega(X) \times t] + \gamma \cdot t, \tag{12.5}$$

where t is the treatment indicator, with 1 for treated and 0 for untreated subjects. In both models, $\omega(X)$ is the binary signature rule, with 1 and 0 for signature positive and negative subgroups respectively.

For scoring-based methods,

$$\omega(X) = I(s \cdot f(X) > s \cdot c), \tag{12.6}$$

where $f(\cdot)$ can be any function that projects multiple biomarkers into a single composite score, c is a candidate cutoff on the score, and $s = \pm 1$ indicates the direction of the cutoff.

For rule-based methods,

$$\omega(X) = \prod_j^m I\left(s_j X_j \geq s_j c_j\right), \tag{12.7}$$

where c_j is the cutoff on the jth selected marker X_j, $s_j = \pm 1$ indicates the direction of the binary cutoff for the selected marker, and m is the number of selected markers.

In this section, we will discuss several algorithms for constructing $\omega(X)$ with the objective of optimizing the statistical significance level for testing $\beta = 0$ in (1) or (2) based on the score test statistics:

$$S\{\omega(\cdot)\} = U\{\omega(\cdot)\}^2/V\{\omega(\cdot)\}, \tag{12.8}$$

where $U\{\omega(\cdot)\} = \sum_{i=1}^n \partial \ell \{\eta(X_i), y_i\}/\partial \beta$ and $V\{\omega(\cdot)\}$ is the corresponding inverse of the Fisher information marix under the null hypothesis with $\beta = 0$ (Tian and Tibshirani 2011). The specific form of this test statistic depends on the employed working model. For example,

$$U\{\omega(\cdot)\} = n^{-\frac{1}{2}} \sum_{i=1}^n t_i \hat{\omega}(X_i) \left\{ y_i - \frac{e^{\hat{\alpha}+\hat{\gamma} t_i}}{1 + e^{\hat{\alpha}+\hat{\gamma} t_i}} \right\} \tag{12.9}$$

can be used for estimating the predictive signature rule for binary responses, where $\hat{\alpha}$ and $\hat{\gamma}$ are consistent estimators for α and γ respectively, in the absence of the interaction term, and $\hat{\omega}(X)$ is the current estimator of the rule.

12.3.1 Scoring-Based Methods

When multiple candidate biomarkers are available, most of the statistical learning methods first create a scoring system $f(X)$ as a function of multiple biomarkers to estimate a subject-specific outcome (i.e., the estimated endpoints of interest in the prognostic case, and the estimated treatment differences in the predictive case). Based on this scoring system, a desired level (cutoff) c of endpoint measures (in a prognostic case) or treatment differences (in a predictive case) needs to be specified to obtain a subgroup of patients.

Many of the popular subgroup identification methods in the statistical literature belong to this category. In the prognostic case, for example, regression-based methods (GLM, LASSO, MARS, etc.), machine learning based methods (CART, GUIDE, Random Forest, SVM and Neural Network, etc.) are among those popular algorithms (Hastie et al. 2009) to construct the scoring system for predicting the outcome of interest, followed by algorithms of finding (or predefining) a cutoff as the second step in defining the subgroup of interest. In the predictive cases, modifications of the above algorithms that take into account the treatment interaction are developed for predicting the treatment difference. Examples include Interaction

Trees (Su et al. 2008, 2009), Bayesian approaches (Berger et al. 2014), Virtual Twins (Foster et al. 2011), Adaptive Index Model (AIM) (Tian and Tibshirani 2011), and outcome weighted learning related methods (Chen et al. 2017; Delmar et al. 2017; Zhao et al. 2012).

We will present the AIM approach in detail as below, in which $f(X)$ is in the form of additive signature score as the sum of individual binary rules,

$$f(X) = \sum_{j}^{m} I\left(s_j X_j \geq s_j c_j\right),$$

where m predictors are selected from a set of p candidates via a cross-validation procedure. The complete AIM procedure for the purpose of subgroup identification is shown below:

AIM procedure:

Step 1. Begin with $f^{(0)}(X) = 0$, $\Lambda = \{1, \ldots, p\}$, where p is the number of candidate predictors.

Step 2. For $j = 1, 2, \ldots, m$, update

$$f^{(j)}(X) \leftarrow f^{(j-1)}(X) + I\left(s_{h(j)} X_{h(j)} \geq s_{h(j)} c_{h(j)}\right) \text{ and}$$

$$\Lambda_j \leftarrow \Lambda_{j-1} \backslash h(j-1)$$

where index $h(j) \in \Lambda_j$ and $(c_{h(j)}, s_{h(j)})$ are selected to maximize $S\{f^{(j)}(X)\}$.

Step 3. BATTing is applied to the resulting AIM score $f(X) = \sum_{j}^{m} I\left(s_j X_j > s_j c_j\right)$ to construct the final signature rule in the form of $I(s_{AIM} f(X) \geq c_{AIM})$ with estimated direction s_{AIM} and cutoff c_{AIM}.

Due to the randomness of cross-validation, the optimal number of predictors m may be different in each implementation for the same dataset. In order to stabilize the variable selection process (i.e., estimation of m), we propose a Monte Carlo procedure that entails repeating the cross-validation multiple times, estimating the optimal number of predictors, m, each time, and using the median of the m s derived from each cross-validation run as the final optimal number of predictors. We performed simulations (results not reported here due to space limitations) to demonstrate how the proposed Monte Carlo procedure stabilized the estimation of m; and based on the simulation results we recommended the optimal number of Monte Carlo repetition to be 25 to 50.

Remark 12.1 We may use $t - \pi_t$ to replace the treatment indicator t in the working model and the corresponding algorithm such as AIM, where π_t is the proportion of treated patients in the study. In this way, the main effect becomes orthogonal to this "centered treatment indicator"; and thus the search for predictive signature based on the score test would not be confounded by the presence of a prognostic effect. On the other hand, when the prognostic and predictive signatures do share common components, the current algorithm may be more sensitive in estimating such predictive signature rules.

Remark 12.2 All scoring-based methods are two-step optimization procedures. To avoid bias due to over-optimism, one needs to use a nested cross-validation method that implements the complete two-step procedure to evaluate the model performance and select the best scoring system among all competing methods. We will discuss this procedure in detail in Sect. 12.4.

Remark 12.3 For the same reason, the visualization tools introduced in Sect. 12.2 cannot be directly applied to evaluate the relationship across all possible cutoffs and their corresponding performance. Instead, the nested cross-validation methods need to be adapted to generate similar visualization (Zhao et al. 2013).

12.3.2 Rule-Based Methods

In contrast to the two-step procedure implemented by scoring-based methods, rule-based methods aim for direct subgroup search, and usually result in simple decision rules for patient subgroup selection that are more interpretable and convenient in clinical practice. Some examples are RULE-Fit (Friedman and Popescu 2008), PRIM (Chen et al. 2015; Friedman and Fisher 1999), SIDES (Lipkovich and Dmitrienko 2014; Lipkovich et al. 2011), AIM-Rule (Huang et al. 2017), and Sequential-BATTing (Huang et al. 2017).

In this section, we describe in detail two rule-based methods for subgroup identification: (1) Sequential-BATTing, a multivariate extension of the Bootstrapping and Aggregating of Thresholds from Trees (BATTing), and (2) AIM-RULE, a multiplicative rules-based modification of the Adaptive Index Model (AIM). We present these subgroup identification methods under the aforementioned unified framework.

Both methods focus on multiplicative signature rules:

$$\omega(X) = \prod_j^m I\left(s_j X_j \geq s_j c_j\right),$$

where c_j is the cutoff on the jth selected marker X_j, $s_j = \pm 1$ indicates the direction of the binary cutoff for the selected marker, and m is the number of selected markers.

12.3.2.1 Sequential BATTing

Sequential BATTing is designed to derive a binary signature rule of (12.7) in a stepwise manner by extending the BATTing procedure. The resulting signature rule is a multiplicative of predictor-threshold pairs. The details of the algorithm are described in the following steps:

Sequential BATTing procedure

Step 1. $\omega^{(0)}(X) = 1$, $\Lambda = \{1, \ldots, p\}$, where p is the number of candidate predictors.
Step 2. For $j = 1, \cdots, m$, first find c_k and s_k for each X_k, $k \in \Lambda_j$ via BAT-Ting procedure described above and $X_{h(j)}$ is then selected to maximize $S\{\omega^{(j-1)}(X)I(s_h X_h \geq s_h c_h)\}$ with respect to h.

$$\omega^{(j)}(X) \leftarrow \omega^{(j-1)}(X) I\left(s_{h(j)} X_{h(j)} \geq s_{h(j)} c_{h(j)}\right) \text{ and}$$

$$\Lambda_j \leftarrow \Lambda_{j-1} \backslash h(j-1).$$

Step 3. The final signature rule $\omega(X) = \omega^{(j-1)}(X)$, if the likelihood of ratio test statistics of $\omega^{(j)}(X)$ vs $\omega^{(j-1)}(X)$ is not significant at a predefined level of α.

Note that the p-values from the likelihood ratio test in the stopping criteria do not have usual interpretation because of multiplicity inherent in subgroup search; and the α cutoff in the stopping criteria is served as a tuning parameter. Nevertheless, the choice of $\alpha = 0.05$ prevents premature termination of the algorithm and encourages inclusion of potentially informative markers.

12.3.2.2 AIM-Rule

Note that step 2 of the AIM algorithm implicitly ranked the selected predictors in terms of their contribution to the model. This order of the predictor importance, however, is not reflected in the additivity form of the AIM score $f(X)$, because it assigns equal weights to all m predictors. Intuitively, the AIM procedure is very efficient if all m predictors have relatively equal contributions to the model; otherwise, the same AIM score may not imply the same effects for different patients. Here, we proposed another variation of the AIM algorithm called AIM-Rule which uses multiplicative binary rules $\widetilde{f}(X) = \prod_j^{h(k)} I\left(s_j X_j \geq s_j c_j\right)$ as the final signature, where $k = 1, 2, \ldots, m$.

AIM-Rule procedures:

Step 1. Construct the AIM score $f(X) = \prod_j^m I\left(s_j X_j \geq s_j c_j\right)$. Without loss of generality, we assume that the relevant features enter the signature rule in the order of $X_1, X_2, \cdots, X_m$.
Step 2. Construct the final signature rule in the form of

$$\widetilde{f}(X) = \prod_{j=1}^{\widetilde{h}} I\left(s_j X_j \geq s_j c_j\right),$$

where the index $\widetilde{h}$ is selected via BATTing based on the ordered signature rules:

$$\left\{I\left(s_1 X_1 \geq s_1 c_1\right), I\left(s_1 X_1 \geq s_1 c_1\right) I\left(s_2 X_2 \geq s_2 c_2\right), \ldots, \prod_j^m I\left(s_j X_j \geq s_j c_j\right)\right\}.$$

12.4   Model Evaluation: Nested Cross-Validation

Based on the signature rule $\omega(X)$ derived from the above algorithms, patients are stratified into signature positive and signature negative groups. It should be noted that the resubstitution p-values for β associated with the final signature rules and the resubstitution summary measures determined from the same data used to derive the stratification signature may be severely biased because the data has already been explored for deriving the signature. Instead, we advocate determining the p-value of the signature via K-fold cross-validation (CV) and refer to such p-value as Predictive Significance (Chen et al. 2015; Huang et al. 2017). From this CV procedure, we can also estimate the effect size and related summary statistics, along with estimates of predictive/prognostic accuracy. For practical purposes, we recommend $K = 5$. We now describe the procedure for deriving the Predictive Significance and the associated summary measures. First, the dataset is randomly split into K subsets (folds). A signature rule is then derived from K-1 folds from one of the algorithms. This signature rule is then applied to the left-out fold, resulting in the assignment of a signature positive or signature negative label for each patient in this fold. This procedure is repeated for each of the other K-1 folds by leaving them out one at a time, resulting in a signature positive or signature negative label for each patient in the entire dataset. All signature positive and negative patients in the entire dataset are then analyzed together and a p-value for β is calculated; we refer to this p-value as CV p-value. Due to the variability in random splitting of the entire dataset, this K-fold CV procedure is repeated multiple times (e.g., 100 times), and the median of the CV p-values across these CV iterations is used as an estimate of the Predictive Significance. Note that the CV p-value preserves the error of falsely claiming a signature when there is no true signature, as demonstrated in the simulation section. Therefore it can be used to conclude that no signature is found if the effect of interest is greater than a pre-specified significance level (i.e., 0.05). In addition to p-values, we can use the same procedure to calculate the CV version of relevant summary statistics (e.g., response rate, median survival time, restricted mean survival time (Tian et al. 2014), sensitivity, specificity, etc.) and point estimates of the treatment effect in each subgroup (odds ratio, hazard ratios, etc.).

Note that this cross-validation procedure evaluates the predictive performance only after aggregating the predictions from all the left-out folds, which is an important difference compared to the more traditional/common approaches that evaluate the predictive performance of each fold separately. The proposed approach is in the same spirit of the pre-validation scheme proposed in Tibshirani and Efron (2002), as well as the cross-validated Kaplan-Meier curves proposed by Simon (2013) and Simon et al. (2011). The proposed cross-validation procedure preserves the sample size of the original training-set, which is particularly important for the subgroup identification algorithms where we evaluate the p-values for testing $\beta = 0$, and also for more reliable estimation of summary statistics and point estimates – this is especially critical when the training data set is not large, as is often the case in Phase-I and Phase-II clinical trials.

The Predictive Significance and related CV version of summary statistics calculated in the aforementioned process help reduce the bias of the resubstitution performance assessment of the final signature and provide a realistic estimation of predictive performance of the recovered signature derived from a method. It is of note that a signature performance estimate based on cross-validation would be of more practical use and more likely to be replicated in a future prospective trial, enriching the patient population based on such signature.

12.5 Optimizing Long-Term Treatment Strategy: An Example of Subgroup Identification Leading to Label Inclusion

On July 30, 2015, the European Medicines Agency approved the use of originator adalimumab (Humira®, AbbVie) 40 mg every-week dosing for the treatment of adults with active moderate to severe hidradenitis suppurativa (HS) who have failed to respond to conventional systemic HS treatments. Humira® was the first, and up to today the only, medication approved for HS in the European Union. The long-term treatment strategy included in the European Medicines Agency Summary of Produce Characteristics for Humira was the result of subgroup identification which provides patients the opportunity to benefit from this treatment beyond 12 weeks, the originally planned primary analysis point. The U.S. Food and Drug Admiration (FDA) approval later (September 10, 2015) did not limit the treatment duration to 12 weeks either.

The sections below summarize the background, methods, and the results of the subgroup analysis that led to the label inclusions of longer-term treatment strategy.

12.5.1 Motivation and Background

Evaluating the benefit–risk profile in a randomized withdrawal setting has become commonplace, required by regulatory agencies for the approval of long-term treatment in a chronic disease setting, and by payers for economic consideration. Various challenges were faced in the clinical development program of adalimumab in the treatment of hidradenitis suppurativa (HS), making a well-powered randomized withdrawal trial unfeasible.

HS is a serious, painful, systemic, chronic skin disease which may persist for decades (Jemec 2012; Revuz 2009; Shlyankevich et al. 2014). Inflammatory skin lesions, including abscesses, fistulas, and nodules, may exhibit purulent, malodorous drainage, and develop tunnels (sinus tracts) (Lipsker et al. 2016) and scarring as disease severity increases (Jemec 2012; Kurzen et al. 2008).

The adalimumab clinical development program included two double-blind, placebo-controlled pivotal studies; both were powered for the primary endpoint at the end of the initial 12-week double blind period. Per agreement with FDA, a subsequent 24-week randomized withdrawal period was included in each study as exploratory, and the outcome of this period would not impact the approvability, for the following reasons:

1. Adalimumab is a fully human, IgG1 monoclonal antibody specific for TNF-α. While immunologic abnormalities have been hypothesized to have a causal role in the disease (Jemec et al. 1996) as significant elevations in levels of the tumor necrosis factor α (TNF-α) have been detected in HS lesions (van der Zee et al. 2011), the unknown nature of responses made pre-specification of dosing strategy for long-term treatment extremely challenging. Randomized trials of other anti-TNF-α agents (infliximab (Grant et al. 2010) and etanercept (Adams et al. 2010)) for the treatment of HS had failed to show a significant benefit.
2. HS lesions may flare, resolve and recur in different body areas; therefore, a high degree of disease fluctuation was expected, and the categorization of long-term treatment was difficult.
3. A well-powered randomized withdrawal study requires a large study size, which was not practical because of the low prevalence of HS. Consequently, FDA has granted adalimumab Orphan Drug designation for the treatment of moderate to severe HS.

12.5.2   Method

PIONEER I and II are two pivotal studies similar in design and in enrollment criteria. Each study had two placebo-controlled, double-blind periods.

- Period A: patients were randomized 1:1 to adalimumab 40 mg weekly dosing (adalimumab weekly dosing) or placebo. The primary endpoint was the proportion of patients achieving Hidradenitis Suppurativa Clinical Response (HiSCR) (Kimball et al. 2016b), which is defined as a $\geq 50\%$ reduction in inflammatory lesion count (sum of abscesses and inflammatory nodules, AN count), and no increase in abscesses or draining fistulas in HS when compared with baseline as a meaningful clinical endpoint for HS treatment.
- Period B: adalimumab-treated patients continuing to Period B were re-randomized at week 12 to adalimumab weekly dosing, adalimumab every-other-week dosing, or matching placebo in a 1:1:1 ratio; Week-12 HiSCR status was included as a stratification factor. Placebo patients were reassigned to adalimumab weekly dosing in PIONEER I or remained on placebo in PIONEER II. Patients who lost response or had worsening or absence of improvement in Period B (defined in Fig. 12.6) were allowed to enter the open label extension study (OLE). All patients were treated in a blinded fashion. Randomization and blinding details have been published (Kimball et al. 2016a).

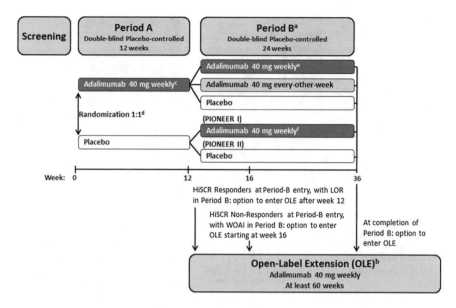

Fig. 12.6 Study design. [a]Week-12 HiSCR Responders through Period B to week 36 or until loss of response (loss of 50% of the AN count improvement gained between baseline and week 12), and Week-12 HiSCR Non-Responders continued Period B to at least week 26 (and up to week 36). [b]Patients could enter the multi-center, 60-week, phase-3 OLE trial (evaluated long-term safety, tolerability, and efficacy of adalimumab for patients with moderate-to-severe HS), if: (1) they completed Period B of their respective PIONEER trial, (2) achieved HiSCR at entry to Period B of their respective PIONEER trial and then experienced a loss of response (LOR), or (3) did not achieve HiSCR at the entry of Period B and then experienced worsening or absence of improvement (WOAI) (greater or equal to the baseline AN count on two consecutive visits after week 12, occurring at least 14 days apart). [c]Starting at week 4 after 160 mg (week 0), 80 mg (week 2). [d]Stratified by baseline Hurley Stage II versus III (PIONEER I & II) & baseline concomitant antibiotic use (PIONEER II). [e]Re-randomization for patients treated with adalimumab in Period A was stratified by Week-12 HiSCR status at entry into Period B, and by baseline Hurley Stage II versus III. [f]40 mg starting at week 16 after 160 mg (week 12), 80 mg (week 14). Abbreviations: HiSCR, Hidradenitis Suppurativa Clinical Response; AN, abscesses and inflammatory nodules; OLE, open-label extension; HS, hidradenitis suppurativa; LOR, loss of response; WOAI, worsening or absence of improvement

This subgroup identification utilized the integrated data from the two studies to ascertain the most clinically appropriate patient group receiving continuous adalimumab weekly dosing over the longer term versus adalimumab discontinuation. The analysis population comprised patients who were re-randomized to either continuation of adalimumab weekly dosing or withdrawal from adalimumab (placebo) in Period B after initial treatment of adalimumab weekly dosing for 12 weeks. The primary endpoint was the proportion of patients achieving HiSCR at the end of Period B. Safety profile was evaluated as well.

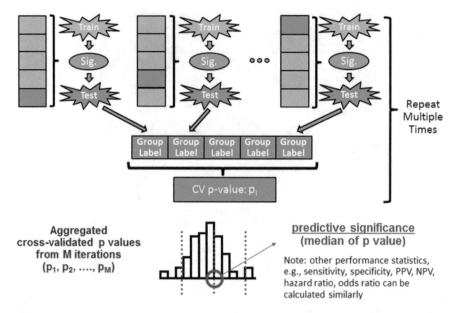

Fig. 12.7 Predictive significance by cross-validation (CV)

Several subgroup identification algorithms were implemented for this subgroup identification exercise, which included: Sequential-BATTing, AIM, AIM-Rule, PRIM, VG (Virtual Twins plus GUIDE) and SIDES.

Candidate variables included: baseline BMI and stratification factors (baseline Hurley Stage and continuation of baseline antibiotics), patients' response status at the end of Period A (HiSCR status, change and percent change from baseline in AN count).

The final subpopulation was proposed by comparing cross-validation performance of these candidate methods via a rigorous statistical framework, as introduced in Sect. 12.4 and shown in Fig. 12.7, which demonstrated that the subgroup identified from Sequential-BATTing is optimal.

The analysis of continued adalimumab weekly dosing vs withdrawal from adalimumab in each population was performed by CMH, adjusting for baseline Hurley Stage and Week-12 HiSCR status. Missing data (including early escape to OLE due to loss of responses or worsening of disease) were handled by non-responder imputation (NRI) (Fig. 12.8).

12.5.3 Result

A total of 199 patients (99 continued adalimumab weekly dosing, 100 withdrawal from adalimumab weekly dosing) in Period B were included for the subgroup

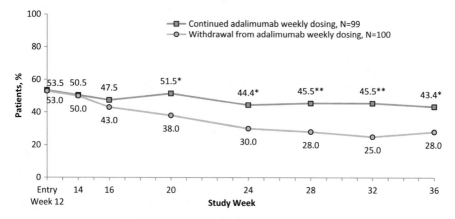

Fig. 12.8 Proportion of patients achieving HiSCR by visit (all patients). *, **, ***: statistically significant at 0.05, 0.01, and 0.005 level

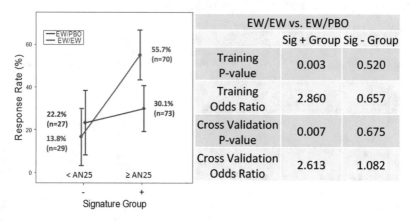

Fig. 12.9 Results from sequential BATTing method. EW/PBO: withdrawal from adalimumab weekly dosing; EW/EW: continuation with adalimumab weekly dosing; AN25: at least 25% reduction in total AN count

identification. The overall HiSCR rates are presented in (Gulliver et al. 2017). The identified signature-positive subgroup comprised patients achieving at least 25% reduction in AN count ($\geq$AN25) after the initial 12 weeks of treatment, named PRR population (*P*artial *R*esponders and HiSCR *R*esponders). The subgroup results are presented in Figs. 12.9, 12.10, and 12.11. The safety profile in the PRR population was similar to the overall population.

The results of these analyses were included in the EU Summary of Product Characteristics (SmPC), Canada Product Monograph, etc.

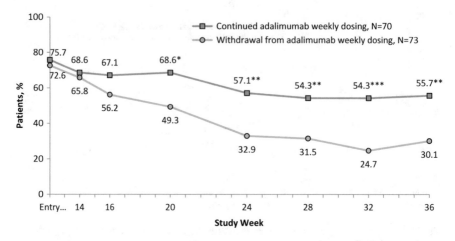

Fig. 12.10 Proportion of patients achieving HiSCR (signature-positive: PRR population). *, **, ***: statistically significant at 0.05, 0.01, and 0.005 level

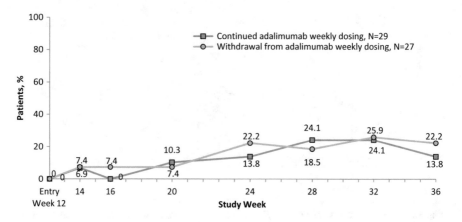

Fig. 12.11 Proportion of patients achieving HiSCR (signature-negative: non-PRR population)

EMA SmPC:

In patients with at least a partial response to Humira 40 mg weekly at Week 12, the HiSCR rate at Week 36 was higher in patients who continued weekly Humira than in patients in whom dosing frequency was reduced to every other week, or in whom treatment was withdrawn.

Canada Product Monograph:

In patients with at least a partial response ($\geq 25\%$ improvement in AN count) to HUMIRA 40 mg weekly at Week 12, the proportion of patients achieving HiSCR at Week 24 was 57.1% in HUMIRA 40 mg weekly, 51.4% in HUMIRA 40 mg every other week and 32.9% in the placebo group. The corresponding proportion at Week 36 was 55.7% in HUMIRA 40 mg weekly, 40.0% in HUMIRA 40 mg every other week and 30.1%.

The identified PRR population was also used in the cost-effectiveness modeling and provided a scientific rationale to payers, allowing the majority of the patients to receive long-term treatment of adalimumab weekly dosing.

12.6 Discussion

Subgroup identification strategies can be applied to various stages of drug development. The considerations when initiating exploratory subgroup identification activities include study design, clinical scenario, and practical considerations. The subgroup identification activities are not only applicable in searching for potential subgroups of treatment responders from a failed trial, but also applicable in identifying subgroups of patients who may experience different outcomes (efficacy, safety, etc.) in a successful trial to further enhance the optimal treatment strategies (dosage, treatment duration, etc.) with a more favorable benefit–risk profile.

There are many discussions in the statistical literature addressing the issue of controlling "type I error" for the subgroup identification exercise. It is worth noting that the retrospective subgroup identification is not a hypothesis testing strategy; instead, like all the statistical learning practices, it is aimed at developing a predictive classifier as accurate as possible, and using a rigorous internal validation paradigm to obtain an unbiased estimate of the statistic of interest (e.g., treatment effect in the signature positive group) resulting from the developed classifier. When exercising subgroup identifications, there is no "one-size-fits-all" method (it is usually difficult to predetermine which method outperforms others due to the difference in dataset structure under the specific problem); hence we recommend considering a variety of methods for identifying patient subgroups in any given dataset. When an independent validation/test dataset is not available, it is particularly important that the performance of the derived signatures from different algorithms needs to be evaluated with careful application of a cross-validation approach, such as the nested cross-validation described in Sect. 12.4 of this chapter. Finally, an independent validation/test dataset (preferably from a similar and proper design) would be ideal to confirm and validate the subgroup finding.

Due to the exploratory nature of the retrospective subgroup identification, one must carefully interpret the results according to the level of evidence generated (Simon et al. 2009), especially when the result is intended for regulatory approval and change of clinical practice. For this purpose, a successful subgroup identification exercise requires cross-disciplinary collaborations from the trial design to the interactions with regulatory authorities. For example, the study needs to be designed appropriately to allow the identification of subgroups. In the adalimumab HS example, with the consideration of the natural disease fluctuation and the unknown response time course, the clinical development program did not follow the traditional randomized withdrawal trial design which would have re-randomized only the initial HiSCR responders; the re-randomization in the HS program included all patients entering Period B, which made the subgroup identification possible. Last but not least, the success of convincing regulatory agencies, payers, and physicians often relies on supporting evidence beyond statistical modeling. For example, the candidate biomarkers feeding to the algorithms must be clinically meaningful; the subgroup signature needs to be biologically plausible and easy to identify and evaluate by physicians or patients; and the effect size needs to be clinically meaningful. In addition, the size of the signature positive subgroup also needs to be considered. The success of the Humira HS example was also influenced by the unmet medical need and impact of the adalimumab treatment.

References

Adams DR, Yankura JA, Fogelberg AC, Anderson BE (2010) Treatment of hidradenitis suppurativa with etanercept injection. Arch Dermatol 146(5):501–504. https://doi.org/10.1001/archdermatol.2010.72

Basu NN, Ingham S, Hodson J, Lalloo F, Bulman M, Howell A, Evans DG (2015) Risk of contralateral breast cancer in BRCA1 and BRCA2 mutation carriers: a 30-year semi-prospective analysis. Fam Cancer 14(4):531–538. https://doi.org/10.1007/s10689-015-9825-9

Berger JO, Wang X, Shen L (2014) A Bayesian approach to subgroup identification. J Biopharm Stat 24(1):110–129. https://doi.org/10.1080/10543406.2013.856026

Bonetti M, Gelber RD (2004) Patterns of treatment effects in subsets of patients in clinical trials. Biostatistics 5(3):465–481. https://doi.org/10.1093/biostatistics/5.3.465

Breiman L (1996) Bagging predictors. Mach Learn 24:123–140. https://doi.org/10.1007/BF00058655

Cantor SB, Sun CC, Tortolero-Luna G, Richards-Kortum R, Follen M (1999) A comparison of C/B ratios from studies using receiver operating characteristic curve analysis. J Clin Epidemiol 52(9):885–892

Chen G, Zhong H, Belousov A, Devanarayan V (2015) A PRIM approach to predictive-signature development for patient stratification. Stat Med 34(2):317–342. https://doi.org/10.1002/sim.6343

Chen S, Tian L, Cai T, Yu M (2017) A general statistical framework for subgroup identification and comparative treatment scoring. Biometrics 73(4):1199–1209. https://doi.org/10.1111/biom.12676

Delmar P, Irl C, Tian L (2017) Innovative methods for the identification of predictive biomarker signatures in oncology: application to bevacizumab. Contemp Clin Trials Commun 5:107–115. https://doi.org/10.1016/j.conctc.2017.01.007

Ferreiros ER, Boissonnet CP, Pizarro R, Merletti PF, Corrado G, Cagide A, Bazzino OO (1999) Independent prognostic value of elevated C-reactive protein in unstable angina. Circulation 100(19):1958–1963

Foster JC, Taylor JM, Ruberg SJ (2011) Subgroup identification from randomized clinical trial data. Stat Med 30(24):2867–2880. https://doi.org/10.1002/sim.4322

Friedman JH, Fisher NI (1999) Bump hunting in high-dimensional data. Stat Comput 9(2):123–143. https://doi.org/10.1023/A:1008894516817

Friedman JH, Popescu BE (2008) Predictive learning via rule ensembles. Ann Appl Stat 2(3):916–954. https://doi.org/10.1214/07-AOAS148

Garon EB, Rizvi NA, Hui R, Leighl N, Balmanoukian AS, Eder JP et al (2015) Pembrolizumab for the treatment of non-small-cell lung cancer. N Engl J Med 372(21):2018–2028. https://doi.org/10.1056/NEJMoa1501824

Grant A, Gonzalez T, Montgomery MO, Cardenas V, Kerdel FA (2010) Infliximab therapy for patients with moderate to severe hidradenitis suppurativa: a randomized, double-blind, placebo-controlled crossover trial. J Am Acad Dermatol 62(2):205–217. https://doi.org/10.1016/j.jaad.2009.06.050

Group F-N B W (2016) BEST (Biomarkers, EndpointS, and other Tools) Resource. Food and Drug Administration (US), Silver Spring

Gulliver W, Okun MM, Martorell A, Geng Z, Huang X, Tang Q, Gu Y (2017) Therapeutic response guided dosing strategy to optimize long-term adalimumab treatment in patients with hidradenitis suppurativa: integrated results from the PIONEER phase 3 trials. J Am Acad Dermatol 76(6):AB259. https://doi.org/10.1016/j.jaad.2017.04.1007

Hastie T, Tibshirani R, Friedman J (2009) The elements of statistical learning: data mining, inference, and prediction, 2nd edn. Springer-Verlag, New York

Hellmann MD, Ciuleanu TE, Pluzanski A, Lee JS, Otterson GA, Audigier-Valette C et al (2018) Nivolumab plus Ipilimumab in lung cancer with a high tumor mutational burden. N Engl J Med 378(22):2093–2104. https://doi.org/10.1056/NEJMoa1801946

Huang X, Sun Y, Trow P, Chatterjee S, Chakravartty A, Tian L, Devanarayan V (2017) Patient subgroup identification for clinical drug development. Stat Med 36(9):1414–1428. https://doi.org/10.1002/sim.7236

Jemec GB (2012) Clinical practice. Hidradenitis suppurativa. N Engl J Med 366(2):158–164. https://doi.org/10.1056/NEJMcp1014163

Jemec GB, Heidenheim M, Nielsen NH (1996) The prevalence of hidradenitis suppurativa and its potential precursor lesions. J Am Acad Dermatol 35(2 Pt 1):191–194

Kimball AB, Okun MM, Williams DA, Gottlieb AB, Papp KA, Zouboulis CC et al (2016a) Two Phase 3 trials of adalimumab for hidradenitis suppurativa. N Engl J Med 375(5):422–434. https://doi.org/10.1056/NEJMoa1504370

Kimball AB, Sobell JM, Zouboulis CC, Gu Y, Williams DA, Sundaram M et al (2016b) HiSCR (Hidradenitis Suppurativa Clinical Response): a novel clinical endpoint to evaluate therapeutic outcomes in patients with hidradenitis suppurativa from the placebo-controlled portion of a phase 2 adalimumab study. J Eur Acad Dermatol Venereol: JEADV 30(6):989–994. https://doi.org/10.1111/jdv.13216

Kurzen H, Kurokawa I, Jemec GB, Emtestam L, Sellheyer K, Giamarellos-Bourboulis EJ et al (2008) What causes hidradenitis suppurativa? Exp Dermatol 17(5):455–472. https://doi.org/10.1111/j.1600-0625.2008.00712_1.x

Ledermann J, Harter P, Gourley C, Friedlander M, Vergote I, Rustin G et al (2012) Olaparib maintenance therapy in platinum-sensitive relapsed ovarian cancer. N Engl J Med 366(15):1382–1392. https://doi.org/10.1056/NEJMoa1105535

Lipkovich I, Dmitrienko A (2014) Strategies for identifying predictive biomarkers and subgroups with enhanced treatment effect in clinical trials using SIDES. J Biopharm Stat 24(1):130–153. https://doi.org/10.1080/10543406.2013.856024

Lipkovich I, Dmitrienko A, Denne J, Enas G (2011) Subgroup identification based on differential effect search–a recursive partitioning method for establishing response to treatment in patient subpopulations. Stat Med 30(21):2601–2621. https://doi.org/10.1002/sim.4289

Lipsker D, Severac F, Freysz M, Sauleau E, Boer J, Emtestam L et al (2016) The ABC of hidradenitis suppurativa: a validated glossary on how to name lesions. Dermatology 232(2):137–142. https://doi.org/10.1159/000443878

Liu X (2012) Classification accuracy and cut point selection. Stat Med 31(23):2676–2686. https://doi.org/10.1002/sim.4509

Mamounas EP, Tang G, Fisher B, Paik S, Shak S, Costantino JP et al (2010) Association between the 21-gene recurrence score assay and risk of locoregional recurrence in node-negative, estrogen receptor-positive breast cancer: results from NSABP B-14 and NSABP B-20. J Clin Oncol 28(10):1677–1683. https://doi.org/10.1200/JCO.2009.23.7610

McNeil BJ, Keller E, Adelstein SJ (1975) Primer on certain elements of medical decision making. N Engl J Med 293(5):211–215. https://doi.org/10.1056/NEJM197507312930501

Metz CE (1978) Basic principles of ROC analysis. Semin Nucl Med 8(4):283–298

Paik S, Shak S, Tang G, Kim C, Baker J, Cronin M et al (2004) A multigene assay to predict recurrence of tamoxifen-treated, node-negative breast cancer. N Engl J Med 351(27):2817–2826. https://doi.org/10.1056/NEJMoa041588

Revuz J (2009) Hidradenitis suppurativa. J Eur Acad Dermatol Venereol: JEADV 23(9):985–998. https://doi.org/10.1111/j.1468-3083.2009.03356.x

Shlyankevich J, Chen AJ, Kim GE, Kimball AB (2014) Hidradenitis suppurativa is a systemic disease with substantial comorbidity burden: a chart-verified case-control analysis. J Am Acad Dermatol 71(6):1144–1150. https://doi.org/10.1016/j.jaad.2014.09.012

Simon RM (2013) Genomic clinical trials and predictive medicine, 1st edn. Cambridge University Press

Simon RM, Paik S, Hayes DF (2009) Use of archived specimens in evaluation of prognostic and predictive biomarkers. J Natl Cancer Inst 101(21):1446–1452. https://doi.org/10.1093/jnci/djp335

Simon RM, Subramanian J, Li M-C, Menezes S (2011) Using cross-validation to evaluate predictive accuracy of survival risk classifiers based on high-dimensional data. Brief Bioinform 12:203–214. https://doi.org/10.1093/bib/bbr001

Stilgenbauer S, Eichhorst B, Schetelig J, Coutre S, Seymour JF, Munir T et al (2016) Venetoclax in relapsed or refractory chronic lymphocytic leukaemia with 17p deletion: a multicentre, open-label, phase 2 study. Lancet Oncol 17(6):768–778. https://doi.org/10.1016/S1470-2045(16)30019-5

Su X, Zhou T, Yan X, Fan J, Yang S (2008) Interaction trees with censored survival data. Int J Biostat 4(1). https://doi.org/10.2202/1557-4679.1071

Su X, Tsai C-L, Wang H, Nickerson DM, Li B (2009) Subgroup analysis via recursive partitioning. J Mach Learn Res 10:141–158

Tian L, Tibshirani R (2011) Adaptive index models for marker-based risk stratification. Biostatistics 12:68–86. https://doi.org/10.1093/biostatistics/kxq047

Tian L, Zhao L, Wei LJ (2014) Predicting the restricted mean event time with the subject's baseline covariates in survival analysis. Biostatistics (Oxford, England) 15:222–233. https://doi.org/10.1093/biostatistics/kxt050

Tibshirani R, Efron B (2002) Pre-validation and inference in microarrays. Stat Appl Genet Mol Biol 1. https://doi.org/10.2202/1544-6115.1000

van der Zee HH, de Ruiter L, van den Broecke DG, Dik WA, Laman JD, Prens EP (2011) Elevated levels of tumour necrosis factor (TNF)-alpha, interleukin (IL)-1beta and IL-10 in hidradenitis suppurativa skin: a rationale for targeting TNF-alpha and IL-1beta. Br J Dermatol 164(6):1292–1298. https://doi.org/10.1111/j.1365-2133.2011.10254.x

Youden WJ (1950) Index for rating diagnostic tests. Cancer 3(1):32–35

Zhao Y, Zeng D, Rush AJ, Kosorok MR (2012) Estimating individualized treatment rules using outcome weighted learning. J Am Stat Assoc 107(449):1106–1118. https://doi.org/10.1080/01621459.2012.695674

Zhao L, Tian L, Cai T, Claggett B, Wei LJ (2013) Effectively selecting a target population for a future comparative study. J Am Stat Assoc 108(502):527–539. https://doi.org/10.1080/01621459.2013.770705

Zweig MH, Campbell G (1993) Receiver-operating characteristic (ROC) plots: a fundamental evaluation tool in clinical medicine. Clin Chem 39(4):561–577

Chapter 13
Statistical Learning Methods for Optimizing Dynamic Treatment Regimes in Subgroup Identification

Yuan Chen, Ying Liu, Donglin Zeng, and Yuanjia Wang

Abstract Many statistical learning methods have been developed to optimize multistage dynamic treatment regimes (DTRs) and identify subgroups that most benefit from DTRs using data from sequential multiple assignment randomized trials (SMARTs) and for observational studies. These methods include regression-based Q-learning and classification-based outcome-weighted learning. For the latter, a variety of loss functions can be considered for classification, such as hinge loss, ramp loss, binomial deviance loss, and squared loss. Furthermore, data augmentation can be used to further improve learning performance. In this chapter, we describe the development of an R-package, namely "DTRlearn2", to incorporate the methods from Q-learning and outcome-weighted learning to be widely used for medical clinical trials and public health observational studies. We illustrate the new R package via application to data from a 2-stage ADHD study. We compare the performance of different learning methods that are obtained from the package. The analysis reveals that children with medication history can benefit from starting treatment of behavioral modification.

Y. Chen
Department of Biostatistics, Mailman School of Public Health, Columbia University, New York, NY, USA
e-mail: yc3281@cumc.columbia.edu

Y. Liu
Department of Psychiatry, Columbia University Irving Medical Center, New York, NY, USA
e-mail: ying.liu@nyspi.columbia.edu

D. Zeng (✉)
Department of Biostatistics, Gillings School of Global Public Health, University of North Carolina at Chapel Hill, Chapel Hill, NC, USA
e-mail: dzeng@email.unc.edu

Y. Wang
Department of Biostatistics, Mailman School of Public Health, Columbia University, New York, NY, USA

Department of Psychiatry, Columbia University, New York, NY, USA
e-mail: yw2016@cumc.columbia.edu

© Springer Nature Switzerland AG 2020
N. Ting et al. (eds.), *Design and Analysis of Subgroups with Biopharmaceutical Applications*, Emerging Topics in Statistics and Biostatistics,
https://doi.org/10.1007/978-3-030-40105-4_13

13.1   Introduction

Heterogeneous treatment responses are commonly observed in patients. Thus, a universal treatment strategy may not be ideal, and tailored treatments adapting to individual characteristics could improve treatment response rates. Dynamic treatment regimes (DTRs, Lavori and Dawson 2000), also known as adaptive treatment strategies (Lavori and Dawson 2000), multi-stage treatment strategies (Thall et al. 2002; Thall and Wathen 2005), or treatment policies (Lunceford et al. 2002; Wahed and Tsiatis 2006), adaptively assign treatments based on patient's intermediate responses and health history over time. DTRs provide useful tools for managing patient's long-term health conditions and for facilitating personalized medicine.

DTRs can be inferred from data collected in Sequential Multiple Assignment Randomization Trials (SMARTs), in which randomization is implemented at each treatment stage. Recently, there have been numerous methods developed to estimate optimal DTRs using SMARTs. In particular, Q-learning, first proposed by Watkins (1989) and Qian and Murphy (2011), models the conditional expectation of the so-called Q-functions in order to infer the best treatment at each stage. Alternatively, A-learning (Murphy 2003) and G-computation (Lavori and Dawson 2004; Moodie et al. 2007) models the contrasts of the Q-functions between two treatments.

Without modeling Q-function but making an analogy between learning optimal DTRs and weighted classification, outcome-weighted learning (O-learning, Zhao et al. 2012, 2015) were developed to estimate DTRs through support vector machines (SVMs). Later on, Liu et al. (2014, 2018) proposed an augmented version of outcome-weighted learning to further improve the efficiency based on fitted Q-functions and derived residuals after removing the average treatment effects. Other non-parametric and machine learning methods exist, for example tree-based methods (Su et al. 2009; Foster et al. 2011; Laber and Zhao 2015; Qiu and Wang 2019) to identify subgroups of people who can benefit more from a certain treatment.

In this paper, we develop software for Q-learning and a variety of outcome-weighted learning methods for DTR estimation using data from SMARTs. We also discuss extensions to handle observational studies. We develop a comprehensive R package, namely "DTRlearn2", to implement these methods. For outcome-weighted learning, we include different loss functions such as hinge loss, ramp loss, binomial deviance loss, and squared loss. The package is illustrated via implementation to both simulated and real data. We also demonstrate how the learned DTRs can be applied to future patients.

13.2   Conceptual Framework

We consider estimating K-stage DTRs from a K-stage SMART. For $k = 1, 2, \ldots, K$, we denote A_k as the assigned treatment at stage k taking values in $\{-1, 1\}$, and R_k as a reward outcome variable (e.g., treatment response measured by reduction of symptoms) observed post the kth stage treatment. Without loss of generality, we assume a higher reward is more desirable. Let X_k be a vector of subject-specific variables collected at stage k just before treatment. A patient's health history information (feature variables) prior to treatment at stage k, denoted by H_k, can be defined recursively as $H_k = (H_{k-1}, A_{k-1}, R_{k-1}, X_k)$ with $H_1 = X_1$. Note that H_k can also be constructed flexibly to allow interaction terms, for example, $H_k = (H_{k-1}, A_{k-1}, R_{k-1}, X_k, H_{k-1}A_{k-1}, R_{k-1}A_{k-1}, X_kA_{k-1})$.

A DTR is a sequence of decision functions, $\mathscr{D} = (\mathscr{D}_1, \mathscr{D}_2, \ldots, \mathscr{D}_K)$, where D_k maps H_k to the domain of A_k, i.e., $\{-1, 1\}$. Our goal is to identify the optimal DTR, defined as D^*, that yields the highest expected total reward if all patients in the population follow the DTR. That is,

$$\mathscr{D}^* = \mathrm{argmax}_{\mathscr{D}} E_{\mathscr{D}}[R_1 + R_2 + \ldots + R_K],$$

where $E_{\mathscr{D}}$ is the expectation under $A_k = D_k(H_k)$. The expectation on the right-hand side is defined as the value function associated with $\mathscr{D}$, denoted by $\mathscr{V}(\mathscr{D})$.

13.2.1   Q-learning

In Q-learning, a key concept is called Q-function, which is defined at each stage as the expected outcome if patients are treated optimally at the current stage and all future stages. Let $Q_k(H_k)$ be the Q-function at stage k. An important property of the Q-function is the following Bellman-type equation:

$$Q_k(H_k) = \max_a E[R_k + Q_{k+1}(H_{k+1})|H_k, A_k = a].$$

Thus, the optimal treatment for a patient with H_k at stage k is the treatment maximizing the expectation on the right-hand side of the above equation.

Motivated by this property, Q-learning proceeds by estimating Q-functions and optimal DTRs through a backward algorithm. Starting from stage K, we estimate the conditional function

$$E[R_K|H_K, A_K],$$

and denote the optimal treatment at stage K as

$$\mathscr{D}_K^*(H_K) = \mathrm{argmax}_a E[R_K|H_K, A_K = a].$$

The optimal conditional expected outcome at stage K can be represented as

$$Q_K(H_K) = E[R_K|H_K, A_K = \mathscr{D}_K^*(H_K)], \tag{13.1}$$

At stage $(k-1)$, $k = K, \ldots, 2$, given $Q_k(H_k)$, we estimate the conditional function

$$E[R_{k-1} + Q_k(H_k)|H_{k-1}, A_{k-1}].$$

The optimal treatment at stage $(k-1)$ is the one that maximizes the expected conditional cumulative outcome from stage $k-1$ to K,

$$\mathscr{D}_{k-1}^*(H_{k-1}) = \text{argmax}_a E[R_{k-1} + Q_k(H_k)|H_{k-1}, A_{k-1} = a].$$

The expected optimal outcome increment from stage $k-1$ to K under the optimal treatment from stage $k-1$ to K can be represented as

$$Q_{k-1}(H_{k-1}) = E[R_{k-1} + Q_k(H_k)|H_{k-1}, A_{k-1} = \mathscr{D}_{k-1}^*(H_{k-1})]. \tag{13.2}$$

13.2.2 Outcome-Weighted Learning

Different from Q-learning, outcome-weighted learning estimates optimal treatment rules via directly maximizing the value function $\mathscr{V}(\mathscr{D})$. If we let $\pi_k(a, h_k)$ denote the treatment assignment probability at stage k in a SMART, i.e., $P(A_k = a|H_k = h_k)$, then this value function can be shown (Qian and Murphy 2011) to be equivalent to

$$\mathscr{V}(\mathscr{D}) = E_{\mathscr{D}}\left[\sum_{k=1}^{K} R_k\right] = E\left[\frac{\Pi_{k=1}^{K} I(A_k = \mathscr{D}_k(H_k))(\sum_{k=1}^{K} R_k)}{\Pi_{k=1}^{K} \pi_k(A_k, H_k)}\right].$$

Hence, the goal is to estimate a DTR that maximizes the value function. Outcome-weighted learning transforms this optimization as a weighted classification problem, where A_k are class labels, H_k are feature variables, and reward outcomes along with treatment assignment probabilities are subject-specific weights.

13.2.2.1 Outcome-Weighted Learning Without Augmentation

Motivated by the observation of the connection between the value maximization and the weighted classification, a general framework of outcome-weighted learning can be described as follows. Starting from stage K, we minimize

$$E\left[\frac{R_K}{\pi_K(A_K, H_K)} I\big(A_K \neq \mathscr{D}_K(H_K)\big)\right]$$

with respect to $\mathscr{D}_K$ to learn the optimal treatment rule $\mathscr{D}_K^*$. Based on Liu et al. (2014, 2018), this minimization is equivalent to minimizing

$$E\left[\frac{|R_K - s_K(H_K)|}{\pi_K(A_K, H_K)} I\big[A_K \operatorname{sign}\big(R_K - s_K(H_K)\big) \neq \mathscr{D}_K(H_K)\big]\right], \tag{13.3}$$

where we subtract the total reward by an arbitrary function $s_K(H_K)$ of feature variables H_k. The rationale is that removal of the main effects independent of treatment should not affect treatment-decision while taking residuals can significantly reduce the variability of weights to improve algorithm performance. Furthermore, the weights in the latter are non-negative so many learning algorithms are applicable. Liu et al. (2014, 2018) theoretically prove the efficiency gain of taking residuals and equivalence of the above two equations.

At stage $(k-1), k = K, \ldots, 2$, we minimize

$$E\left[M_k^K \frac{|\sum_{j=k-1}^K R_j - s_{k-1}(H_{k-1})|}{\prod_{j=k-1}^K \pi_j(A_j, H_j)}\right.$$

$$\left. I\left[A_{k-1} \operatorname{sign}\left(\sum_{j=k-1}^K R_j - s_{k-1}(H_{k-1})\right) \neq \mathscr{D}_{k-1}(H_{k-1})\right]\right] \tag{13.4}$$

with respect to $\mathscr{D}_{k-1}$, where $M_k^K = I(A_k = \mathscr{D}_k^*, \ldots, A_K = \mathscr{D}_K^*)$, that is, the assigned treatment being the optimal treatment from stage k to K.

As remark, the algorithms in Zhao et al. (2012, 2015) are the special cases of the above algorithm if we choose $s_k(H_k)$ to be the minimal value of the cumulative reward.

13.2.2.2 Augmented Outcome-Weighted Learning (AOL)

In (13.3), we only utilize those subjects whose treatments are optimal from stage k to K when learning the rule at stage $(k-1)$. This can result in much information loss. Later, Liu et al. (2014, 2018) proposed augmented outcome-weighted learning (AOL) which utilizes partial information from all subjects to improve efficiency while preserving validity of DTRs. More specifically, at stage k, AOL constructs an augmented outcome G_k with the following form

$$G_k = \frac{M_k^K \sum_{j=k}^K R_j}{\prod_{j=k}^K \pi_j} - \sum_{j=k}^K \left[\frac{M_k^{j-1}}{\prod_{l=k}^{j-1} \pi_j} \left(\frac{I(A_j = \mathscr{D}_j^*(H_j))}{\pi_j} - 1\right) m_{kj}(H_j)\right], \tag{13.5}$$

where $M_k^{j-1} = I(A_1 = \mathscr{D}_1^*, \ldots, A_{j-1} = \mathscr{D}_{j-1}^*, A_j \neq \mathscr{D}_j^*)$ for $j > k$, $M_k^{k-1} = 1$, and $m_{kj}(H_j)$ is the expected outcome increment under optimal treatments from stage k up to stage $(j-1)$. That is,

$$
m_{kj}(H_j) = E[Q_k|H_j, M_k^{j-1} = 1]
$$
$$
= R_k + \ldots + R_{j-1} + E[Q_j|H_j, A_k = \mathscr{D}_k^*(H_k), \ldots, A_{j-1} = \mathscr{D}_{j-1}^*(H_{j-1})].
$$
(13.6)

With the augmented outcome G_k, we can estimate the DTR at stage $(k-1)$ by minimizing

$$
E\left[\frac{|R_{k-1}+G_k-s_{k-1}(H_{k-1})|}{\pi_{k-1}(A_{k-1},H_{k-1})} \right.
$$
$$
\left. I\left[A_{k-1} \operatorname{sign}\left(R_{k-1} + G_k - s_{k-1}(H_{k-1})\right) \neq \mathscr{D}_{k-1}(H_{k-1})\right]\right]. \quad (13.7)
$$

Liu et al. (2014, 2018) showed that the augmentation in AOL can reduce the stochastic variability of DTRs asymptotically and demonstrated significant improvement over outcome-weighted learning without augmentation. In addition, they also showed that AOL is robust when the regression models in the augmentation terms are misspecified.

13.2.2.3 Surrogate Loss Functions in Outcome-Weighted Learning

As we have seen in the previous sections, learning the optimal rule at each stage is reduced to the problem of minimizing a weighted misclassification error. Since 0–1 loss is not continuous, surrogate loss functions are used to replace 0–1 loss as commonly done in most of supervised learning methods.

At stage k, let $\mathscr{D}_k(H_k) = \operatorname{sign}(f(H_k))$, where $f(.)$ is the decision function at stage k, $k = 1, 2, \ldots, K$. Denote y_k as the class label, and w_k as the weight in the classification problem. For example, at stage K in outcome-weighted learning with or without augmentation,

$$
y_K = A_K \operatorname{sign}(R_K - s_K(H_K)), \quad w_K = \frac{|R_K - s_K(H_K)|}{\pi_K(A_K, H_K)}.
$$

We consider the following surrogate loss functions:

- *Hinge loss*: Hinge loss is the loss function used in support vector machine (SVM). It takes the form of $L(y_k, f(H_k)) = [1 - y_k f(H_k)]_+$.
- *Ramp loss*: Ramp loss provides a tight bound on the misclassification rate (Collobert et al. 2006; Keshet and McAllester 2011). The function can be expressed as

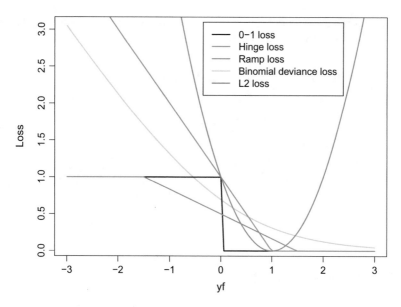

Fig. 13.1 Loss functions, plotted as a function of yf. y: class label, which takes value in $\{-1, 1\}$; f: the decision function

$$L_s(y_k, f(H_k)) = I\left(y_k f(H_k) \leq -\frac{s}{2}\right) + \left(\frac{1}{2} - \frac{y_k f(H_k)}{s}\right) I\left(-\frac{s}{2} < y_k f(H_k) < \frac{s}{2}\right),$$

$$(13.8)$$

where s is a positive slope parameter.

- *Binomial deviance loss*: Binomial deviance loss is minimized in logistic regression, where

$$L(y_k, f(H_k)) = \log\left(1 + \exp\left[-y_{ki} f_k(H_{ik})\right]\right).$$

- *Squared or L_2 loss*: L_2 loss is minimized in least squares estimation, where

$$L(y_k, f(H_k)) = (y_{ik} - f_k(H_{ik}))^2.$$

The above loss functions are plotted in Fig. 13.1 and they all lead to Fisher consistent DTRs in theory (Neykov et al. 2016).

13.2.3   Evaluation of DTRs

Benefit function, proposed in Qiu et al. (2018), is the difference in the value function between those who follow the DTR and those who follow the opposite treatment assignment to the DTRs. They have shown that a single-stage DTR is optimal if and only if its corresponding benefit function is non-negative for any subset of the population. We extend the benefit function to the multi-stage by defining the benefit as

$$\delta(\mathscr{D}) = E_{\mathscr{D}}\left[\sum_{k=1}^{K} R_k\right] - E_{\bar{\mathscr{D}}}\left[\sum_{k=1}^{K} R_k\right]$$

$$= E\left[\frac{\Pi_{k=1}^{K} I\left[A_k = \mathscr{D}_k(H_k)\right]\left(\sum_{k=1}^{K} R_k\right)}{\Pi_{k=1}^{K} \pi_k(A_k, H_k)}\right]$$

$$- E\left[\frac{\left(1 - \Pi_{k=1}^{K} I\left[A_k = \mathscr{D}_k(H_k)\right]\right)\left(\sum_{k=1}^{K} R_k\right)}{\Pi_{k=1}^{K} \pi_k(A_k, H_k)}\right].$$

This quantity can be useful to evaluate the overall benefit gain of $\mathscr{D}$.

13.3 Estimation and Algorithm

Based on the framework in Sect. 13.2, we estimate K-stage DTRs from a K-stage SMART with n subjects. We denote A_{ik} as the treatment assignment for subject i at stage k, R_{ik} as the observed outcome post the kth stage treatment for subject i, and H_{ik} as subject i's health history information prior to the treatment at stage k, for $k = 1, 2, \ldots, K, i = 1, 2, \ldots, n$.

13.3.1 Q-learning

With backward learning, starting from stage K, we estimate the conditional function $E[R_K | H_K, A_K]$ via least squares, which minimizes

$$n^{-1} \sum_{i=1}^{n} \left(R_{iK} - \left[\beta_{0K} + H_{iK}^T \beta_{1K} + \beta_{2K} A_{iK} + (H_{iK} A_{iK})^T \beta_{3K}\right]\right)^2.$$

Lasso penalty can be added for variable selection, by minimizing

$$n^{-1} \sum_{i=1}^{n} \left(R_{iK} - \left[\beta_{0K} + H_{iK}^T \beta_{1K} + \beta_{2K} A_{iK} + (H_{iK} A_{iK})^T \beta_{3K}\right]\right)^2$$

$$+ \lambda \left(\sum_{j=1}^{p_{1K}} |\beta_{1Kj}| + \sum_{j=1}^{p_{2K}} |\beta_{2Kj}| + \sum_{j=1}^{p_{3K}} |\beta_{3Kj}| \right),$$

where p_{1K}, p_{2K}, and p_{3K} is the dimension of β_{1K}, β_{2K}, and β_{3K}.

Based on the estimated coefficients $\widehat{\beta}_{0K}, \widehat{\beta}_{1K}, \widehat{\beta}_{2K}, \widehat{\beta}_{3K}$, the estimated outcome $\widehat{R}_{iK}$ under treatment $A_K = 1$ and $A_K = -1$ can be computed by

$$\widehat{R}_{iK}(H_{iK}, A_K) = \widehat{\beta}_{0K} + H_{iK}^T \widehat{\beta}_{1K} + \widehat{\beta}_{2K} A_K + (H_{iK} A_K)^T \widehat{\beta}_{3K}.$$

The estimated treatment rule $\widehat{\mathscr{D}}_K$ is set to be the treatment that yields a higher estimated outcome, that is

$$\widehat{\mathscr{D}}_K(H_{iK}) = \mathrm{argmax}_a \widehat{R}_{iK}(H_{iK}, A_K = a) = \mathrm{sign}(\widehat{\beta}_{2K} + H_{iK}^T \widehat{\beta}_{3K}).$$

The estimated optimal Q-function at stage K for subject i can then be represented as

$$\widehat{Q}_{iK}(H_{iK}) = \widehat{\beta}_{0K} + H_{iK}^T \widehat{\beta}_{1K} + \widehat{\beta}_{2K} \widehat{D}_K(H_{iK}) + (H_{iK} \widehat{D}_K(H_{iK}))^T \widehat{\beta}_{3K}. \quad (13.9)$$

At stage $(k-1)$, $k = K, \ldots, 2$, with $\widehat{Q}_{ik}$ estimated from stage k, we estimate the conditional function $E[R_{k-1} + Q_k | H_{k-1}, A_{k-1}]$ via least squares, which minimizes

$$\sum_{i=1}^{n} \left(R_{i,k-1} + \widehat{Q}_{ik} - \left[\beta_{0,k-1} + H_{i,k-1}^T \beta_{1,k-1} + \beta_{2,k-1} A_{i,k-1} + (H_{i,k-1} A_{i,k-1})^T \beta_{3,k-1} \right] \right)^2.$$

Lasso penalty can be added for variable selection.

Define $\widetilde{R}_{i,k-1} = R_{i,k-1} + \widehat{Q}_{ik}$, then the estimated outcome $\widehat{\widetilde{R}}_{i,k-1}$ under treatment $A_{k-1} = 1$ and $A_{k-1} = -1$ can be computed with

$$\widehat{\widetilde{R}}_{i,k-1}(H_{i,k-1}, A_{k-1}) = \widehat{\beta}_{0,k-1} + H_{i,k-1}^T \widehat{\beta}_{1,k-1} + \widehat{\beta}_{2,k-1} A_{k-1} + (H_{i,k-1} A_{k-1})^T \widehat{\beta}_{3,k-1}.$$

Similar to stage K, the estimated optimal treatment is the one that yields a higher estimated outcome, that is

$$\widehat{\mathscr{D}}_{k-1}(H_{i,k-1}) = \mathrm{argmax}_a \widehat{\widetilde{R}}_{i,k-1}(H_{i,k-1}, A_{i,k-1} = a) = \mathrm{sign}(\widehat{\beta}_{2,k-1} + H_{i,k-1}^T \widehat{\beta}_{3,k-1}).$$

Then the optimal Q-function at stage $k - 1$ can be estimated by

$$\widehat{Q}_{i,k-1}(H_{i,k-1}) = \widehat{\beta}_{0,k-1} + H_{i,k-1}^T \widehat{\beta}_{1,k-1} + \widehat{\beta}_{2,k-1} \widehat{D}_{k-1}(H_{i,k-1})$$
$$+ (H_{i,k-1} \widehat{D}_{k-1}(H_{i,k-1}))^T \widehat{\beta}_{3,k-1}. \quad (13.10)$$

If no tailoring variable is selected at stage k under lasso penalty, i.e., $\widehat{\beta}_{2,k} + H_{i,k}^T \widehat{\beta}_{3,k} = 0$, we randomly assign a treatment (1 or -1) with equal probability.

13.3.2 Outcome-Weighted Learning

For learning the sequence of optimal treatments for K stages, we use backward algorithm starting from the last stage with index K. We denote the decision function learned at stage K as $\widehat{f}_K(H_K)$, and the estimated DTR at stage K as $\widehat{\mathscr{D}}_K = \text{sign}(\widehat{f}_K(H_K))$. At stage $(k - 1), k = K, K - 1, \ldots, 2$, we denote $\widehat{M}_k^K = I(A_k = \widehat{\mathscr{D}}_k, \ldots, A_K = \widehat{\mathscr{D}}_K)$, and use $\widehat{M}_k^K$ to approximate M_k^K in (13.4).

With augmented outcomes, at stage $(k - 1), k = K, K - 1, \ldots, 2$, similarly, we use $\widehat{D}_j(H_j)$ to approximate $D_j^*(H_j)$, and take $\widehat{M}_k^{j-1} = I(A_1 = \widehat{\mathscr{D}}_1, \ldots, A_{j-1} = \widehat{\mathscr{D}}_{j-1}, A_j \neq \widehat{\mathscr{D}}_j)$ to approximate M_k^{j-1} in (13.5). In constructing $m_{kj}(H_j)$, we estimate the last expectation term in (13.6) by the estimated optimal Q-function $\widehat{Q}_j$ at stage j from Q-learning ((13.9), (13.10)). The functions $s_k(H_k), k = 1, 2, \ldots, K$ in ((13.3), (13.4), (13.7)) are constructed by least squares with H_k being the predictors.

The decision functions $\widehat{f}_k, k = K, K - 1, \ldots, 1$, are estimated by solving the following minimization problems under different loss functions. We adopt the notation of w_k and y_k from Sect. 13.2.2.3 as the weight and the class label for the weighted classification problem at stage k. Note that y_k takes value in $\{-1, 1\}$, $k = K, K - 1, \ldots, 1$.

13.3.2.1 Under SVM Hinge Loss

Under the hinge loss in the SVM framework, at each stage k, we minimize

$$Cn^{-1} \sum_{i=1}^{n} w_{ik}[1 - y_{ik} f_k(H_{ik})]_+ + \frac{1}{2}\|f_k\|^2, \tag{13.11}$$

where C is a regularization parameter, and $\|f\|$ is the norm in a reproducing kernel Hilbert space (RKHS). Linear kernel and radial basis function (RBF) kernel (also known as Gaussian kernel) are implemented under the hinge loss. Linear kernel on two samples $\mathbf{x}$ and $\mathbf{x}'$ is defined as

$$K(\mathbf{x}, \mathbf{x}') = \langle \mathbf{x}, \mathbf{x}' \rangle, \tag{13.12}$$

where $\langle \cdot, \cdot \rangle$ is the inner product. The RBF kernel on two samples $\mathbf{x}$ and $\mathbf{x}'$ is defined as

$$K(\mathbf{x}, \mathbf{x}') = \exp\left(-\frac{\|\mathbf{x} - \mathbf{x}'\|^2}{2\sigma^2}\right),$$

where $\|\cdot\|$ is the Euclidean distance and σ is a non-negative tunning parameter.

The solution to the minimization problem in (13.11) can be obtained by solving the Lagrangian dual problem via quadratic programming. The estimated decision function at stage k can be represented as

$$\widehat{f_k}(H_k) = \widehat{\beta}_{0k} + \sum_{i=1}^{n} \widehat{\alpha}_{ik} \, y_{ik} K(H_k, H_{ik}),$$

where $\widehat{\alpha}_{ik}, i = 1, 2, \ldots, n$ is the solution to the dual problem at stage k. With linear kernel, the estimated decision function can also be represented as

$$\widehat{f_k}(H_k) = \widehat{\beta}_{0k} + H_k^T \widehat{\beta}_k.$$

The estimated probability that treatment 1 (vs. -1) is the optimal treatment at stage k can be calculated by Platt scaling as

$$\frac{e^{\widehat{f_k}(H_k)}}{1 + e^{\widehat{f_k}(H_k)}}. \tag{13.13}$$

13.3.2.2 Under SVM Ramp Loss

We implement the linear kernel under ramp loss in the SVM framework. At each stage k, we minimize

$$Cn^{-1} \sum_{i=1}^{n} w_{ik} \, L_s(y_{ik}, f_k(H_{ik})) + \frac{1}{2} ||\beta_k||^2 \tag{13.14}$$

over β_k, where $||\beta_k||$ is the Euclidean norm of β_k, $L_s(\cdot)$ is the hinge loss function defined in (13.8), C is a regularization parameter, and the decision function has the form $f_k(H_{ik}) = \beta_{0k} + H_{ik}^T \beta_k$.

Since ramp loss is a non-convex loss function, we solve this optimization problem by the difference of two convex functions algorithm (DCA) (Tao et al. 1996). Basically, we express the ramp loss function as the difference of two convex functions and iteratively solve the Lagrangian dual problem via quadratic programming. To improve convergence, initial values of β_k is set to be the estimated $\widehat{\beta}_k$ from SVM hinge loss with linear kernel. The convergence criterion at each stage is set as the change in the loss function (13.14) to be less than 10^{-5}. The maximum number of iterations to run at each stage is set to be 20.

The estimated decision function at stage k can be represented as

$$\widehat{f_k}(H_k) = \widehat{\beta}_{0k} + \sum_{i=1}^{n} \widehat{\alpha}_{ik} \, y_{ik} K(H_k, H_{ik}), \quad \text{or} \quad \widehat{f_k}(H_k) = \widehat{\beta}_{0k} + H_k^T \widehat{\beta}_k,$$

where $\widehat{\alpha}_{ik}, i = 1, 2, \ldots, n$ is the solution to the Lagrangian dual problem at stage k, and the linear kernel $K(\cdot, \cdot)$ is defined in (13.12). The estimated probability that treatment 1 (vs. -1) is the optimal treatment at stage k can be calculated by Platt scaling similar to (13.13).

13.3.2.3   Under Binomial Deviance Loss

Binomial deviance loss is minimized in logistic regression. We construct the decision function as $f_k(H_k) = \beta_{0k} + H_k^T \beta_k$ at each stage k, and minimize

$$n^{-1} \sum_{i=1}^{n} w_{ik} \log \left(1 + \exp[-y_{ik} \, f_k(H_{ik})] \right),$$

Lasso penalty can be added to conduct variable selection by minimizing

$$n^{-1} \sum_{i=1}^{n} w_{ik} \log \left(1 + \exp[-y_{ik} \, f_k(H_{ik})] \right) + \lambda \sum_{j=1}^{p_k} |\beta_{kj}|,$$

where p_k is the dimension of β_k. The estimated decision function at stage k is

$$\widehat{f}_k(H_k) = \widehat{\beta}_{0k} + H_k^T \widehat{\beta}_k,$$

and the estimated probability that treatment 1 (vs. -1) is the optimal treatment can be represented as (13.13).

13.3.2.4   Under L_2 Loss

L_2 loss is minimized in the least squares estimation. We construct the decision function as $f_k(H_k) = \beta_{0k} + H_k^T \beta_k$ at each stage k, and minimize

$$n^{-1} \sum_{i=1}^{n} w_{ik} (y_{ik} - f_k(H_{ik}))^2$$

over β_k. Lasso penalty can be added for variable selection by minimizing

$$n^{-1} \sum_{i=1}^{n} w_{ik} (y_{ik} - f_k(H_{ik}))^2 + \lambda \sum_{j=1}^{p_k} |\beta_{kj}|$$

where p_k is the dimension of β_k. The estimated decision function at stage k is

$$\widehat{f}_k(H_k) = \widehat{\beta}_{0k} + H_k^T \widehat{\beta}_k$$

and the estimated probability that treatment 1 (vs. -1) is the optimal treatment at stage k is (13.13).

13.3.2.5 Tuning Parameters

When estimating DTR under SVM hinge loss, there is a regularization parameter C and an additional tuning parameter σ if RBF kernel is used. This spread parameter σ can also be chosen using a heuristic method, for example, $\sigma = 1/d_m^2$ where d_m is the median pairwise Euclidean distance defined as median $\{||X_i - X_j|| : A_i \neq A_j\}$ (Wang et al. 2018). Under SVM ramp loss with linear kernel, there is a regularization parameter C and a slope parameter s. We consider all of them as tuning parameters. We choose the values of the tuning parameters such that the learned DTR has the best value function. At each stage, we perform m-fold cross validation for all possible combinations of the tuning parameters over a given range, and choose the set of values that yields the highest empirical value function on the validation sets.

13.3.3 Handling Observational Study Data

Treatment assignment probability at stage k, $P(A_k = a | H_k = h_k)$, is defined as the probability of being assigned to the observed treatment a given subject's health history up to stage k. The treatment assignment probability at stage k, denoted as $\pi_k(a, h_k)$, plays a role in constructing the weight w_k, $k = 1, 2, \ldots, K$.

In order to infer DTR with causal interpretation, we need three conditions—(C1) the stable unit treatment value assumption (SUTVA), that is the potential outcome at stage k, $R_k(a_1, \ldots, a_k)$, equals the observed R_k under $A_1 = a_1, \ldots, A_k = a_k$, $k = 1, 2, \ldots, K$; (C2) the sequential ignorability assumption or no unobserved confounder assumption, that is at each stage k the treatment A_k is assigned independently of the potential future outcomes conditional on patient's history information up to this stage H_k; (C3) the positive assumption that $P(A_k = a | H_k = h_k) > 0, \forall a$. (C2) and (C3) are guaranteed in SMARTs, and the treatment assignment probabilities are known by design in randomized trials which can be directly used in DTR estimation.

For observational studies, under assumptions (C1)–(C3), we can still estimate valid DTRs. A propensity score (Rosenbaum and Rubin 1983), the probability of receiving the treatment, can be estimated for each individual, through logistic regression at each stage k with H_k being the covariates. Lasso penalty can be applied when the dimension of H_k is high. This propensity score $\widehat{\pi}_k(a, h_k)$ can be used in the DTR estimation procedure for observational study.

13.3.4 DTR Evaluation and Future Use

13.3.4.1 Empirical Value Function and Benefit Function

The empirical value function and benefit function can be calculated after the estimation of DTR by

$$\widehat{\mathscr{V}}(\mathscr{D}) = n^{-1} \sum_{i=1}^{n} \left[\frac{\Pi_{k=1}^{K} I(A_{ik} = \widehat{\mathscr{D}}_k(H_{ik}))(\sum_{k=1}^{K} R_{ik})}{\Pi_{k=1}^{K} \pi_{ik}(A_{ik}, H_{ik})} \right],$$

$$\widehat{\delta}(\mathscr{D}) = n^{-1} \sum_{i=1}^{n} \left[\frac{\Pi_{k=1}^{K} I(A_{ik} = \widehat{\mathscr{D}}_k(H_{ik}))(\sum_{k=1}^{K} R_{ik})}{\Pi_{k=1}^{K} \pi_{ik}(A_{ik}, H_{ik})} \right.$$

$$\left. - \frac{\left(1 - \Pi_{k=1}^{K} I[A_{ik} = \widehat{\mathscr{D}}_k(H_{ik})]\right)\left(\sum_{k=1}^{K} R_{ik}\right)}{\Pi_{k=1}^{K} \pi_{ik}(A_{ik}, H_{ik})} \right].$$

13.3.4.2 Apply the Learned DTR to an Independent Sample

After estimating the DTR from the training data, we can apply the fitted DTR $\widehat{\mathscr{D}} = (\widehat{\mathscr{D}}_1, \widehat{\mathscr{D}}_2, \ldots, \widehat{\mathscr{D}}_K)$ to independent patients where the feature variables H_k, $k = 1, 2, .., K$ are the same as the training sample. If only partial feature variables are observed in the new sample, for example, H_k is observed from stage 1 up to stage j, $j \leq K$, treatment recommendations for the new sample from stage 1 to stage j can be provided by applying $(\widehat{\mathscr{D}}_1, \ldots, \widehat{\mathscr{D}}_j)$ to $(H_1, \ldots, H_j)$, respectively. If all the information $(H_k, A_k, R_k, \pi_k), k = 1, 2, \ldots K$, in the new sample has been observed, one can evaluate the performance of the learned DTR $\widehat{\mathscr{D}}$ on the new sample, and compute the value function and benefit function.

13.4 Software and Illustrations

Models in Sect. 13.3 are implemented in the R package "DTRlearn2".

13.4.1 DTR Estimation

13.4.1.1 Q-learning

The function for estimating DTRs using Q-learning is ql (). The function argument is given by

```
ql (H, AA, RR, K, pi='estimated', lasso=TRUE, m=4)
```

- K is the number of stages.
- AA and RR are the observed treatment assignments and reward outcomes respectively at the K stages for all subjects in the sample. They can be a vector if K is 1, or a list of K vectors corresponding to the K stages.
- H is the subject history information before treatment at each stage. It can be a vector or a matrix when only baseline information is used in estimating the DTR; otherwise, it would be a list of length K, representing the history information corresponding to the K stages. A patient's history information prior to the treatment at stage k can be constructed recursively as $H_k = (H_{k-1}, A_{k-1}, R_{k-1}, X_k)$ with $H_1 = X_1$, where X_k is subject-specific variables collected at stage k before the treatment, A_k is the treatment at stage k, and R_k is the outcome observed post the treatment at stage k. Higher order or interaction terms can also be easily incorporated in H_k, e.g., $H_k = (H_{k-1}, A_{k-1}, R_{k-1}, X_k, H_{k-1}A_{k-1}, R_{k-1}A_{k-1}, X_kA_{k-1})$.
- pi is the treatment assignment probabilities at the K stages for all subjects in the sample. It can be a user specified input if the treatment assignment probabilities are known. It is a vector if K=1, or a list of K vectors corresponding to the K stages. The default is pi="estimated", that is we estimate the treatment assignment probabilities based on lasso-penalized logistic regressions. See Sect. 13.3.3 for more details.
- lasso specifies whether to add lasso penalty at each stage when fitting the model. Lasso penalty is encouraged especially when the dimension of H_k is high. The default is lasso=TRUE.
- m is the number of folds in the m-fold cross validation. It is used when lasso=TRUE is specified. The default is m=4.

After fitting the function ql(), a list of results is returned as an object. It contains the following attributes

- stage1 consists of the stage 1 outputs;
- stage2 consists of the stage 2 outputs;
  ...
- stageK consists of the stage K outputs;
- valuefun is the overall empirical value function under the estimated DTR;
- benefit is the overall empirical benefit function under the estimated DTR;
- pi is the treatment assignment probabilities of the observed treatments for each subject at the K stages. It is a list of K vectors. If pi="estimated" is specified as input, the estimated treatment assignment probabilities from lasso-penalized logistic regressions will be returned.
  And in each stagek outputs, $k = 1, 2, \ldots, K$, we have the following results
- stagek$co is the estimated coefficients of $(1, H_k, A_k, H_kA_k)$, the variables in the model at stage k;
- stagek$treatment is the estimated optimal treatments at stage k for each subject in the sample;
- stagek$Q is the estimated optimal outcome increments from stage k to K (the estimated optimal Q-functions at stage k) for each subject in the sample.

13.4.1.2 Outcome-Weighted Learning

The function for estimating DTRs using outcome-weighted learning methods is `owl()`. The function argument is given by

```
owl (H, AA, RR, n, K, pi='estimated', res.lasso=TRUE,
loss='hinge', kernel='linear', augment=TRUE, c=2.^(-2:
2), sigma=c(0.03,0.05,0.07),s=2.^(-2:2),m=4)
```

- n is the sample size, and K is the number of stages.
- AA and RR are the observed treatment assignments and reward outcomes at the K stages for all subjects in the sample. They can be a vector if K is 1, or a list of K vectors corresponding to the K stages.
- H is the subject history information before treatment at each stage. It can be a vector or a matrix when only baseline information is used in estimating the DTR; otherwise, it would be a list of length K, representing the history information corresponding to the K stages. A patient's history information prior to the treatment at stage k can be constructed recursively as $H_k = (H_{k-1}, A_{k-1}, R_{k-1}, X_k)$ with $H_1 = X_1$, where X_k is subject-specific variables collected at stage k before the treatment, A_k is the treatment at stage k, and R_k is the outcome observed post the treatment at stage k. Higher order or interaction terms can also be easily incorporated in H_k, e.g., $H_k = (H_{k-1}, A_{k-1}, R_{k-1}, X_k, H_{k-1}A_{k-1}, R_{k-1}A_{k-1}, X_kA_{k-1})$. Please standardize all the variables in H to have mean 0 and standard deviation of 1 before using H as the input.
- pi is the treatment assignment probabilities at the K stages for all subjects in the sample. It can be a user specified input if the treatment assignment probabilities are known. pi is a vector if K=1 or a list of K vectors corresponding to the K stages. The default is pi="estimated", that is we estimate the treatment assignment probabilities based on lasso-penalized logistic regressions. See Sect. 13.3.3 for more details.
- res.lasso specifies whether or not to use lasso penalty in fitting $s_k(H_k)$, the least squares to acquire the residuals in constructing the weights at stage k, $k = 1, 2, \ldots, K$. Lasso penalty is encouraged especially when the dimension of H_k is high. The default is res.lasso=TRUE.
- loss specifies which loss function to use for the weighted classification problem at each stage. The options are "hinge", "ramp", "logit", "logit.lasso", "l2", "l2.lasso". "hinge" and "ramp" are for the SVM hinge loss and SVM ramp loss. "logit" and "logit.lasso" are for the binomial deviance loss used in the logistic regression, where lasso penalty is applied under "logit.lasso". "l2" and "l2.lasso" are for the L_2 or square loss, where lasso penalty is applied under "l2.lasso". The default is loss="hinge".
- kernel specifies which kernel function to use under SVM hinge loss and SVM ramp loss, i.e., when loss="hinge" or loss="ramp". "linear" and

"rbf" kernel are implemented under SVM hinge loss, and "linear" kernel is implemented under SVM ramp loss. The default is kernel="linear".

- augment specifies whether or not to use augmented outcomes at each stage. Augmentation is recommended when there are multiple stages with a small sample size. The default is augment=TRUE.
- c is a vector which specifies the values of the regularization parameter C for tuning under SVM hinge loss or SVM ramp loss. The default is c=c(0.25,0.5,1,2,4). In practice, a wider range of c can be specified based on the data.
- sigma is a vector which specifies the values of the parameter σ in the RBF kernel for tuning under SVM hinge loss, i.e., when loss="hinge" and kernel="rbf". The default is sigma=c(0.03,0.05,0.07). In practice, a wider range of sigma can be specified based on the data.
- s is a vector which specifies the values of the slope parameter s in the SVM ramp loss for tuning, i.e., when loss="ramp" and kernel="linear". The default is c=c(0.25,0.5,1,2,4). In practice, a wider range of s can be specified based on the data.
- m is the number of folds in the m-fold cross validation. The m-fold cross validation is implemented in selecting the tuning parameters c, sigma or s. It is also used for choosing the tuning parameter for the lasso penalty when res.lasso=T, loss="logit.lasso", or loss="l2.lasso" is specified. The default is m=4.

After fitting the function owl(), a list of results is returned as an object. It contains the following attributes

- stage1 consists of the stage 1 outputs;
- stage2 consists of the stage 2 outputs;

  ...

- stageK consists the stage K outputs;
- valuefun is the overall empirical value function under the estimated DTR;
- benefit is the overall empirical benefit function under the estimated DTR;
- pi is the treatment assignment probabilities of the observed treatments for each subject at the K stages. It is a list of K vectors. If pi='estimated' is specified as input, the estimated treatment assignment probabilities from lasso-penalized logistic regressions will be returned.
- type type of the returned object corresponding to the loss and kernel And in each stagek result, k=1, 2, ..., K, we have the following possible outputs.
- stagek$beta0 is $\widehat{\beta}_{0k}$, the estimated coefficient of the intercept in the decision function.
- stagek$beta is $\widehat{\beta}_k$, the estimated coefficients of H_k in the decision function. It is not returned with RBF kernel under SVM hinge loss.
- stagek$fit is the fitted decision functions for each subject in the sample.

- `stagek$probability` is the estimated probabilities that treatment 1 (vs. -1) is the optimal treatment for each subject in the sample, which is calculated by exp(stagek$fit)/(1 + exp(stagek$fit).
- `stagek$treatment` is the estimated optimal treatments for each subject in the sample.
- `stagek$c` is the best regularization parameter C in SVM hinge loss or SVM ramp loss, chosen from the values specified in c via cross validation.
- `stagek$sigma` is the best parameter σ in the RBF kernel, chosen from the values specified in `sigma` via cross validation.
- `stagek$s` is the best slope parameter s in the ramp loss, chosen from the values specified in s via cross validation.
- `stagek$iter` is the number of iterations conducted under SVM ramp loss.
- `stagek$alpha1` is the solutions to the Lagrangian dual problem under SVM hinge loss or SVM ramp loss. It is used for constructing the decision function on the new sample.
- `stagek$H` is the patient history matrix H_k, which is returned only under SVM hinge loss with RBF kernel. It is used for constructing the RBF kernel on the new sample.

13.4.2 Apply the Estimated DTR to an Independent Sample

The `predict()` function can be applied to the `object` returned after running `ql()` or `owl()` on the training sample, to apply the learned DTR to another independent sample. The new sample should have the same observed subject history variables H_k for $k = 1, 2, .., j$, where $j \leq K$. Then treatments up to stage j can be recommended for subjects in the new sample. To do this, we call `predict()` with argument

```
predict1 = predict(object, H=newH, K=j)
```

To see the outputs, call `predict1$treatment` for the recommended treatments for each subject at the j stages, which is a list of j vectors. The fitted decision functions of the j stages are also available when predict with `owl` objects, by calling `predict1$fit`.

If all the information (H_k, A_k, R_k, π_k), $k = 1, 2, \ldots K$, in the new sample has been observed, we can evaluate the performance of the learned DTR $\widehat{\mathscr{D}}$ on the new sample and see the value and benefit function. To do this, we call `predict` with argument

```
predict2 = predict(object, H=newH, AA=newAA, RR=newRR,
K=K, pi=newpi)
```

To see the outputs, similarly, call `predict2$treatment` and `predict2$fit` for the recommended treatments and decision functions for the K stages. Additionally, the overall empirical value function and benefit function on the new sample applying the learned DTR are stored in the following attributes of the function output `predict2$valuefun` and `predict2$benefit`.

13.4.3 Other Details

There are several packages we used in implementing the functions in the package. `kernlab` is used in constructing the kernels and solving the quadratic programming problem under SVM hinge loss or SVM ramp loss. `glmnet` is used in implementing lasso penalized models. `MASS` is used for generating some of the results.

To reduce the errors occurred due to matrix singularity when solving the quadratic programming problem, we modify the quadratic matrix in the quadratic programming. An eigenvalue decomposition is firstly conducted for the quadratic matrix as $A = V \Lambda V^{-1}$, and those eigenvalues that are smaller than 10^{-5} times of the biggest eigenvalue is truncated to 10^{-5} times the biggest eigenvalue. The modified quadratic matrix is calculated as $\tilde{A} = V \tilde{\Lambda} V^{-1}$, where $\tilde{\Lambda}$ is the diagonal matrix with the modified eigenvalues (Rebonato and Jäckel 2011).

NAs will be returned in the results with an error message printed if too few samples were selected for estimating the DTR. The tuning parameters under which computation issues are encountered in solving the quadratic programming will be skipped in comparison for choosing the best tuning parameters.

13.5   Simulations and Real Data Implementation

13.5.1 Simulations

We simulated a four-stage SMART with baseline variables $X_1, \ldots, X_{30}$ from a multivariate Gaussian distribution. $X_1, \ldots, X_{10}$ had variance 1 and pairwise correlation 0.2; $X_{11}, \ldots, X_{30}$ had mean 0 and were uncorrelated with each other and with $X_1, \ldots, X_{10}$. We assumed that subjects were from 10 latent groups with equal size, and these 10 groups were characterized by the different means of the feature variables $X_1, \ldots, X_{10}$. The means of $X_1, \ldots, X_{10}$ for the 10 groups were generated from $\mathcal{N}(0, 5)$. Each latent group was assumed to have its own optimal treatment sequence, which was generated as

$$A_{kg}^* = 2([g/(2k - 1)] \mod 2) - 1, \quad k = 1, 2, 3, 4, \quad g = 1, 2, \ldots, 10.$$

for group g at stage k. The primary outcome was observed only at the end of the trial, which was generated as

$$R_1 = R_2 = R_3 = 0, \quad R_4 = \sum_{k=1}^{4} A_k A_{kg}^*, \quad k = 1, 2, 3, 4, \quad g = 1, 2, \ldots, 10.$$

where A_k, the assigned treatment group (1 or -1) at stage k, was randomly generated with equal probability.

We simulated a large test set with sample size 20,000 according to this setting. The empirical optimal value function on this test set was 4.02. We then simulated 100 training datasets of sample size 100 with the same means in $X_1, \ldots, X_{10}$ with the test set. Q-learning and different outcome-weighted learning methods were implemented on each training set and the learned rules were applied to the test set. The history information at stage 1 was constructed as $H_1 = (X_1, \ldots, X_{30})$, and at stage k, k = 2, 3, 4, the history information was constructed recursively as $H_k = (H_{k-1}, A_{k-1}, H_{k-1}A_{k-1})$ since there were no intermediate outcomes. The regularization parameter C in the outcome-weighted learning was tuned on the grid of 2^p where p is an integer ranging from -15 to 15; σ in the RBF kernel and s in the ramp loss were tuned on the grid of 2^p where p is an integer ranging from -8 to 8. The running time for 1 iteration with the specified tuning parameters is the following: 0.10 s for Q-learning; for outcome-weighted learning, 0.74 s under binomial deviance loss, 0.76 s under squared loss, 5.5 s under SVM hinge loss with linear kernel, 610 s under SVM hinge loss with RBF kernel, and 492 s under SVM ramp loss with linear kernel when running on a computer with 2.7 GHz Intel Core i5 processor.

The value functions evaluated on the test set with the DTRs learned from 100 simulated training sets are plotted in Fig. 13.2. Augmented outcome-weighted learning under SVM hinge loss with RBF kernel yields the highest value function (mean 1.49, median 1.53) followed by augmented outcome-weighted learning under SVM hinge loss with linear kernel (mean 1.19, median 1.18) and Q-learning (mean 0.92, median of 0.9). In this setting, we didn't assume any parametric decision boundaries, but with decision boundaries implicitly determined by the underlying latent groups with differentiable feature variable means; thus, Q-learning suffers from misspecification of the decision rules. Outcome-weighted learning under SVM hinge loss, especially with RBF kernel, allows more flexible decision boundaries. However, with a small training sample size ($n = 100$), in the backward stages, outcome-weighted learning can suffer from an even smaller sample size at early stages (stage 2 and stage 1). Augmented outcome-weighted learning borrows information from Q-learning and improves the efficiency substantially in this case.

13.5.2   Illustration of the Real Data Implementation

We applied Q-learning and outcome-weighted learning methods to a two-stage SMART mimicking a real world study of children affected by attention deficit hyperactive disorder (ADHD) (Pelham and Fabiano 2008). At the first stage,

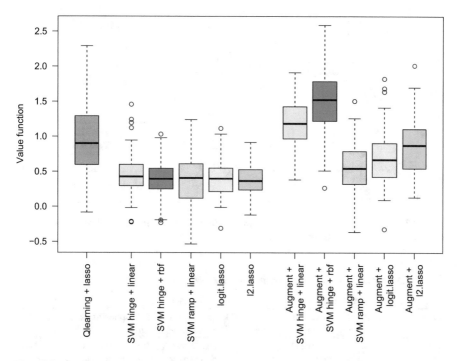

Fig. 13.2 Empirical value functions on the test set with DTRs learned from 100 simulated 4-stage SMARTs of sample size 100 with Q-learning and various outcome-weighted learning methods without or with augmented outcomes

children were randomized to treatment of low-intensity behavioral modification (BMOD) or low-intensity methamphetamine (MED) with equal probability. At second stage, children were randomized to treatment of low-intensity BMOD + low-intensity MED, or high-intensity BMOD with equal probability. The primary outcome of study was participant's school performance score ranging from 1 to 5, which was assessed at the end of the study for all participants. The diagram of the study design is shown in Fig. 13.3. Variables in this ADHD study are listed in Table 13.1.

First load the data, and construct the matrix of the health history information for each individual at the two stages and standardize each variable.

```
library(DTRlearn2)
data(adhd)
attach(adhd)
R> H1 = scale(cbind(o11, o12, o13, o14))
R> H2 = scale(cbind(H1, a1, H1*a1, r,o22,r*a1,o22*a1))
R> colnames(H2)[12] = "r*a1"
R> colnames(H2)[13] = "o22*a1"
```

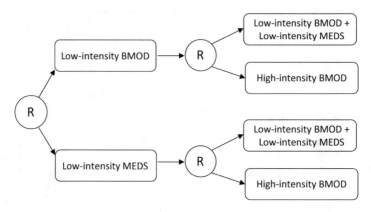

Fig. 13.3 Study design of the two-stage SMART for Children with ADHD. BMOD: behavioral modification; MEDS: methamphetamine

Table 13.1 Variables in the dataset for the two-stage SMART of 150 children with ADHD

Variable name	Variable type	Variable description
O11	Baseline covariate (binary; coded as 0/1)	Diagnosed with ODD (oppositional defiant disorder) before the first-stage intervention
O12	Baseline covariate (continuous)	ADHD score: ADHD symptoms at the end of the previous school year (ranging from 0 to 3, larger values for fewer symptoms)
O13	Baseline covariate (binary; coded as 0/1)	Medication during the previous school year
O14	Baseline covariate (binary; coded as 0/1)	Race: white (coded 1) versus nonwhite (coded 0)
A1	First stage intervention (binary; coded as −1/1)	A1 = −1 for low-intensity MEDS; A1 = 1 for low-intensity BMOD
R	First stage response indicator (binary; coded as 0/1)	R = 0 if participant did not respond to the first stage intervention; R= 1 if he or she responded
O21	Intermediate outcome (postbaseline covariate, continuous)	Number of months until non-response (maximum: 8 months, NA for responders)
O22	Intermediate outcome (post-baseline covariate, binary; coded as 0/1)	Adherence to the stage 1 intervention: 1 for high adherence
A2	Second stage intervention (binary; coded as −1/1)	A2 = −1 for low-intensity BMOD + MEDS; A2 = 1 for high-intensity BMOD
Y	Primary outcome (continuous)	School performance at the end of the school year (ranging from 1 to 5, higher values reflect better performance)

Randomly split the dataset to a training set and a test set with ratio of 2:1.

```
R> n = length(a1)
R> nfold = 3
R> set.seed(1)
R> train_idx = sample(1:n, n/nfold*(nfold-1))
R> ntrain = length(train_idx)
R> ntest = n - length(train_idx)
```

Estimate the DTRs from the training set using Q-learning.

```
R> fit1 = ql(H=list(H1[train_idx,], H2[train_idx,]),
+    AA=list(a1[train_idx], a2[train_idx]),
+    RR=list(rep(0, ntrain), y[train_idx]),
+    pi=list(rep(0.5, ntrain), rep(0.5,ntrain)), K=2,
                                    m=3, lasso=T)
```

Apply the estimated DTRs from Q-learning on the test set, and output the recommended treatments for the two stages as well as the value function of the estimated DTRs evaluated on the test set.

```
R> pred1 = predict(fit1, H=list(H1[-train_idx,],
                                H2[-train_idx,]),
+    AA=list(a1[-train_idx],a2[-train_idx]),
+    RR=list(rep(0, ntest), y[-train_idx]),
+    pi=list(rep(0.5,ntest), rep(0.5,ntest)), K=2)

R> pred1$treatment[[1]]
       [,1]
 [1,]    1
 [2,]   -1
 [3,]    1
 [4,]    1
 [5,]   -1
...


R> pred1$treatment[[2]]
       [,1]
 [1,]    1
 [2,]   -1
 [3,]   -1
 [4,]   -1
 [5,]    1
...


R> pred1$valuefun
[1] 3.44
```

Next, implement outcome-weighted learning with SVM hinge loss under the linear kernel with augmented outcomes. We tune the regularization parameter C on the grid of 2^{-15} to 2^{15}.

```
R> c = 2^(-15:15)
R> set.seed(1)
R> fit2 = owl(H=list(H1[train_idx,], H2[train_idx,]),
+      AA=list(a1[train_idx], a2[train_idx]),
+      RR=list(rep(0, ntrain), y[train_idx]),
+      pi=list(rep(0.5, ntrain), rep(0.5,ntrain)),
+      n=ntrain, K=2, res.lasso = T, loss="hinge",
       kernel="linear",
+      augment=T, c=c, m=3)
```

Apply the estimated DTRs on the test set, and output the recommended treatments for the two stages as well as the value function of the estimated DTRs evaluated on the test set.

```
R> pred2 = predict(fit2, H=list(H1[-train_idx,],
    H2[-train_idx,]),
+      AA=list(a1[-train_idx],a2[-train_idx]),
+      RR=list(rep(0, ntest), y[-train_idx]),
+      pi=list(rep(0.5,ntest), rep(0.5,ntest)), K=2)

R> pred2$treatment[[1]]
         1
  [1,]   1
  [2,]  -1
  [3,]   1
  [4,]  -1
  [5,]  -1
...

R> pred2$treatment[[2]]
         1
  [1,]   1
  [2,]  -1
  [3,]  -1
  [4,]  -1
  [5,]   1
   ...

R> pred2$valuefun
[1] 3.92
```

Outcome-weighted learning yields a higher value function evaluated on the test set. Hence, we fit the estimated DTR though outcome-weighted learning on the entire sample to examine the contribution of tailoring variables by examining the estimated standardized coefficients.

```
R> set.seed(1)
R> fit2_whole = owl (H=list(H1, H2), AA=list(a1,a2),
    RR=list(rep(0, n), y),
+      n=n, K=2, pi=list(rep(0.5, n), rep(0.5,n)),
    res.lasso = T, loss="hinge",
+      kernel="linear", augment=T, c=c)

R> fit2_whole$stage1$beta
              1
o11 -8.633966e-06
o12  5.337246e-05
o13 -8.965257e-01
o14  2.913989e-02

R> fit2_whole$stage2$beta
              1
o11      -3.986267e-05
o12       1.309333e-04
o13       1.206403e-04
o14       2.774660e-05
a1       -4.813819e-05
o11      -2.086035e-05
o12      -1.331149e-05
o13       4.701919e-05
o14       4.035830e-05
r         1.076440e-01
o22       8.855914e-01
r*a1      1.322140e-01
o22*a1   -1.527959e-01
```

The variable with the largest contribution to tailor stage 1 treatments is prior medication (O13), with an estimated coefficient of -0.90 on the standardized variable. This result suggests that children who do not have medication history prior to the trial should start with behavioral modification. The variables with a large contribution to tailor stage 2 treatments are adherence to stage 1 intervention (O22, standardized estimated coefficient 0.88) and O22*A1 (standardized estimated coefficient -0.15). This result suggests that children who start with behavioral modification in the first stage and adhere but do not achieve adequate response should be treated by high-intensity behavioral modification at the second stage. Additionally, children who respond after the first stage of behavioral intervention can benefit more from the augmented behavioral intervention at the second stage (estimated standardized coefficient of $0.11 + 0.13 = 0.24$ based on r and $r * a1$ coefficient estimates).

13.6 Summary

We provide a comprehensive software, "DTRlearn2", to estimate general K-stage DTRs from SMARTs and possibly from observational data with Q-learning and a variety of outcome-weighted learning methods. Penalizations are allowed for variable selection and model regularization. Q-learning can be efficient when the Q-function is correctly specified but it can also suffer from incorrect model specification, while outcome-weighted learning is more robust since it directly optimizes with respect to the treatment rule. With the outcome-weighted learning scheme, different loss functions—SVM hinge loss, SVM ramp loss, binomial deviance loss, and L_2 loss—can be adopted to solve the weighted classification problem at each stage. SVM hinge losses are powerful in learning complex decision rules using the kernel trick, especially under the rbf kernel. SVM ramp loss is more robust to outliers since the loss function is bounded unlike other loss functions. But the SVM methods all require tuning over the hyperparameters. In practice, different methods can be applied to estimate the DTRs, and the method that yields the best value function and benefit function can be determined. Augmentation is allowed for outcome-weighted learning methods to improve efficiency, especially for estimating multi-stage DTRs and when sample size is small. After estimating the DTR on the training sample, the fitted DTR can be easily applied to a new sample for individualized treatment recommendations or DTR evaluation. "DTRlearn2" facilitates these processes.

Acknowledgements This work is supported by U.S. NIH grants NS073671 and GM124104.

References

Collobert R, Sinz F, Weston J, Bottou L (2006) Trading convexity for scalability. In: Proceedings of the 23rd international conference on machine learning. ACM, New York, pp 201–208

Foster JC, Taylor JM, Ruberg SJ (2011) Subgroup identification from randomized clinical trial data. Stat Med 30(24):2867–2880

Keshet J, McAllester DA (2011) Generalization bounds and consistency for latent structural probit and ramp loss. In: Advances in neural information processing systems, pp 2205–2212

Laber E, Zhao Y (2015) Tree-based methods for individualized treatment regimes. Biometrika 102(3):501–514

Lavori PW, Dawson R (2000) A design for testing clinical strategies: biased adaptive within-subject randomization. J R Stat Soc: Ser A (Stat Soc) 163(1):29–38

Lavori PW, Dawson R (2004) Dynamic treatment regimes: practical design considerations. Clin Trials 1(1):9–20

Liu Y, Wang Y, Kosorok MR, Zhao Y, Zeng D (2014) Robust hybrid learning for estimating personalized dynamic treatment regimens. arXiv preprint. arXiv:161102314

Liu Y, Wang Y, Kosorok M, Zhao Y, Zeng D (2018) Augmented outcome-weighted learning for estimating optimal dynamic treatment regimens. Stat Med 37(26):3776–3788

Lunceford JK, Davidian M, Tsiatis AA (2002) Estimation of survival distributions of treatment policies in two-stage randomization designs in clinical trials. Biometrics 58(1):48–57

Moodie EE, Richardson TS, Stephens DA (2007) Demystifying optimal dynamic treatment regimes. Biometrics 63(2):447–455

Murphy SA (2003) Optimal dynamic treatment regimes. J R Stat Soc Ser B (Stat Methodol) 65(2):331–355

Neykov M, Liu JS, Cai T (2016) On the characterization of a class of fisher-consistent loss functions and its application to boosting. J Mach Learn Res 17(70):1–32

Pelham WE, Fabiano GA (2008) Evidence-based psychosocial treatments for attention-deficit/hyperactivity disorder. J Clin Child Adolesc Psychol 37(1):184–214

Qian M, Murphy SA (2011) Performance guarantees for individualized treatment rules. Ann Stat 39(2):1180

Qiu X, Wang Y (2019) Composite interaction tree for simultaneous learning of optimal individualized treatment rules and subgroups. Stat Med 38(14):2632–2651

Qiu X, Zeng D, Wang Y (2018) Estimation and evaluation of linear individualized treatment rules to guarantee performance. Biometrics 72(2):517–528

Rebonato R, Jäckel P (2011) The most general methodology to create a valid correlation matrix for risk management and option pricing purposes. Available at SSRN 1969689

Rosenbaum PR, Rubin DB (1983) The central role of the propensity score in observational studies for causal effects. Biometrika 70(1):41–55

Su X, Tsai CL, Wang H, Nickerson DM, Li B (2009) Subgroup analysis via recursive partitioning. J Mach Learn Res 10(Feb):141–158

Tao PD, Muu LD et al (1996) Numerical solution for optimization over the efficient set by dc optimization algorithms. Oper Res Lett 19(3):117–128

Thall PF, Wathen JK (2005) Covariate-adjusted adaptive randomization in a sarcoma trial with multi-stage treatments. Stat Med 24(13):1947–1964

Thall PF, Sung HG, Estey EH (2002) Selecting therapeutic strategies based on efficacy and death in multicourse clinical trials. J Am Stat Assoc 97(457):29–39

Wahed AS, Tsiatis AA (2006) Semiparametric efficient estimation of survival distributions in two-stage randomisation designs in clinical trials with censored data. Biometrika 93(1):163–177

Wang Y, Fu H, Zeng D (2018) Learning optimal personalized treatment rules in consideration of benefit and risk: with an application to treating type 2 diabetes patients with insulin therapies. J Am Stat Assoc 113(521):1–13

Watkins C (1989) Learning from delayed rewards. Ph.D. thesis, King's College, Cambridge

Zhao Y, Zeng D, Rush AJ, Kosorok MR (2012) Estimating individualized treatment rules using outcome weighted learning. J Am Stat Assoc 107(499):1106–1118

Zhao YQ, Zeng D, Laber EB, Kosorok MR (2015) New statistical learning methods for estimating optimal dynamic treatment regimes. J Am Stat Assoc 110(510):583–598

Part III
General Issues About Subgroup Analysis, Including Regulatory Considerations

Chapter 14
Subgroups in Design and Analysis of Clinical Trials, General Considerations

Li Ming Dong, Heng Li, Ram Tiwari, and Lilly Q. Yue

Abstract Assessing whether different subgroups of clinical trial participants have different treatment benefits plays a crucial role in the interpretation of the clinical trial findings. All analyses of subgroup treatment effects are not conducted for the same purpose. While an exploratory subgroup analysis can be performed after the trial completion to evaluate the heterogeneity/homogeneity of the treatment effects across various subgroups, focused subgroup analysis can also be planned at the trial design stage to establish treatment effect in a specific subpopulation in addition to or in lieu of the overall study population. This chapter discusses general statistical issues and points to consider in subgroup analysis in confirmatory randomized controlled clinical trials for medical products. The chapter covers the following three topics:

1. General issues in subgroup analysis as part of the overall evaluation of a clinical trial;
2. Trial design considerations to establish treatment efficacy in specific subgroup of patients; and
3. Bayesian subgroup analysis.

A summary is provided at the end of the chapter.

As a means of evaluating the safety and effectiveness of medical products, clinical trials enroll targeted patients who are considered most likely to benefit from the treatment under investigation. Nevertheless, the enrolled patients are typically far from being homogeneous. They vary in biological characteristics such as age, sex, race, disease stage, biomarkers etc. It is well-known that there are times when

L. M. Dong (✉) · H. Li · R. Tiwari · L. Q. Yue
The Center for Devices and Radiological Health, U.S. Food and Drug Administration, Silver Spring, MD, USA
e-mail: Liming.Dong@fda.hhs.gov; Heng.Li@fda.hhs.gov; Ram.Tiwari@fda.hhs.gov; Lilly.Yue@fda.hhs.gov

© Springer Nature Switzerland AG 2020
N. Ting et al. (eds.), *Design and Analysis of Subgroups with Biopharmaceutical Applications*, Emerging Topics in Statistics and Biostatistics,
https://doi.org/10.1007/978-3-030-40105-4_14

a medical product performs differently in different subgroups defined by those characteristics. Therefore, in making treatment decisions, physicians as well as patients are often interested in finding out whether differences exist in the benefit-risk profile of a treatment across such subgroups. This knowledge can also help us better understand how the treatment works and may lead to the identification of a better-defined patient population for the treatment. As such, assessing treatment effect in different subsets of trial participants, or subgroup analysis, plays a crucial role in the interpretation of the clinical trial findings and regulatory decision making. Besides subgroups defined by biological characteristics, another kind of subgroups that can influence the performance of a medical product is geographical region, because of regional differences in patient population or in medical practice. The statistical issues discussed herein apply to those subgroups as well.

This chapter covers three topics in subgroup analysis in confirmatory clinical trials: (1) General issues in subgroup analysis as part of the overall evaluation of a clinical trial; (2) Trial design considerations to establish treatment efficacy in specific subgroup of patients; and (3) Bayesian subgroup analysis. The confirmatory clinical trial in this chapter refers to randomized parallel-group comparative clinical trials conducted to confirm the effect of a medical product or intervention. The issues and considerations are described mostly in the context of superiority trials; however, the considerations apply to other types of clinical trials such as non-inferiority trials.

14.1 General Issues in Subgroup Analysis as Part of the Overall Evaluation of a Clinical Trial

For confirmatory clinical trials, after analyses on the overall patient population have been conducted, additional analyses are performed routinely for subgroups based on demographics, disease severity, geographical region and other relevant factors as an integral part of evaluation of trial results. There are two scenarios to consider with a completed clinical trial: (1) The clinical trial met its objectives in the overall study population, in terms of statistical and clinical significance on the key hypotheses regarding primary efficacy/safety endpoints; (2) The trial did not meet its objectives in the overall study population.

14.1.1 Successful Trial for the Overall Study Population

When a clinical trial met its objectives in the study population, such as a significant treatment effect is detected in the overall study population in a superiority trial, observed differences in the primary efficacy/safety endpoints (and in some cases, important secondary endpoints as well) are examined for various subgroups of interest. The objective of the subgroup analysis is typically to gain insight into the level of consistency/heterogeneity of the treatment effect across the subgroups. Relative consistency among the subgroups may provide evidence that the findings are robust over the intended patient population, while signs of heterogeneity may be used to inform clinical practice. However, challenge in interpreting subgroup analysis findings arises when subgroup analyses were not included as part of a pre-specified and multiplicity-adjusted statistical plan. Some observed subgroup treatment effects could be due to true heterogeneity and others are due to chance alone. A key challenge is how to distinguish true heterogeneity from the play of chance, especially when the number of subgroups is large and subgroup sample sizes are small.

In addition to examining observed treatment effects by various subgroups, the consistency/heterogeneity of treatment effect across subgroups can be assessed through statistical testing of treatment-by-subgroup interaction, quantitative or qualitative. With quantitative interactions, the magnitude of the treatment effect may vary across subpopulations, but the subgroup-specific treatment effects are in the same direction. Suspected quantitative interactions usually do not lead to restrictions on the population for which a product can be deemed efficacious; however, variations in observed treatment effect among subgroups need to be reported. With qualitative interactions, the treatment difference is nonzero in at least one subgroup but is zero or goes in the opposite direction in at least one other subgroup. In such a case, considerable concerns arise in the case of superiority trials, as this implies the investigational treatment is no better than or even worse than the control for certain subgroups. It should be noted that in practice, clinical trials usually have low power to detect potentially important treatment-by-subgroup interactions, and failure to detect a significant interaction does not imply the absence of an important interaction. It is also known that observed subgroup difference may be explained by other characteristics, for example, gender difference may be due to difference in body size.

If the treatment effect is consistent across the subgroups of interest, or the treatment effect varies in magnitude across the subgroups but not too much, the subgroup analyses can increase the confidence in the robustness of the study results in the overall patient population, and a conclusion about the overall effectiveness of the treatment could be drawn. When there is a significant treatment-by-subgroup interaction, investigation is needed to explore possible reasons that may account for the heterogeneity of the treatment effects, and conclusions of a favorable benefit-risk profile may be restricted to a certain subgroup of the studied population. It is known that when a large number of subgroup analyses are conducted when the treatment effect is not large, the probability is high that some subgroups will have treatment effect estimates in the opposite direction. In other words, there is a great uncertainty in observed subgroup treatment effects and caution is needed in attempt to interpret the observed subgroup effects.

14.1.2   Failed Trial for the Intended Overall Patient Population

If a study didn't meet its objectives in the intended overall patient population, but some exploratory subgroup analyses identified one or more subgroups in which the investigational therapy is apparently beneficial, it is difficult to distinguish whether the observed treatment benefit is due to true benefit or chance. Alosh et al. (2015) calculated the probability of observing at least one subgroup with a nominal p-value <0.025 for selected numbers of subgroup factors and levels when the true treatment effect in the overall study population and in all the subgroups is zero. For three subgrouping factors with three levels each, the probability of at least one subgroup with a nominal p-value <0.025 when the overall finding was not significant is 0.15; the probability is much higher with large number of subgroup factors commonly seen in clinical trials. Therefore, the subgroup findings should usually be considered as hypothesis generating that need to be confirmed by one or more new studies (Alosh et al. 2015).

When planning confirmatory clinical trials, due to limited information available on the treatment effect for the study population and even less information about treatment effect in subpopulations, the trials typically assume implicitly the consistency of treatment effect across different subpopulations. However, from a scientific perspective, patients with different biological characteristics may respond differently to treatment. As such, subgroup analyses looking for signs of gross deviation from homogeneity assumption in a successful trial is a necessary first step in investigating whether overall findings can be applied to subpopulations with various biological characteristics. For failed trials, subgroup analyses are also conducted to understand the treatment as well as to search for subpopulations that may benefit from the treatment. Despite subgroup-level treatment effect heterogeneity is rarely identifiable reliably in a single clinical trial, report of subgroup analysis results is important. The availability of subgroup data not only provides needed information for the clinicians and patients to make clinical decisions, but also provides information for planning future study or makes future meta-analysis of a specific subgroup possible when data for the same subgroup from other trials are available.

14.2   Trial Design Considerations to Establish Treatment Efficacy in Specific Subgroup of Patients

Clinical trials target patients who are expected to benefit from the treatment under investigation. If there is indication, either from previous studies or based on scientific rationale, that treatment effects may vary substantially across different subgroups, the trial perhaps should be conducted only among the group of the patients who are most likely to benefit from the treatment, because an overall average "treatment effect" obtained from mixing such subgroups presents great difficulty for interpretation.

When there is uncertainty about in which study population the treatment is most effective, or the investigators want to establish the treatment efficacy either in the overall patient population or in a pre-specified target subgroup in one clinical trial, it is possible to design a clinical trial to achieve that goal. In this case, it is critically important to plan at the design stage. The planning needs to consider two aspects. One is the control of overall type I error rate due to the multiplicity issue, the other is power. The two alternative paths to trial success, one for the overall study population and one for the targeted subgroup, give rise to multiplicity issue. Adjustment for this multiplicity is needed to ensure proper control of the type I error rate. The other aspect for consideration is that the study power for both overall study population and the targeted subgroup should be adequate, and a sufficient number of patients in the subgroup are needed for reliable subgroup results. Sample size estimation can be done separately for the overall study population and the targeted subgroup, each with its desired power and expected size of treatment effect. In such trials, randomization stratified by the targeted subgroup and its complementary subgroup is recommended because of the advantage of creating balance within each subgroup. This is especially useful when the goal is to establish treatment efficacy in the targeted subgroup. In addition to statistical considerations, care also needs to be given to whether the estimated sample size for the targeted subgroup would be adequate for safety evaluation of the treatment in the subgroup.

In some situations, trials may need to enroll a higher proportion of target subgroup than its prevalence in the general population (over sampling) to increase power for the target subgroup. For this enrichment design, the estimate and inference of the treatment effect for the entire study population need to adjust for the enrichment sampling scheme. Methods have been proposed for various types of enrichment designs (Zhao et al. 2010; Simon and Simon 2013; Lai et al. 2019).

Although treatment effect in the overall study population and in the targeted subgroup of study population can be estimated and tested statistically, the study findings for the overall patient population should be carefully interpreted. For example, if the treatment effect in the overall study population is mostly driven by the large effect size in the targeted subgroup, i.e., the treatment under investigation has little or even negative effect in the complementary subgroup, it would raise the question whether the treatment effect applies to patients in the complementary subgroup. This is in fact the case of treatment effect by subgroup interaction. To better understand treatment effect for overall patient population, an evaluation of the complementary subgroup should be performed.

14.3 Bayesian Subgroup Analysis

The statistical inference for subgroup analysis that has been discussed so far in this chapter can be viewed as a frequentist approach. Frequentist inference for subgroup-specific treatment effect has limitations in that it forces the analyst into a dichotomy of either solely using overall treatment effect or solely using within-subgroup treatment effect. Bayesian inference can overcome this limitation, by "borrowing" information across subgroups. For example, Simon (2002) proposed a method in which prior distributions are assigned to parameters in such a way that the probability of a qualitative interaction is *a priori* small, thereby making information in the "other" subgroups relevant in estimating treatment effect for a given subgroup.

An approach for Bayesian subgroup analysis that has a larger body of literature is the Bayesian hierarchical modeling. Under this rubric, various methods of estimating treatment effect in subgroups have been proposed using different models (Louis 1984; Dixon and Simon 1991; Gamalo-Siebers et al. 2016; Hsu et al. 2019). The Bayesian estimators from this approach shrink the observed subgroup treatment effect toward an overall treatment effect, resembling the shrinkage estimator under the frequentist framework.

The basic idea and general framework of Bayesian hierarchical modeling can be illustrated using a model based on normal distribution. Following Pennello and Rothman's notation (2018), let y_j be the sample estimate of the treatment effect for subgroup $j, j = 1, 2, \ldots, J$, such as differences in sample average or proportions or log hazard ratios, and $y = (y_1, y_2, \ldots, y_J)$. In the first level of the hierarchical model, y_j given μ_j and σ_j^2, are is assumed to be independently distributed with

$$y_j \mid \mu_j, \sigma_j^2 \sim N\left(\mu_j, \sigma_j^2\right),$$

where μ_j is the treatment effect and σ_j^2 is the variance for subgroup j. In the second level μ_j are specified as independently distributed with

$$\mu_j \sim N\left(\mu_0, \sigma_\mu^2\right)$$

where μ_0 is the mean of the subgroup treatment effects, i.e., the overall treatment effect, and σ_μ^2 is the between-subgroup variance of the subgroup treatment effects. Finally, prior distributions are specified for μ_0 and σ_μ^2, and σ_j^2 are replaced by their sample point estimates. Many proposed methods are analogous to the above hierarchical model although the model settings and priors are different (Louis 1984; Dixon and Simon 1991; Gamalo-Siebers et al. 2016; Hsu et al. 2019). Simulation may be needed because analytical posterior distributions may be difficult to obtain for some choices of priors.

Subgroup analysis with Bayesian hierarchical modeling offers an alternative to frequentist inference in investigating treatment homogeneity/heterogeneity among subpopulations of clinical trial participants, where point estimate of a subgroup treatment effect parameter is obtained using only the data in this subgroup. Often times when large differences between subgroup treatment effects are observed, questions would arise as to whether such differences reflect the true underlying treatment effects or are simply due to randomness. Bayesian approaches provide posterior distribution of the subgroup treatment effect parameters while integrating the treatment effects from other subgroups. The posterior distribution allows the calculation of the probability of extreme subgroup treatment effect. In addition, the Bayes estimates which shrinks the subgroup effect toward the overall treatment effect may be more appealing because the apparently extreme subgroup effects in some subgroups are likely exaggerated especially when many subgroups are examined.

Although a useful tool, Bayesian subgroup analysis with hierarchical models has limitations. A crucial assumption in this approach is the exchangeability of the subgroup effects, implied by the specification that treatment effects in different subgroups are samples from a common probability distribution. This exchangeability assumption needs to hold a priori, i.e., it cannot be checked with the trial data, therefore its validity needs to be considered carefully. For example, in the subgroup analysis by gender, one should ask whether it is a clinically plausible assumption that effects in males and females are exchangeable.

It should be noted that, just as with the frequentist subgroup analysis, when the analysis is not pre-planned, subgroup analysis with Bayesian hierarchical modeling is post hoc in nature and should be considered as exploratory. As discussed earlier, findings from such analyses need to be confirmed by one or more new studies.

14.4 Summary

Subgroup analyses are routinely performed in randomized confirmatory clinical trials to examine treatment effect in subgroups of study populations. Estimation or testing of treatment effect in a specific subpopulation of the trial can be achieved with validity in an appropriately designed trial taking into consideration type I error control and adequate statistical power. An exploratory subgroup analysis can be conducted to evaluate the heterogeneity/homogeneity of the overall treatment effect across various subgroups. While the observed consistency of treatment effects across subgroups provide confidence in the estimated overall treatment effect, the observed heterogeneity of subgroup treatment effects, whether under a frequentist or a Bayesian paradigm, needs to be carefully investigated, and sometimes generates hypotheses for new trials.

Even if subgroup-level treatment effect heterogeneity cannot be ascertained reliably in a clinical trial, report of subgroup analysis results is important in not only providing useful information to the clinicians and patients for making clinical decisions, but also in informing the design of future studies.

References

Alosh M, Fritsch K, Huque M et al (2015) Statistical considerations on subgroup analysis in clinical trials. Stat Biopharm Res 7(4):286–303
Dixon DO, Simon R (1991) Bayesian subset analysis. Biometrics 47:871–881

Gamalo-Siebers M, Tiwari R, LaVange L (2016) Flexible shrinkage estimation of subgroup effects through Dirichlet process priors. J Biopharm Stat 26(6):1040–1055

Hsu Y, Zalkikar J, Tiwari RC (2019) Hierarchical Bayes approach for subgroup analysis. Stat Methods Med Res 28(1):275–288. http://journals.sagepub.com/doi/pdf/10.1177/0962280217721782

Lai TL, Lavori PW, Tsang KW (2019) Adaptive enrichment designs for confirmatory trials. Stat Med 38:613–624. https://doi.org/10.1002/sim.7946

Louis TA (1984) Estimating a population of parameter values using Bayes and empirical Bayes methods. J Am Stat Assoc 79:393–398

Pennello G, Rothmann M (2018) Chapter 10 Bayesian subgroup analysis with hierarchical models. In: Peace KE et al (eds) Biopharmaceutical applied statistics symposium, ICSA book series in statistics. Springer Nature Singapore Pte Ltd, Singapore, pp 175–192. https://doi.org/10.1007/978-981-10-7826-2_10

Simon N, Simon R (2013) Adaptive enrichment designs for clinical trials. Biostatistics 14(4):613–625

Simon R (2002) Bayesian subset analysis: application to studying treatment-by-gender interactions. Stat Med 21:2909–2916

Zhao YD, Dmitrienko A, Tamura R (2010) Design and analysis considerations in clinical trials with a sensitive subpopulation. Stat Biopharm Res 2:72–83

Chapter 15
Subgroup Analysis: A View from Industry

Oliver N. Keene and Daniel J. Bratton

Abstract Subgroup analysis in clinical trials for regulatory and reimbursement purposes can be confirmatory or exploratory in nature. Confirmatory subgroup analysis requires pre-specification of the proposed analysis and appropriate control of the type I error rate. Exploratory subgroup analysis is a feature of Phase III clinical trials. Examination of the results by sex, age and race is required by FDA and submissions for regulatory approval typically involve numerous further analyses by baseline characteristics such as disease severity. For efficacy these exploratory analyses are often directed at providing reassurance that the overall estimated treatment effect translates into benefit for each of the subgroups and for safety to investigate the existence of signals in more vulnerable subgroups. For reimbursement purposes, extensive analysis is required to try to identify those groups experiencing most benefit and for whom the medicine is therefore most cost-effective.

Exploratory subgroup analyses present a major challenge in interpretation due to the large number of subgroups examined. Effect sizes can vary largely from the overall treatment effect estimate and even be in opposite directions due to chance alone. The commonly used statistical methods to assess consistency of effect all have limitations. There is an important role for statistical modelling and an increasing interest from industry in Bayesian shrinkage techniques which balance emphasis on a specific observed differential subgroup effect with the overall treatment effect.

When planning and designing confirmatory trials of new medicines, discussion and agreement with regulatory and reimbursement authorities on the population is exceptionally valuable. Pre-identification of a small number of important biologically plausible subgroups which require exploration is helpful for interpretation.

O. N. Keene (✉) · D. J. Bratton
GlaxoSmithKline Research and Development, Middlesex, UK
e-mail: oliver.n.keene@gsk.com

© Springer Nature Switzerland AG 2020   309
N. Ting et al. (eds.), *Design and Analysis of Subgroups with Biopharmaceutical Applications*, Emerging Topics in Statistics and Biostatistics,
https://doi.org/10.1007/978-3-030-40105-4_15

15.1 Introduction

The classic design of a company-sponsored late-stage trial is directed at providing a single overall estimate of the effect of a medicine on the primary endpoint. Many important stakeholders find a single summary of response on an endpoint to be incomplete. Patients want to know if this average effect will apply to them with their own individual set of baseline characteristics which will vary among patients studied in the clinical trial. Physicians are concerned with identifying those patients for whom the medicine will be more effective or less effective. Payers only want to pay for a medicine for patient groups where the medicine is cost-effective.

The need for subgroup analyses is therefore unavoidable for late stage clinical trials performed by the pharmaceutical industry. They are regularly requested by practising physicians seeking to understand the results of the trial in the context of the diversity of the patients who consult them.

In a regulatory setting, the FDA require summaries of efficacy and safety by demographic subgroups (FDA 2015) and for a multi-regional trial an evaluation of consistency of treatment effects across regions is required by ICH E17 (ICH 2017). In a reimbursement setting, the Institute for Quality and Efficiency in Health Care (IQWiG) in Germany requires analysis by sex, age, country and disease severity for all patient relevant endpoints, including safety endpoints as well as efficacy endpoints (IQWiG 2017). These requirements are independent of an a priori expectation that a particular subgroup will experience a different treatment effect. As well as these mandated subgroups, further subgroup analyses are also frequently requested by regulatory and reimbursement agencies to assess consistency of treatment effects.

In the next sections we start by defining what is meant by a subgroup effect. We then review the key issue of multiplicity. Later sections describe analysis methods that go beyond the simple approaches of separate analysis of subgroups according to a specific characteristic and interaction tests.

15.2 Defining a Subgroup Effect

It is important to define what is meant by a subgroup effect as this terminology can have different interpretations. Subgroups can be dichotomous (e.g. male/female), categorical (e.g. region), ordered categorical (e.g. disease score at baseline) or based on a continuous measure (e.g. age). For subgroups defined by a continuous measure, patients are often categorised based on values lying within specific cut-points. A more powerful method of evaluation is often to retain the continuous scale and use a modelling approach (this is discussed later in the chapter).

Subgroup effects considered in this chapter are defined by baseline characteristics measured prior to treatment. Analysis based on differentiating patients according to a post-randomisation measurement can be misleading, because a particular treatment effect may influence classification to the subgroup (Yusuf et al. 1991).

Table 15.1 Illustrative example of importance of scale of measurement for subgroup effects

Baseline event rate	Placebo event rate	Active event rate	Absolute reduction	Percentage reduction (%)
0	0.8	0.6	0.2	25
1	1.2	0.9	0.3	25
2 or more	2.0	1.5	0.5	25

There are two keys aspects to describing a subgroup effect for a typical phase III superiority study. Firstly, the baseline characteristic may affect the outcome regardless of treatment and therefore be a prognostic variable. In modelling terms, this would be a main effect. For example, severe patients may have poorer outcomes than milder patients. Secondly, a baseline characteristic may influence the effect of active treatment compared to placebo and therefore be a predictive variable. In modelling terms this would be an interaction of the treatment effect with the variable. The same covariate can be both prognostic and predictive; it is examination of potential predictive variables that is the focus of this chapter.

Importantly, whether a differential treatment effect exists may depend on the scale of measurement used (Keene 1995). For instance, consider the example below shown in Table 15.1. The outcome variable is the number of events during treatment, and this has been split according to number of events in the previous year. For those with a baseline event rate of 0 per year, the event rate after randomisation is 0.8 on placebo compared to 0.6 on treatment, a reduction of 25%. The same percentage reduction of 25% applies to those with two or more events in the previous year. However, some may consider absolute reductions as more clinically relevant; these are very different, a reduction of 0.5 events/year for this group compared to 0.2 events/year for the group with 0 events at baseline. Model based analysis of event rates such as the negative binomial model (Keene et al. 2007) express treatment effects in terms of relative reductions and therefore there would be no statistical interaction. However, for a payer, there may be more willingness to fund a medicine that reduces event rates by 0.5 events a year than one that reduces events by 0.2 events.

15.3   Multiplicity in Subgroup Analysis

The major difficulty when interpreting subgroup analysis is that subgroup differences in treatment effect can arise by chance and it is exceptionally hard to identify what is a true difference. While there is a general acknowledgement that results from small subgroups are unreliable, unfortunately results from analyses of larger subgroups of patients are often interpreted as the true results for that group of patients, ignoring the fact that it is likely that some groups will show bigger or smaller differences simply by chance. While multiplicity issues can also arise in clinical trials from other sources such as multiple endpoints, the issue is particularly difficult for subgroups.

In the classic illustration of the problem, the ISIS-2 authors (ISIS 1988) examined the outcome by astrological birth sign. While the overall results showed an impressively positive effect for aspirin on mortality, for patients born under Gemini or Libra there was a small observed increase in mortality.

For trials performed by the pharmaceutical industry, prior specification of a subgroup, combined with an appropriate strategy to strongly control the type I error rate is required if a claim of efficacy in a subgroup is to be approved (EMA 2002). If the effect of a treatment is expected to be stronger in a subgroup compared to the complimentary subgroup, then studying this subgroup alone is an option. However, where the subgroup is defined by a biomarker, there is a desire from regulatory authorities to understand the effect in both the biomarker positive and biomarker negative patients. The FDA guideline on enrichment designs (FDA 2019) suggests that the type I error rate for the study be shared between a test conducted using only the enriched subpopulation and a test conducted using the entire population. In this case, it will be beneficial to increase sample size in the group where greater efficacy is expected. Simple strategies for this sharing include Bonferroni adjustment of p-values or hierarchical testing but increases in power can be obtained using strategies that take advantage of the correlation between the test statistics for analysis of a subgroup and analysis of the whole population (Song and Chi 2007).

Historically, there have been concerns about inferring efficacy in a post-hoc subgroup in trials where the overall effect was not positive. However, the current emphasis in regulatory and reimbursement submissions is on showing that specific subgroups derive benefit from the medicine in the presence of a positive effect overall. Regulators seek assurance that effects are consistent across subgroups and payers seek to restrict access to medicines to those groups where the benefit is strongest.

Li et al. (2007) investigated the probability of observing negative subgroup results when the treatment effect is positive and homogeneous across subgroups. Negative here is defined as an effect size in the opposite direction to the overall result. They show that if a trial with 90% power to detect an overall effect and total sample size of 338 is divided into five equally sized subgroups, the probability of observing at least one negative subgroup result is 32%. Each subgroup in this case has more than 65 patients.

The number of different subgroups that are typically examined in a confirmatory clinical trial of a new medicine is extensive and this can create challenges in interpretation. For an integrated summary of effectiveness, the FDA guideline (FDA 2015) includes the following list of subpopulations to be considered: age, sex, race, disease severity, prior treatment, concomitant illness, concomitant drugs, alcohol, tobacco, body weight and renal or hepatic functional impairment. While some of these subpopulations may not be applicable to a specific medicine, most will be and there will be a perceived need to split the data into subgroups according to multiple criteria. The greater the number of subgroup analyses performed and the smaller the resulting subgroups are, the higher chance that there will be subgroups with seemingly no benefit or potential harm from treatment. This issue of selection bias is recognised in the European regulatory guideline on subgroup analysis (EMA 2019) which states: "Not only might the play of chance impact the estimated effect, but it is tempting to focus on subgroups with extreme effects".

For submission for reimbursement in Germany, the Institute for Quality and Efficiency in Health Care (IQWiG) requires analysis by sex, age, country and disease severity for all patient relevant endpoints, including safety endpoints as well as efficacy endpoints (IQWiG 2017). These analyses are usually performed in the population for whom reimbursement is sought, often already a subgroup of the trial population. This requirement can typically lead to an excessively large number of subgroup analyses (e.g. 5 characteristics × 20 endpoints = 100 subgroup analyses) and can involve very small sample size in some analyses. The credibility of the analysis produced for IQWIG submissions has therefore been questioned (Ruof et al. 2014) and the value of this exhaustive exercise in the determination of the cost-effectiveness of the medicine is unclear.

When assessing whether observed differences across levels of a subgroup represent a true difference, it is possible to use checklists such as that provided by Sun et al. (2010). In practice, discussion often focuses on biological rationale (Hemmings 2014; Pocock et al. 2002). Unfortunately, biological plausibility is a somewhat elusive concept as most subgroup analyses have a degree of plausibility and therefore it is helpful to plan subgroup analysis in advance of unblinding of the trial. One possibility is to divide proposals for subgroup analysis into whether (a) a differential effect is anticipated, (b) a differential effect is biologically plausible but not anticipated and (c) observed differential effects are hypothesis generating (Dane et al. 2019). The weight given to the observed findings could then depend on which category the subgroup analysis was assigned to as well as the overall number of subgroup analyses performed.

Replication across endpoints and across two or more trials strengthens the support for a hypothesis of a different effect in a specific subgroup. In particular because of regression to the mean, treatment effects from exploratory subgroup analyses that show the biggest differential often cannot be reproduced.

15.4 Statistical Methods

The next sections describe commonly used statistical methods for investigating exploratory subgroup effects. This is not an exhaustive list and other methods are available. The focus is on methods that explore treatment by a single covariate, which is a problem in many practical cases in analysis of data from clinical trials.

15.4.1 Separate Analysis by Subgroup

Separate analysis by subgroup can be performed by either using an entirely separate analysis for the specific subgroup or via a model of the complete data with a treatment interaction term for the subgroup under investigation.

Graphical representation of subgroup analyses is a key component in facilitating the interpretation of subgroup analyses. Forest plots, displaying treatment effect estimates for each subgroup along with the associated confidence interval, is one of the most common displays used.

Interpretation of such forest plots however is not straightforward. For example, it is not possible to draw valid inferences about consistency of effect by comparing the individual subgroup p-values or by assessing whether the CIs in the forest plot cross the line of no difference. A significant difference in one subgroup but not the other is not necessarily evidence of a significant difference between the subgroups.

When performing subgroup analysis, it is common to classify a continuous variable such as age into categories and to analyse each subgroup separately. A key choice then is the number and location of the cut-points used to define the categories. Usually these might be dictated by clinical relevance. For example, it is often necessary to define body mass index (BMI) subgroups by <18.5, 18.5–<25, 25–<30 and ≥ 30 kg/m^2 as these are commonly used in clinical practice to identify underweight, normal, overweight and obese patients respectively. Where possible, it is helpful to state the cut-points prior to unblinding as different choices of cut-points can result in different estimates of treatment effect (Royston et al. 2006).

However, pre-specifying cut-points is not without issues. In some cases, there might be insufficient data in a particular pre-defined subgroup to allow estimation of a treatment effect. In such cases the subgroup could be combined with a neighbouring group but then the analysis can lose some of its value in estimating treatment differences in groups of interest or even miss a true interaction. A potential solution to this problem is to define subgroups by quantiles of the observed covariate distribution (e.g. quartiles) to help ensure that there will be sufficient data within each subgroup. Although such an approach might help to identify associations between the treatment effect and the covariate, the chosen subgroups may not have an easy clinical interpretation.

15.4.2 Interaction Tests

The classical approach to assessing consistency of effects across subgroups is to perform an interaction test. The focus of interest here is the contrast between the effects in the different subgroups, rather than examining a specific subgroup in isolation. For a factor with multiple levels such as region, a global test of any difference across all categories can be performed or a test for a specific category vs. the rest of the population. The ICH E9 guideline indicates in section 5.7 that interaction testing is the first step in undertaking subgroup analyses (ICH 1999).

However, in practice, simple significance tests for interaction are on their own of limited value when investigating subgroup differences. Firstly, as interaction tests are tests of significance, they have an associated fixed type I error rate. If this is fixed at 5%, then even if there are no true differences among subgroups, 5% of the tests will be expected to be significant suggesting a differential subgroup effect. Because of the low power of interaction tests, tests at the 10% or 20% level have been suggested (Hemmings 2014). In these cases, even more false positive results are to be expected.

Secondly, they have low power to detect heterogeneity. For example, in the simple case of a continuous endpoint with two equal sized groups, the variance of the interaction contrast is four times the variance of the overall treatment difference. This implies that only unlikely large interaction effects can be detected with any certainty.

Absence of statistically significant interactions does not imply consistency of the treatment effect in the studied population since absence of statistical significance cannot be taken to imply equality or consistency. To require only absence of statistical significance in an interaction test, or only directional consistency, would not be sufficiently sensitive filters to detect differences of potential interest.

The need to go beyond simple interaction tests is recognised in the CHMP guideline on subgroup analysis (EMA 2019) which states that "The sole reporting of an isolated p-value from a test for interaction is an inadequate basis for decision making". The guideline recommends including estimates of the size of the interaction contrasts with associated confidence intervals to show what differences a trial can reliably detect.

15.4.3   Stepwise Regression

When subgroups are assessed individually, the analysis does not account for potential imbalances in known effect-modifiers between groups. Multivariable analysis of an endpoint including various subgroups of interest and their interaction with treatment may be required to determine whether the effects observed within a subgroup are partially or wholly affected by other factors. In addition, a modelling approach allows for correlation between covariates instead of examining these in isolation.

Selection methods based on stepwise regression can be a useful exploratory approach for this to help determine the most influential factors on treatment effect. Various approaches are available for such an analysis, including forward, backward and stepwise selection (Royston and Sauerbrei 2009).

In backward selection, all subgroups of interest and their interactions with treatment are included in a model and the term with the largest p-value is removed if it is above a specified significance threshold (e.g. $p = 0.1$). The process continues iteratively removing the term with the largest p-value above the significance threshold at successive steps until all remaining covariates are significant at this level. Alternatively, selection methods based on information criteria (e.g. Akaike's information criterion) or penalised likelihood could be used rather than p-values for the individual covariates. Main effects should only be removed if its interaction with treatment has also been removed at a previous step. Forward selection is essentially the opposite to this, with terms being added separately to a model and retaining that with the smallest p-value below some specified threshold for the next step of the process. This is repeated until no covariates are significant when added to the model. In this case interaction terms should only be added to the model if the corresponding main effect was added at a previous iteration. Stepwise selection is a combination of the two approaches, testing variables for inclusion or exclusion at each step and allowing previously included or excluded variables to be removed or reincluded respectively. Ideally all methods will result in the same final model, but this is not always the case.

Results from such models should be deemed to be exploratory in nature since the selection procedure will tend to lead to an over-estimation of the effects of the selected covariates and, as in the case of separate analyses of subgroups, type I error rate is not strictly controlled. However, they are a useful tool for hypothesis generation or building prediction models.

When control of type I error is required, then potential methods are reviewed in Dane et al. (2019), Ballarini et al. (2018) and Thomas and Bornkamp (2017). Dane at al describe resampling methods and Balletini et al use penalised regression with a Lasso-type penalty as a model selection and estimation technique. Thomas and Bornkamp include model averaging in addition to resampling and Lasso methods. However, absence of statistical significance does not imply that the effects are the same in each subgroup and in a response to the Dane et al. article, Hemmings and Koch (2019) argue that "power should be prioritised over type I error where the objective is to generate signals for further inspection".

15.4.4   Fractional Polynomial Modelling Approaches with Continuous Covariates

As stated above, it is common to analyse a continuous variable by classifying the variable into categories. Clear disadvantages of this approach are the loss of information (Altman and Royston 2006; Royston et al. 2006) and the assumption that patients close to a cut-point will have different responses when these are likely to be similar. While these subgroup analyses provide treatment effect estimates within a narrower range of the baseline covariate than in the overall study, they do not necessarily adequately estimate the effect of treatment for a particular value of that covariate which might be more useful to an individual patient.

A more informative approach is to create a statistical model of the outcome by treatment as a function of the covariate (Keene and Garrett 2014). For instance, the covariate of interest can be entered into the model as a continuous linear term along with its interaction with treatment. Such a model allows treatment differences to be estimated at particular values of the covariate of interest rather than in groups. A resulting plot of the estimated treatment difference versus the covariate can potentially show in more detail how the treatment effect varies over the range of the covariate than a forest plot of subgroup effects focusing on specific categories.

However, if the relationship between treatment efficacy and the covariate of interest is non-linear then a model where the prognostic and predictive effects of a covariate are represented by linear terms may fit the data poorly. For instance, if a treatment has lower efficacy in patients who are underweight (<18.5 kg/m^2) and overweight (≥ 25 kg/m^2) compared to patients in the 'normal' range, then the association between efficacy and BMI is non-linear. In such a case a linear model will miss such a result whereas the subgroup analysis is more likely to demonstrate this interaction. Other transformations of the covariate could be assessed (e.g. log transformation or adding a quadratic term) but while some transformations might fit the data better and more closely align with subgroup estimates, there may be more appropriate functions to use.

Fractional polynomial models (FPs) offer the flexibility to identify non-linear treatment-covariate interactions (Royston and Altman 1994). In the FP framework, various transformations of a covariate are assessed and the model which describes the data best is selected. Transformations of the covariate of interest X that are assessed are of the form X^p, where p is chosen from a set S of eight powers: $S = \{-2, -1, -0.5, 0, 0.5, 1, 2, 3\}$. Here $p = 0$ indicates a log transformation of X. Each transformed covariate is assessed individually and that which maximises the likelihood of the model is used to assess the treatment interaction (Royston and Sauerbrei 2004). The model can include other covariates, for instance those pre-specified in the primary analysis, or a multivariable fractional polynomial (MFP) algorithm can be applied prior to modelling the treatment interaction to determine the most influential prognostic factors for the outcome and their best fitting forms (Royston and Sauerbrei 2009).

An FP model containing a single transformation of the covariate X is referred to as an FP1 model. To increase the flexibility of the modelling procedure, two transformations of the covariate can be entered into the model using powers from the same set S, so the model contains terms X^p and X^q where $p, q \in S$. This is referred to as an FP2 model. If $p = q$ then this is referred to as a repeated-powers model and one of the terms is replaced by $X^p \log(X)$. Unlike an FP1 model, including two transformations of X allows non-monotonic functions to be fitted thus greatly increasing the flexibility of the modelling. Examples of FP1 and FP2 functions are shown in Fig. 15.1. More than two transformations of the covariate could be used, but such models do not greatly increase the flexibility of the modelling procedure over and above FP2 models, can greatly increase the time taken to find the best fitting model, and may lead to overfitting.

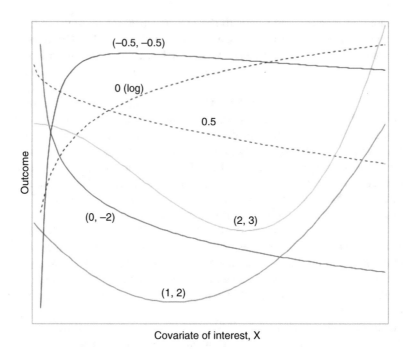

Fig. 15.1 Examples of FP1 (dashed lines) and FP2 (solid lines) functions

It is important to check the results of the FP modelling, particularly if it indicates a treatment-covariate interaction. Should there be an interaction, then this is also likely to be indicated by a subgroup analysis. Therefore, estimating treatment effects within a number of subgroups, for instance defined by quartiles or quintiles, can show whether there is agreement between the two approaches. Disagreement should be a signal of caution and investigated as it could be an artefact of the modelling— for instance due to influential outliers of the covariate which are less likely to affect a subgroup analysis.

Although FP modelling has several advantages over subgroup analysis, it is not without some potential pitfalls. FPs can behave strangely at the tails of the covariate, particularly close to 0 when negative powers are used. However, given that tails contain little data and that the CIs for the treatment effect line are likely to be wide, the plot of the treatment interaction can simply be truncated so that only the middle 90 or 95% of the distribution of the covariate are presented. There are also issues with scaling and ensuring that the covariate is strictly positive prior to modelling, but suitable solutions are available (Royston and Sauerbrei 2007).

An example of the value of a modelling approach is provided by the METREO and METREX trials of mepolizumab in patients with COPD (Pavord et al. 2017). These two randomised, placebo-controlled, double-blind, parallel group trials compared mepolizumab (100 mg in METREX, 100 or 300 mg in METREO) with placebo, given every 4 weeks for 52 weeks in patients with COPD who had a history of moderate or severe exacerbations while taking inhaled triple maintenance therapy. The trials were funded by GlaxoSmithKline (ClinicalTrials.gov numbers: METREO: NCT02105961, METREX: NCT02105948). The primary variable was the rate of moderate/severe exacerbations and analysis was performed using a negative binomial generalised linear model with a log link function (Keene et al. 2007).

The key covariate of interest was the screening blood eosinophil count. A pre-specified meta-analysis of the two studies was performed to examine the result of the studies by subgroups defined by categories of screening blood eosinophil count and the results are shown in Fig. 15.2. The estimated exacerbation rate reduction in patients with a screening eosinophil count between 300 and <500 cells/μL is 18%, however, some patients are likely to fare better than others within this category and so it is not clear for example what the estimated treatment effect is for a patient with say an eosinophil count of 400 cells/μL. In addition, the subgroup analysis implies a cliff effect at the cut-offs whereby two similar values of eosinophils correspond to markedly different treatment effect estimates. In this example a patient with a screening eosinophil count of 499 cells/μL and another with 500 cells/μL are estimated to achieve a 18% and 33% reduction in exacerbations, respectively, when there is a negligible difference between the two eosinophil values.

The relationship between exacerbation rate reduction with mepolizumab and screening eosinophil count has been analysed using fractional polynomial modelling and the results are shown in Fig. 15.3. Here the best fitting model was an FP2

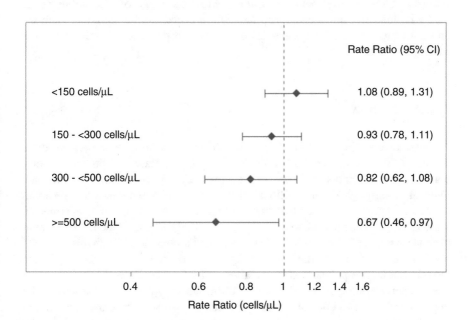

Fig. 15.2 Rate of moderate/severe exacerbations by screening blood eosinophil count: METREO/METREX trials

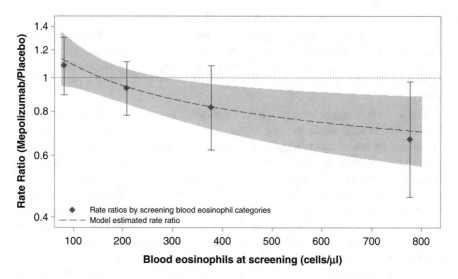

Fig. 15.3 Fractional polynomial modelling of exacerbations by screening eosinophils

function with repeated powers of (−0.5, −0.5). Estimates from the analysis based on the categories in Fig. 15.2 are overlaid on the FP plot against the mean eosinophil level in each category. The most obvious difference is that FP modelling estimates a smooth treatment effect curve across the range of eosinophils rather than a biologically implausible step-function obtained from a subgroup analysis, thus allowing more accurate estimates of treatment efficacy to be made at specific eosinophil values.

15.4.5 Splines

An alternative method to model treatment interactions with a continuous covariate is using splines. With splines and unlike FPs, the covariate is subdivided at cut-points defined as 'knots' and then separate regression curves are modelled in each segment using polynomial functions. These piecewise polynomials are anchored at the knots in such a way that the resulting curve is smooth and continuous. Various approaches are available for spline modelling but one of the more common methods is restricted cubic splines (Durrleman and Simon 1989). With this approach polynomial functions are fitted in each segment. A cubic function is used as this is the smallest degree polynomial which allows an inflection. Since cubic splines are likely to behave poorly at the tails due to lack of data, the splines are 'restricted' to be linear outside the two boundary knots. This can give an advantage over FPs, which as mentioned above can behave erratically in the lower tail particularly if values of the covariate are close to 0. Similarly, since functions are estimated in intervals of the covariate, splines may be less prone to outliers of the covariate compared to FPs.

An obvious additional step for splines is the need to specify the number and location of the knots, much like categorization in subgroup analysis. The choice of the number of knots can depend on the sample size and the prior belief in how 'undulating' the relationship is between efficacy and the covariate. Too many knots can lead to overfitting while too few can impede the flexibility of the modelling and thus might miss a true non-linear association. Authors have suggested using between 3 and 5 knots depending on sample size (Harrell 2001; Croxford 2016). For the location of the knots, Harrell (2001) has suggested particular quantiles depending on the number of knots to ensure that there is sufficient data within each interval to estimate the cubic function. For instance, for three knots Harrell recommends using the 10th, 50th, and 90th percentiles of the covariate, while for five knots use the 5th, 27.5th, 50th, 72.5th and 95th percentiles.

Despite the above guidance, the choice of knots can affect the resulting curve and so restricted-cubic splines can suffer from similar issues to subgroup analysis of the covariate. It is therefore important to pre-specify the knots where possible. Alternatively, penalized splines use many knots but discourage overfitting by restricting model complexity based on some penalty parameter (Eilers and Marx 1996). For instance, one option is to choose the spline which minimises the AIC (Binder et al. 2013). Penalised splines therefore avoid the need to specify the number and location of the knots, and hence some of the potential pitfalls of restricted cubic splines.

With these approaches, unlike FP modelling, there is currently no suitable procedure for simultaneously selecting functional forms and variables in a multivariable procedure (Binder et al. 2013). Binder et al. (2013) in their comparison of splines and FPs, concluded that for large sample sizes, the two methods often estimated similar curves, while for moderate sample sizes, FPs tended to outperform splines and were easier to implement.

Restricted cubic spline models were applied to the mepolizumab trial described above to model the efficacy of mepolizumab versus placebo on exacerbations by screening blood eosinophil count. Figure 15.4 shows the resulting curves for a spline with three knots and another with five knots using the percentiles as suggested by Harrell (2001) above and compares these curves to the best fitting FP2 function presented in Fig. 15.3. The 3-knot spline resulted in estimates curve very close the FP2 function while the 5-knot spline was more variable, likely due to fewer data points between knots. This demonstrates the need to carefully pre-specify the number of knots up-front, as the estimated curve can be sensitive to this choice.

15.4.6 Shrinkage Methods

As discussed above, when there is no true difference in efficacy between subgroups, spurious interactions can arise. This is especially the case if many subgroups are assessed, or if a specific subgroup contains a large number of levels. Subgroups including a small amount of data are particularly susceptible to showing a difference

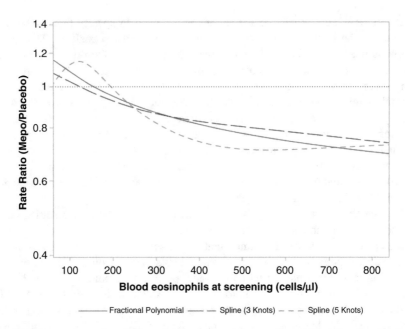

Fig. 15.4 Comparison of modelling of exacerbations by screening eosinophils using fractional polynomials with two powers and splines with three and five knots

to the complementary group due to the higher variability. Although the estimate in any one subgroup does not have a statistical bias in isolation, focusing on the specific result for that subgroup ignores relevant information from other groups.

Shrinkage methods are a technique to incorporate this information and move subgroup estimates toward the overall effect. They also increase the precision of the estimates by borrowing information across subgroups. Various shrinkage methods are available, including Empirical Bayes and Bayesian Hierarchical modelling. In the Empirical Bayes approach (Quan et al. 2013) the treatment effect d_i is first estimated in each subgroup i using data in that subgroup only. The subgroup estimates are then combined in a random-effects meta-analysis to obtain an estimate of the overall treatment effect, d, and the level of heterogeneity between the subgroup estimates as measured by the between-subgroup variability, τ^2. Subgroup estimates are then moved toward d by taking a weighted average of the original estimate d_i and d with weights w_i and $(1 - w_i)$ respectively where $w_i = \tau^2/(\tau^2 + s_i^2)$ and s_i^2 is the estimated variance of within-subgroup effect d_i. The result is that the original subgroup estimates are shrunk towards the overall effect, with this shrinkage being larger the higher the variability between estimates.

Another approach, Bayesian hierarchical modelling (Spiegelhalter et al. 1999), assumes that the effect in each subgroup d_i is a random quantity drawn from some common distribution centred around the overall treatment effect, d, i.e. $d_i \sim N(d, \tau^2)$. The subgroup effects are assumed to be exchangeable in that there is no reason a priori to believe that the effect in one group will be different from another. Prior

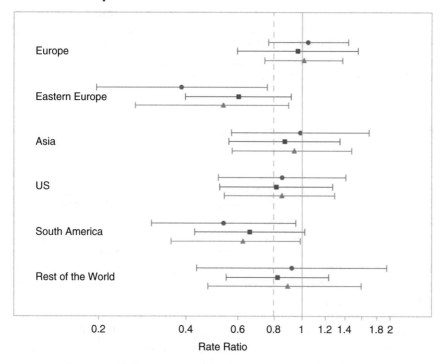

Fig. 15.5 Rate of moderate/severe exacerbations by region: METREO trial

distributions are placed on the random effect and the hyperparameters d and τ^2 to then estimate posterior distributions for the d_i and corresponding credible intervals to provide shrunken estimates of the subgroup effects.

Figure 15.5 shows a subgroup analysis of exacerbation rates by region for one of the example trials (METREO) described above. In the standard analysis using data within each subgroup separately, there appears to be a more beneficial effect of treatment in the Eastern Europe region compared to other regions, and the effect looks more favourable than the overall rate ratio of ~0.80 but confidence intervals are wide. The Empirical Bayes estimates are somewhat closer to the overall effect and the confidence intervals of most estimates are also narrower due to the borrowing of additional information from other regions. The Bayesian hierarchical analysis estimates are slightly closer to those from the original analysis, and CIs also have similar width. Thus, shrinkage techniques can incorporate prior scepticism about observing large positive or negative effects in subgroups which are unlikely to be true.

The above approach is useful primarily for evaluating a specific covariate as each patient needs to be included in a single category i.e. subgroups must be disjoint. If there is interest in assessing multiple subgroups simultaneously then patients need to be split by the covariates of interest (e.g. male European smokers). This is likely to lead to groups containing few patients, thus affecting the stability of the model. Instead, approaches involving model averaging of subgroup-specific models can be used (Bornkamp et al. 2017). Subgroups are assessed in individual models and the model averaging applies shrinkage across all groups.

15.4.7 Bayesian Dynamic Borrowing

One novel technique which may become increasingly useful in evaluation of subgroup effects is Bayesian dynamic borrowing (Schmidli et al. 2014; Gamalo-Siebers et al. 2017). As described above, it is often required to show evidence of effect in a subgroup alongside an overall positive effect. A separate analysis of the subgroup in question is limited by sample size and does not take account of the information on the effects of treatment in the complementary subgroup. A Bayesian statistical approach is one natural quantitative method to explicitly borrow information from the complementary subgroup to provide inferences on the subgroup under evaluation.

The approach works as follows. A robust mixture prior is constructed as a weighted combination of an informative prior and a non-informative prior. The results from the complementary subgroup are used for the informative prior for the response in the subgroup of interest. The non-informative prior consists of a weak prior distribution centred on a mean of zero, reflecting no relevance of results from the complementary subgroup. This weighted combination of priors allows for dynamic borrowing of prior information; the analysis learns how much of the complementary subgroup prior information to borrow based on the consistency between the subgroup of interest and the complementary subgroup.

The prior weight, w, assigned to the informative prior component represents the prior degree of confidence in the similarity of the two subgroups. At lower prior weights the mixture prior presents a heavier tailed distribution with more prior weight being applied to the non-informative weak prior component. When the mixture prior is combined with the observed efficacy data, w is updated using Bayes theorem according to how consistent the data in the subgroup are with the complementary subgroup; the stronger the evidence of consistency, the greater the increase in the posterior weight (w^*) relative to the prior weight (w). Conversely, when there is prior-data conflict, w^* will be lower than w and will tend to zero as evidence of conflict increases, so that the informative prior is down-weighted and posterior inference is based almost entirely on the observed data in the subgroup.

To assess the strength of prior belief in the consistency assumption required to show efficacy in the subgroup, a tipping point analysis can be carried out to identify how much prior weight (w) needs to be placed on the complimentary subgroup component of the robust mixture prior for the estimate of efficacy in the subgroup of interest to show statistically significant evidence of treatment benefit (in a Bayesian framework, this corresponds to a posterior probability that there is a treatment benefit of greater than 97.5%).

Subgroups that may be suitable for use of this dynamic borrowing approach include those subgroups of specific regulatory interest e.g. sex, race, region. For example, in a trial which includes both paediatric and adult subjects, there may be insufficient paediatric subjects to show statistical significance if this subgroup is analysed separately. A Bayesian dynamic borrowing approach of the adult data would allow assessment of the degree of belief needed that adult efficacy applied to paediatrics in order to conclude that there was evidence of efficacy in the paediatric subgroup.

15.4.8 Partitioning Methods

When there are more than a few pre-defined covariates, e.g. when there are multiple biomarkers under consideration, selection methods based on stepwise regression approaches become increasingly problematic. If there is interest in investigating complex models which go beyond evaluating relationships between treatment and a single covariate then stepwise regression may not be feasible due to the substantial number of potential two-way and three-way covariate interactions (Ruberg and Shen 2015). If there is more than one continuous covariate under evaluation, then a cut-point approach may be needed for the additional continuous variable and this brings the disadvantages described above.

Cluster analysis approaches group patients rather than examine covariates in series. They aim to identify subgroups of patients whose responses are more similar (in some sense) to each other than to those in other groups and the output is a classification tree. Historically cluster analysis has sometimes been performed with the aim of finding subgroups where the p-value for the difference between treatments is maximised, but such approaches have poor reproducibility. A more promising method is the SIDES (Subgroup Identification based on Differential Effect Search) method described by Lipkovich et al. (2011) and by Lipkovich and Dmitrienko (2014).

SIDES is a recursive partitioning method to establish response to treatment in patient subpopulations. The idea is to build a collection of subgroups by recursively partitioning a database into two subgroups at each parent group, such that the treatment effect within one of the two subgroups is maximised compared with the other subgroup. The process of data splitting continues until a predefined stopping condition has been satisfied.

An alternative approach to identify subgroups of patients with enhanced benefit is the virtual twins method described by Foster et al. (2011). The procedure works by first building a model to predict the response on treatment and control for each patient. Each patient comprises a set of 'twins' who differ only by the treatment they receive. This can be done by applying a random forest to each treatment group and then using the forest for a patient's opposite treatment to predict their response on that treatment (Foster et al. 2011). Random forests are particularly useful for this step as they exhibit low bias and prediction variance while avoiding overfitting, despite potentially dealing with a large number of covariates (Lipkovich et al. 2017).

The predicted within-patient treatment differences are then taken as observed values and used as the outcome for the subsequent subgroup identification step which uses a regression tree (or classification tree if the differences are dichotomised) to find a small number of strongly associated covariates. These are used to identify a subgroup of patients with a predicted treatment contrast greater than some clinically relevant threshold. For instance, if an asthma trial estimated the effect of treatment on FEV_1 to be 50 mL which might not be clinically relevant in many cases, then the procedure could be used to identify a subgroup likely to achieve a value more worthwhile, such as 100 mL. The enhanced treatment effect is then estimated as the difference between the effect in the subgroup and the overall population effect. Since the naïve estimate of this will be over-optimistic because the subgroup was estimated from the same data, Foster et al. (2011) describe a bias-corrected bootstrap procedure to obtain a better estimate of the effect.

Concerns can arise that clustering algorithms such as SIDES and the virtual twins method may over-fit the available data. In order to mitigate these concerns, a common practice is to divide the data into independent training and validation datasets. It is important to ensure that the training and test data sets are balanced with respect to the treatment variable and all prespecified categorical covariates (Lipkovich et al. 2011). A treatment effect identified based on the training set is considered to be confirmed if the effect is demonstrated in the validation data set.

These methods may require large sample sizes and/or large enhanced treatment effects to identify and confirm subgroups (Foster et al. 2011). If sample size is limited, it may not be practical to divide the dataset into training and validation datasets with a separate trial required to confirm findings.

A key disadvantage of the SIDES and Virtual Twins approach is the partitioning of continuous variables above and below a specific cut-point. As described above, this implies a cliff-edge effect at the cut-point which is biologically implausible.

Machine learning approaches combine different classification trees into ensembles of trees. There is no simple output showing how patients are classified; rather multiple trees are pooled in various combinations. These methods are primarily directed at prediction of response using a large number of input variables rather than at scientific understanding of which specific baseline characteristics predict response.

15.5 Discussion

In the case of a confirmatory trial for regulatory purposes, it could be argued that the burden of proof to establish an effect in each heterogeneous subgroup is with the trial sponsor. In particular, examination of results by sex and race is increasingly emphasised e.g. there are calls for efficacy to be established separately for both women and men. Many diseases are more prevalent in one sex rather than another; for example, trials of severe asthma have recruited a majority of female participants while COPD trials reflect the historically greater incidence of smoking among men. Depending on where trials are conducted, there is likely to be imbalance in the numbers of patients across the potential classifications of race and there can be confounding of race and region which may make it difficult to disentangle medical practice from race. Small numbers of patients in a specific race category leads to large variability as reflected in wide confidence intervals for the observed effect. Going forward there is likely to be an increasing need for recruitment to trials to reflect a greater diversity in the groups studied even if this does not reflect the relative prevalence of the disease being studied and to have larger sample sizes to allow appropriate assessment of effects in subgroups defined by sex and race. One novel approach that may be helpful is Bayesian dynamic borrowing which quantifies the degree of belief needed from the complementary subgroup to confirm efficacy in the subgroup under evaluation.

Exploratory subgroup analyses are a major scientific and statistical challenge (Peto 2011) and because of multiplicity issues it is hard to identify true quantitative interactions. Subgroup analysis should depend on the heterogeneity of the population and there should be fewer requirements for these analyses when the overall population is targeted (Keene and Garrett 2014).

Formal methods for defining consistency of effect are problematic. Tests of interaction are of limited value as they do not formally provide evidence for a lack of effect, although more emphasis could be placed on estimation of the interaction effect to direct a more rational approach to assessing consistency. Methods of subgroup analysis which strongly control type I error may be able to conclude a lack of evidence for differential effects but may not identify potentially clinical relevant differences in treatment effect. Bayesian shrinkage estimates can be helpful in the interpretation of differential subgroup effects as they balance the overall effect with that observed in the particular subgroup.

A modelling approach can be enlightening in identifying covariates which predict both the absolute level of outcome and the extent to which the treatment effect is modified in that subgroup. Newer methods such as the SIDES method allow consideration of multiple covariates and the interrelationships of these covariates on treatment effect. However, for continuous variables these methods employ a partitioning (cut-point approach).

Fractional polynomial modelling and splines allow a broad range of relationships between a continuous baseline characteristic and outcome and can show treatment interactions in greater clarity compared with categorisation of the covariate. These models of outcomes against a specific covariate avoid imposition of arbitrary cut-offs for continuous variables and can determine cut-offs for treatment based on the clinical relevance of the treatment effect observed. Thus, a modelling analysis is arguably more aligned to a stratified medicine paradigm where a specific expected treatment effect can be estimated more accurately for an individual based on their value for the covariate. Prediction intervals for an individual patient will nonetheless be wide as models summarise results of a trial over a range of values.

The key issue in subgroup analysis is whether heterogeneity can reasonably be assumed. When designing a clinical trial, it is usual to assume that a common effect size holds for all patient groups. If there is a scientific rationale for heterogeneous effects across subgroups defined by a specific characteristic, then it may be necessary to show effects of treatment separately in each subgroup which implies large increases in sample size for trials. Grouin et al. (2005) for example states: "If substantial heterogeneity of the treatment effect across subgroups is suspected at the design stage, then the whole basis of the trial is undermined."

The conundrum of subgroup analysis is therefore that consistency of effect has to be assumed at some level. The trial population is already a subgroup of possible patients who could be treated. Within that trial population, subgroups can be defined based on a specific characteristic. Analysis of this specific subgroup represents a combined effect across all other characteristics. Analysis of subgroups of subgroups is possible in theory, but in practice sample size quickly becomes very small.

In conclusion therefore, the desire for individualised medicine is never likely to be completely satisfied by examination of clinical trial data which by its nature only recruits a limited number of individuals. In general, only broad statements regarding effects of individual characteristics is likely to be possible.

References

Altman DG, Royston P (2006) The cost of dichotomising continuous variables. BMJ 332(7549):1080

Ballarini NM, Rosenkranz GK, Jaki T, König F, Posch M (2018) Subgroup identification in clinical trials via the predicted individual treatment effect. PLoS One 13(10):e0205971. https://doi.org/10.1371/journal.pone.0205971

Binder H, Sauerbrei W, Royston P (2013) Comparison between splines and fractional polynomials for multivariable model building with continuous covariates: a simulation study with continuous response. Stat Med 32(13):2262–2277

Bornkamp B, Ohlssen D, Magnusson BP, Schmidli H (2017) Model averaging for treatment effect estimation in subgroups. Pharm Stat 16(2):133–142

Croxford R (2016) Restricted cubic spline regression: a brief introduction. In: SAS proceedings. Institute for Clinical Evaluative Sciences, Toronto, p 5621

Dane A, Spencer A, Rosenkranz G, Lipkovich I, Parke T, on behalf of the PSI/EFSPI Working Group on Subgroup Analysis (2019) Subgroup analysis and interpretation for phase 3 confirmatory trials: white paper of the EFSPI/PSI working group on subgroup analysis. Pharm Stat 18:126–139. https://doi.org/10.1002/pst.1919

Durrleman S, Simon R (1989) Flexible regression models with cubic splines. Stat Med 8(5):551–561

Eilers PHC, Marx BD (1996) Flexible smoothing with B-splines and penalties. Stat Sci 11(2):89–121

European Medicines Agency (EMA) (2002) Points to consider on multiplicity issues in clinical trials. CPMP/EWP/908/99. Available at http://www.ema.europa.eu/docs/en_GB/document_library/Scientific_guideline/2009/09/WC500003640.pdf. Accessed Mar 2019

European Medicines Agency (EMA) (2019) Guideline on the investigation of subgroups in confirmatory clinical trials. https://www.ema.europa.eu/en/documents/scientific-guideline/guideline-investigation-subgroups-confirmatory-clinical-trials_en.pdf. Accessed Mar 2019

FDA (October 2015) Integrated summary of effectiveness, guidance for industry. Available at https://www.fda.gov/downloads/drugs/guidances/ucm079803.pdf. Accessed Mar 2019

FDA (March 2019) Enrichment strategies for clinical trials to support determination of effectiveness of human drugs and biological products, guidance for industry. Available at https://www.fda.gov/Drugs/GuidanceComplianceRegulatoryInformation/Guidances/UCM332181. Accessed Mar 2019

Foster JC, Taylor JM, Ruberg SJ (2011) Subgroup identification from randomized clinical trial data. Stat Med 30(24):2867–2880

Gamalo-Siebers M, Savic J, Basu C et al (2017) Statistical modeling for Bayesian extrapolation of adult clinical trial information in pediatric drug evaluation. Pharm Stat 10(1002):1807

Grouin JM, Coste M, Lewis J (2005) Subgroup analyses in randomized clinical trials: statistical and regulatory issues. J Biopharm Stat 15(5):869–882

Harrell FE (2001) Regression modelling strategies: with applications to linear models, logistic regression, and survival analysis. Springer-Verlag New York, Inc., New York

Hemmings R (2014) An overview of statistical and regulatory issues in the planning, analysis, and interpretation of subgroup analyses in confirmatory clinical trials. J Biopharm Stat 24(1):4–18

Hemmings R, Koch A (2019) Commentary on: subgroup analysis and interpretation for phase 3 confirmatory trials: white paper of the EFSPI/PSI working group on subgroup analysis by Dane, Spencer, Rosenkranz, Lipkovich, and Parke. Pharm Stat 18:140–144. https://doi.org/10.1002/pst.1935144

International Conference on Harmonisation (ICH) (1999) Statistical principles for clinical trials. Stat Med 18:1905–1942

International Conference on Harmonisation (ICH) (November 2017) E17: General principles for planning and design of multi-regional clinical trials. https://www.ich.org/fileadmin/Public_Web_Site/ICH_Products/Guidelines/Efficacy/E17/E17EWG_Step4_2017_1116.pdf. Accessed Mar 2019

IQWiG (July 2017). General methods, version 5. https://www.iqwig.de/en/methods/methods-paper.3020.html. Accessed Mar 2019

ISIS-2 (1988) Randomised trial of intravenous streptokinase, oral aspirin, both, or neither among 17,187 cases of suspected acute myocardial infarction: ISIS-2. ISIS-2 (second international study of infarct survival) collaborative group. Lancet 2(8607):349–360

Keene ON (1995) The log transformation is special. Stat Med 14(8):811–819

Keene ON, Garrett AD (2014) Subgroups: time to go back to basic statistical principles? J Biopharm Stat 24(1):58–71

Keene ON, Jones MRK, Lane PW, Anderson J (2007) Analysis of exacerbation rates in asthma and chronic obstructive pulmonary disease: example from the TRISTAN study. Pharm Stat 6:89–97

Li Z, Chuang-Stein C, Hoseyni C (2007) The probability of observing negative subgroup results when the treatment effect is positive and homogeneous across all subgroups. Drug Inf J 41(1):47–56

Lipkovich I, Dmitrienko A (2014) Strategies for identifying predictive biomarkers and subgroups with enhanced treatment effect in clinical trials using SIDES. J Biopharm Stat 24(1):130–153

Lipkovich I, Dmitrienko A, Denne J, Enas G (2011) Subgroup identification based on differential effect search (SIDES): a recursive partitioning method for establishing response to treatment in patient subpopulations. Stat Med 30(21):2601–2621

Lipkovich I, Dmitrienko A, B D'Agostino R Sr (2017) Tutorial in biostatistics: data-driven subgroup identification and analysis in clinical trials. Stat Med 36(1):136–196

Pavord ID, Chanez P, Criner GJ, Kerstjens HA, Korn S, Lugogo N, Martinot JB, Sagara H, Albers FC, Bradford ES, Harris SS, Mayer B, Rubin DB, Yancey SW, Sciruba FC (2017) Mepolizumab for eosinophilic chronic obstructive pulmonary disease. N Engl J Med 377(17):1613–1629

Peto R (2011) Current misconception 3: that subgroup-specific trial mortality results often provide a good basis for individualising patient care. Br J Cancer 104(7):1057–1058

Pocock SJ, Assmann SE, Enos LE, Kasten LE (2002) Subgroup analysis, covariate adjustment and baseline comparisons in clinical trial reporting: current practice and problems. Stat Med 21(19):2917–2930

Quan H, Li M, Shih WJ, Ouyang SP, Chen J, Zhang J, Zhao PL (2013) Empirical shrinkage estimator for consistency assessment of treatment effects in multi-regional clinical trials. Stat Med 32(10):1691–1706

Royston P, Altman DG (1994) Regression using fractional polynomials of continuous covariates: parsimonious parametric modelling. Appl Stat 43:429–467

Royston P, Sauerbrei W (2004) A new approach to modelling interactions between treatment and continuous covariates in clinical trials by using fractional polynomials. Stat Med 23(16):2509–2525

Royston P, Sauerbrei W (2007) Improving the robustness of fractional polynomial models by preliminary covariate transformation: a pragmatic approach. Comput Stat Data Anal 51(9):4240–4253

Royston P, Sauerbrei W (2009) Multivariable model-building. Wiley, Hoboken

Royston P, Altman DG, Sauerbrei W (2006) Dichotomizing continuous predictors in multiple regression: a bad idea. Stat Med 25(1):127–141

Ruberg SJ, Shen L (2015) Personalized medicine: four perspectives of tailored medicine. Stat Biopharm Res 7(3):214–229

Ruof J, Dintsios CM, Schwartz FW (2014) Questioning patient subgroups for benefit assessment: challenging the German Gemeinsamer Bundesausschuss approach. Value Health 17(4):307–309

Schmidli H, Gsteiger S, Roychoudhury S, O'Hagan A, Spiegelhalter D, Neuenschwander B (2014) Robust meta-analytic-predictive priors in clinical trials with historical control information. Biometrics 70(4):1023–1032

Song Y, Chi GY (2007) A method for testing a prespecified subgroup in clinical trials. Stat Med 26:3535–3549

Spiegelhalter DJ, Myles JP, Jones DR, Abrams KR (1999) An introduction to Bayesian methods in health technology assessment. BMJ 319:508–512

Sun X, Briel M, Walter SD, Guyatt GH (2010) Is a subgroup effect believable? Updating criteria to evaluate the credibility of subgroup analyses. BMJ 340:850–854

Thomas M, Bornkamp B (2017) Comparing approaches to treatment effect estimation for subgroups in clinical trials. Stat Biopharm Res 9(2):160–171

Yusuf S, Wittes J, Probstfield J, Tyroler HA (1991) Analysis and interpretation of treatment effects in subgroups of patients in randomized clinical trials. JAMA 266(1):93–98

Chapter 16
Subgroup Analysis from Bayesian Perspectives

Yang Liu, Lijiang Geng, Xiaojing Wang, Donghui Zhang, and Ming-Hui Chen

Abstract Identifying the sub-population structures along with the tailored treatments for all groups plays a critical rule for assigning the best available treatment to an individual patient. Subgroup analysis, a key to develop personalized medicine, becomes increasingly important over the past decade. Besides frequentist methods, there are a spectrum of methods developed from Bayesian perspectives to identify subgroups. In this chapter, we provide a comprehensive overview of Bayesian methods and discuss their properties. We further examine empirical performance of the two Bayesian methods via simulation studies and a real data analysis.

16.1 Introduction

In order to provide the best available treatment for individual patients, it is critical to examine whether heterogeneous treatment effect exists among the patient population. Many exploratory methods are developed in the literature to identify subgroups. Among them there are a variety of frequentist approaches, for instance, recursive tree based methods such as Interaction Trees (Su et al. 2009), Virtual Twins (Foster et al. 2011), Subgroup Identification based on Differential Effect Search (SIDES) (Lipkovich et al. 2011), Qualitative Interaction Trees (Dusseldorp and Van Mechelen 2014) and Generalized Unbiased Interaction Detection and Estimation (GUIDE) (Loh et al. 2015). Some optimization-oriented optimal treatment regime methodologies (Zhao et al. 2012, 2015; Tian et al. 2014; Chen et al. 2017) are also developed within the frequentist framework.

Y. Liu · L. Geng · X. Wang · M.-H. Chen (✉)
Department of Statistics, University of Connecticut, Storrs, CT, USA
e-mail: ming-hui.chen@uconn.edu

D. Zhang
Global Biostatistics and Programming, Sanofi US, Bridgewater, NJ, USA

© Springer Nature Switzerland AG 2020   331
N. Ting et al. (eds.), *Design and Analysis of Subgroups with Biopharmaceutical Applications*, Emerging Topics in Statistics and Biostatistics,
https://doi.org/10.1007/978-3-030-40105-4_16

Meanwhile, many Bayesian methods are proposed from different perspectives to . identify subgroups. In Sect. 16.2, we give an overview of some recently developed Bayesian methods for subgroup analysis. Simulation studies are conducted in Sect. 16.3. Section 16.4 presents a real data analysis. We conclude this chapter with a brief discussion in Sect. 16.5.

16.2 Bayesian Subgroup Analysis Methods

In subgroup analysis, a nonparametric mean structure $E(Y|X, trt) = g(X, trt)$ is often considered for the data Y, where $g(\cdot)$ is a multivariate function representing an underlying mechanism of the signal, trt indicates the treatment option, and X is a vector of potential covariates used to identify subgroups. In a commonly investigated scenario, there are two treatment options, placebo or treatment, i.e., $trt = 0$ or $trt = 1$. Then, the difference of treatment effects $\Delta(X)$ between these two options can be defined as

$$\Delta(X) = E(Y \mid X, trt = 1) - E(Y \mid X, trt = 0) = g(X, 1) - g(X, 0).$$

Therefore, we can equivalently model the nonparametric mean structure as

$$E(Y \mid X, trt) = g(X, 0) + \Delta(X)trt = a(X) + \Delta(X)trt. \tag{16.2.1}$$

The first term $a(X)$ in Eq. (16.2.1) is usually referred as the *prognostic effect*, since it affects the response at the same amount regardless of the treatment assignment. $\Delta(X)$ is often called the *predictive effect* or *predictive subgroup effect*, as $\Delta(X)trt$ affects the response differently under the different treatment assignment trt.

Tracing back to the literature in the twentieth century, Dixon and Simon (1991) proposed a linear regression model

$$E(Y \mid X, trt) = \mu + \tau trt + X\beta + \gamma Xtrt \tag{16.2.2}$$

with the first-order term γX serving as $\Delta(X)$, and a linear function of X serving as the prognostic effect $a(X)$, assuming the covariate X has two possible values. The parameters $(\mu, \tau, \beta, \gamma)$ are estimated using a Bayesian approach. Jones et al. (2011) extended the previous linear regression framework of Dixon and Simon (1991) by allowing second-order and third-order interaction terms for the predictive effects. These two methods are not directly applicable when there are other types of covariates, and may not work well when there are a large number of candidate variables.

Many other Bayesian subgroup analysis methods have been proposed from various perspectives. Below we introduce several recently developed Bayesian methods grouped by their similarity.

16.2.1 Tree-Based Bayesian Subgroup Analysis Methods

There are a few Bayesian subgroup analysis approaches which are linked to tree structures. The advantage of a tree structure is that it can handle interactions and nonlinear relationships between covariates and responses in an implicit way.

Berger et al. (2014) used a tree-splitting process to construct the treatment (subgroup) submodels, i.e., $\Delta(X)$ and baseline (prognostic) submodels, i.e., $a(X)$, which simultaneously incorporate the predictive effects and prognostic effects in the modeling. The tree-splitting process is randomly bisecting the covariate space recursively and leads to an allowable partition of the entire population arising from terminal nodes of a tree based on covariate splits, with possible zero treatment or baseline effects. There are several key steps in stochastically splitting a tree: (1) randomly select an ordering of covariates for splitting; (2) randomly determine the existence of a zero effect node at each level, and then randomly choose one of the nodes at that level to be the zero effect, which is a terminal node; (3) randomly decide non-zero effect nodes at each level to be further split by the corresponding covariate at that level; if not it becomes a terminal node. The detailed elaboration of the tree constructions is discussed in Wang (2012). The advantage of this tree splitting process is the elimination of possible partitions of the entire population without scientific meaning in comparison of treatment or baseline effects, which dramatically reduces the total number of models considered in the model space for the outcome.

The simplest way to model the outcome is to combine the treatment and baseline submodels with additive effects. Then, the model space for the outcome Y includes all possible distinct combinations of these two submodels. Next, the prior probabilities of the outcome models are assigned according to the stochastic scheme to generate trees. Once the prior specification is complete, the Bayesian model average techniques are utilized for subgroup analysis and, as a byproduct, the yielded results provide individual probabilities of treatment effect that might be useful for personalized medicine.

Here, we briefly discuss their main idea of defining an outcome model and specifying the priors. Let Ω denote the set of covariates in the study. Let X_{ij} be the j-th binary covariate for the i-th person, where $j \in \Omega$ and $i = 1, \cdots, n$. If we allow at most one covariate to split the treatment submodel, we are going to have five different types of models, i.e., $S_i^{1,0} = 0$, $S_i^{2,0} = trt_i \mu_2$, $S_i^{3,j} = trt_i \mu_{3j} \mathbf{1}_{\{X_{ij}=0\}} + trt_i \mu'_{3j} \mathbf{1}_{\{X_{ij}=1\}}$, $S_i^{4,j} = trt_i \mu_{4j} \mathbf{1}_{\{X_{ij}=0\}}$, $S_i^{5,j} = trt_i \mu_{5j} \mathbf{1}_{\{X_{ij}=1\}}$, where μ_2 is the mean overall treatment effect (if present), $\mu_{3j}, \mu'_{3j}, \mu_{4j}$ and μ_{5j} are the potential treatment (predictive) effects in the subgroups defined by the covariate j, trt_i is the treatment indicator, and $\mathbf{1}_{\{\cdot\}}$ is the indicator function. Similarly, there are two possible types of baseline submodels via splitting one factor. That is, $B_i^{1,0} = \mu_1$ and $B_i^{2,k} = \mu_1 + \beta_k \mathbf{1}_{\{X_{ik}=0\}}$, where μ_1 is the overall mean and β_k is the mean baseline effect for covariate $k \in \Omega$.

Then, the outcome model in Berger et al. (2014) is

$$Y_i = S_i^{h,j} + B_i^{\ell,k} + \epsilon_i, \quad \epsilon_i \sim \mathcal{N}(0, \sigma^2), \tag{16.2.3}$$

$i = 1, \cdots, n$, $h = 1, \cdots, 5$, $\ell = 1, 2$ and $j, k \in \{0, \Omega\}$. Let m be the number of covariates considered, then the total number of models for at most one covariate splitting is $2 + 5m + 3m^2$, which is a huge reduction from 2^{m+1} possible models when m is large.

The method developed in Berger et al. (2014) automatically takes account of multiplicity adjustment in the prior specification for the model space. The prior probability is computable via specifying three interpretable prior inputs, which are: (1) specifying the prior probability that an individual has no treatment (predictive) effect and no baseline (prognostic) effect, respectively; (2) assigning relative effect odds for a covariate i has an effect compared to the first covariate; (3) defining the ratio of the sum of the prior probabilities of the submodels with $i - 1$ split and the sum of the prior probabilities of the submodels with i splits. An advantage for this prior specification is that the experts can easily incorporate pre-experimental preference to specific subgroups. See Section 3 of Berger et al. (2014) for more details of computing the prior probability for each outcome model based on the three interpretable inputs.

Once the prior specification for the outcome model and the unknown parameters in the model is complete, then we can summarize the posterior quantity we are interested in. In Berger et al. (2014), they summarized the posterior quantity of interest using the Bayesian model averaging idea. Two interesting posterior summaries discussed in their paper are:

1. *Individual Treatment Effects*: first, the probability for an individual to have treatment effects is given by $P_i = \sum_{\mathcal{M}_\kappa \in \mathcal{M}} P(\mathcal{M}_\kappa \mid Y_1, \cdots, Y_n) \mathbf{1}_{\{\mu_{i\kappa} \neq 0\}}$, for any $i = 1, \cdots, n$, where $\mathcal{M}$ denotes the entire model space for the outcome model, $\mathcal{M}_\kappa$ is a specific outcome model in the model space, $\mu_{i\kappa}$ is the subgroup treatment effect associated with the ith individual in the given model $\mathcal{M}_\kappa$ and $\bar{\mu}_{i\kappa}$ is the posterior mean of $\mu_{i\kappa}$. Then, the *individual treatment effect size* for each individual is defined as weighted average of $\bar{\mu}_{i\kappa}$, i.e., $\Lambda_i = \sum_\kappa P(\mathcal{M}_\kappa \mid Y_1, \cdots, Y_n) \bar{\mu}_{i,\kappa} \mathbf{1}_{\{\mu_{i,\kappa} \neq 0\}} / P_i$.

2. *Subgroup Treatment Effects*: based on individual posterior probability for the treatment effects, the posterior probability of a nonzero treatment effect for Subgroup j (denoted as S_j) is defined as an average of P_i over the subgroups that individual j belongs to (using the symbol $\{\#i \in S_j\}$, i.e., $Q_j = \sum_{i \in S_j} P_i / \{\#i \in S_j\}$. Similarly, the subgroup treatment effect size for S_j is calculated via $\Delta_j = \sum_{i \in S_j} P_i \Lambda_i / \sum_{i \in S_j} P_i$.

The Bayesian approach described in Berger et al. (2014) can be generally extended to allow more than one covariate used in splitting. However, when more than two covariates are utilized in tree-splitting process, the total number of models that we need to consider will be increasing and the model enumeration scheme in Berger et al. (2014) becomes impossible.

Sivaganesan et al. (2017) restricted the scope from searching for subgroup effects among all possible subgroups, to searching for subgroup effects among only a few pre-determined candidate groups. More specifically, the authors focus on identifying subgroup effects related to certain pre-specified covariates and shapes of subgroups. Any center regions in the covariate space will be excluded from consideration, for instance, a subgroup defined as $\{a < X_1 < b, c < X_2 < d\}$ would be excluded. For any subgroup A, the amount by which its predictive effect $\Delta(A) = E(Y|X \in A, trt = 1) - E(Y|X \in A, trt = 0)$ exceeds the predictive effects of entire patient population $\Delta(C) = E(Y|trt = 1) - E(Y|trt = 0)$, that is, $\delta(A) = \Delta(A) - \Delta(C)$, is used as the primary measure for identifying any potential enhanced subgroup effects. The author defined a utility function to compare potential subgroups:

$$U(A) = \begin{cases} \frac{[(|A|-N)_+]^d}{(1+c)^{nvar(A)-1}}[\delta(A) - T_s], & \text{if } \emptyset \subset A \subset C, \\ 0, & A = \emptyset, \end{cases} \tag{16.2.4}$$

where $|A|$ is the number of observations in A, and $nvar(A)$ is the number of covariates used to define A, $\{x\}_+ = \max(0, x)$, N is the pre-specified minimum subgroup size, $c, d > 0$ are constants to control the "reward" for the subgroup size and the "penalty" for complex subgroups, respectively. T_s is the minimum threshold for $\delta(A)$ which corresponds to the clinically meaningful effect magnitude. Bayesian Additive Regression Trees (BART) (Chipman et al. 2010) approach is used to fit the response Y on the combined covariate space (X, trt) as a nonparametric function, to get the predicted value of $\delta(A)$ for each subgroup A. Subgroups with larger positive expected utility are preferred. Since the candidate subgroups are pre-specified, the process of exploring from the entire covariate space is omitted, which makes this approach differ from many other exploratory subgroup analysis methods.

Zhao et al. (2018) proposed another BART-based subgroup analysis approach to identify important biomarkers. They modeled the predictive effects $\Delta(X)$ with a single tree for better interpretability, and impose an additive tree structure on the prognostic effects $a(X)$ to enhance model fitting. Such an additive tree structure allows more flexibility for the prognostic effect comparing to the commonly assumed linear structures in Dixon and Simon (1991), Jones et al. (2011), and Schnell et al. (2016), which may lead to better estimation performance for the predictive effects $\Delta(X)$ at the same time. However, the computation time will also increase quickly when sample size and number of candidate variables get larger.

Similar to BART, the posterior sampling procedure is carried out using Bayesian backfitting algorithm (Hastie et al. 2000). The posterior probability that a biomarker served as a splitting variable in the predictive tree will be used to determine whether any covariate has notable predictive subgroup effect. In order to reduce "type I error", i.e., claiming irrelevant covariates as predictive biomarkers, no specific subgroup will be declared when such posterior probabilities for all biomarkers are less than a certain threshold. Several simulation scenarios with at most two biomarkers are considered in the paper, and the estimated probabilities for the true predictive biomarker(s) to rank as the top two predictive variables are reported. The method seems to identify the predictive biomarkers well when there is only one predictive biomarker, despite presence of some prognostic effects. Meanwhile it appears to be underpowered when there are two predictive biomarkers in the model, especially for the purpose of identifying both predictive variables as the top two candidate biomarkers.

16.2.2 ANOVA-Based Bayesian Subgroup Analysis Methods

Sivaganesan et al. (2011) developed a Bayesian approach from model selection perspective by considering each covariate separately and constructed the model space by enumerating the possible cases for different levels of treatment-subgroup effects. First denote M_{00} and M_{10} the overall null and the overall effect model, representing no treatment effect and homogeneous treatment effect in the whole population, and the model space of "overall effect", i.e., the model space of no treatment-subgroup interaction models, is $\mathcal{M}_0 = \{M_{00}, M_{01}\}$. Then for each covariate, define models in the model space by introducing the cluster membership indicator $\boldsymbol{\gamma} = (\gamma_1, \ldots, \gamma_S)$, where the elements in $\boldsymbol{\gamma}$ range from 0 to number of distinct non-zero treatment-subgroup effects and represent the order of appearances of distinct treatment-subgroup effects, and S is the number of levels of the covariate. To demonstrate this setting more clearly, Table 16.1 shows an example of models defined by a covariate of two levels.

Use the zero-enriched Polya urn scheme as the probability distribution on the model space $\mathcal{M}_{\mathcal{X}}$. After getting the posterior model probabilities, the authors proposed a decision-making algorithm, comparing the posteriors of models in $\mathcal{M}_0$ with the models defined by each covariate to determine whether notable subgroup effects should be reported. In the algorithm, two threshold values c_0 and c_1 are used for comparing model posterior probabilities. The model selected is the most likely model, and also beats the overall null model M_{00} and the overall effect model M_{01} as its posterior probability odds exceeding c_0 and c_1. Therefore, c_0 represents the threshold for the posterior probability odds of the overall or a subgroup effect model against the overall null model, and c_1 represents the threshold for the posterior probability odds of a subgroup effect model against the overall effect model. When

Table 16.1 Example of model space $\mathcal{M}_{\mathcal{X}}$ defined by covariate X

Model index	$\boldsymbol{\gamma}$	Treatment-subgroup effects
M_0	$(0, 0)$	$\Delta(X = 0) = \Delta(X = 1) = 0$
M_1	$(1, 0)$	$\Delta(X = 0) \neq 0, \ \Delta(X = 1) = 0$
M_2	$(0, 1)$	$\Delta(X = 0) = 0, \ \Delta(X = 1) \neq 0$
M_3	$(1, 1)$	$\Delta(X = 0) = \Delta(X = 1) \neq 0$
M_4	$(1, 2)$	$\Delta(X = 0) \neq \Delta(X = 1), \ \Delta(X = 0), \ \Delta(X = 1) \neq 0$

X has two levels

no subgroup or overall effect models satisfy the comparing conditions, the overall null model will be selected at last. A characteristic of this algorithm is that, when selecting subgroup models, it only compares models within the model space of each covariate, and in the end reports either models in $\mathcal{M}_0$, or one or more subgroup models defined by different covariates. Therefore, this method cannot discover subgroups defined by interactions of multiple covariates, unless data transformation is done. However, an advantage of this method is that it does not only discover subgroups, but also detect orders of subgroup effect sizes.

Liu et al. (2017) extended Sivaganesan et al. (2011) by considering two variables at a time and enumerated all possible situations for the mean levels to construct the model space. The authors elaborated on the case that there are two covariates of interest and each has two levels, which are specified a priori by the investigators. Similar to the decision algorithm introduced in Sivaganesan et al. (2011), a stepwise procedure is adopted based on posterior model probabilities to determine potential subgroup effects. The model space grows quickly when more covariates are considered and/or there are more than two levels for each covariate.

Both of these two ANOVA-based methods do not model prognostic effects as a function of the covariates, and the results may be biased when there exist covariate-dependent prognostic effects.

16.2.3 Other Types of Bayesian Subgroup Analysis Methods

Schnell et al. (2016) also used a linear combination of the covariates to model both prognostic effects and predictive effects similar to Dixon and Simon (1991). Denote the predictive effects as $\Delta(x) = x'\gamma$ for any covariate vector x, and define the beneficial subgroup as $B_\gamma = \{x : \Delta(x) > \delta, \delta > 0\}$ for a pre-specified threshold δ. This method aims to find a credible subgroup pair (D, S) satisfying $D \subseteq B_\gamma \subseteq S$, where D, defined as the "exclusive credible subgroup", is the region such that the posterior probability of $\Delta(x) > \delta$ for all $x \in D$ is no less than $1 - \alpha$. The "inclusive credible subgroup" S is defined as the region such that the posterior probability of including all x, s.t. $\Delta(x) > \delta$ for all patients in S is no less than $1 - \alpha$. The highest posterior density method is applied to find the $1 - \alpha$ credible region G_α for the posterior distribution of γ. Therefore (D, S) can be constructed as: $D = \{x : x'\gamma > \delta$ for all $\gamma \in G_\alpha\}$, and $S^c = \{x : x'\gamma \leq \delta$ for all $\gamma \in G_\alpha\}$. There are two other ways of constructing (D, S) discussed in the paper.

This approach may work well when the dimension of the parameter space is low, while the computational costs increase quickly when the number of candidate covariates increases. When the dimension of parameter space is high, it is also difficult to interpret the credible subgroup pair (D, S) and characterize the patient population within it.

Gu et al. (2013) applied a two-stage Bayesian lasso approach to time-to-event responses and also used the first-order terms of X to model the predictive effects. Three different treatment options are considered. In the first stage, linear combinations of main effects, overall treatment effects and first-order treatment-covariate interactions are considered to model the predictive effects, and shrinkage priors are specified on the parameters, and a distance-based criterion is implemented to help screen the unimportant biomarkers. In the second stage, all the biomarkers retained after the first stage will be included in the model, and the Bayesian adaptive lasso approach is deployed to perform further biomarker selection. The authors only considered the case when true predictive effects are linear structured in the simulation study, and the robustness of this method remains unknown when the predictive effect model is misspecified.

This method does not consider higher-order treatment subgroup interactions and the variable selection step does not extend further to split point selections to identify a potential subgroup such as $\{X_1 > 0.5\}$. Also, the sure screening property for the first stage has not been established yet for this method, we shall be wary of the fact that certain important biomarkers may be missed since the variables excluded from the first stage will never enter the second stage.

16.3 Simulation Studies

In this section, we carry out simulation studies to examine the empirical performance of two aforementioned methods (Berger et al. 2014; Sivaganesan et al. 2011). Both methods consider subgroups defined by one binary variable. We focus on testing of the scenarios listed below

(a) $y_i = 2 + \epsilon_i$,
(b) $y_i = 2 + 2trt_i + \epsilon_i$,
(c) $y_i = 2 + 2trt_i I(X_{i1} = 0) + \epsilon_i$,
(d) $y_i = 2 + 2trt_i I(X_{i1} = 0) + I(X_{i1} = 0) + 2I(X_{i2} = 0) - 3I(X_{i3} = 0) + \epsilon_i$,
(e) $y_i = 2 + 2trt_i \{I(X_{i1} = 0) + I(X_{i2} = 0)\} + \epsilon_i$,
(f) $y_i = 2 + 2trt_i \{I(X_{i1} = 0) - I(X_{i2} = 0)\} + \epsilon_i$,

where y_i is the i-th univariate response. The treatment variable $trt_i \overset{i.i.d}{\sim}$ Bernoulli(0.5). Ten independent binary covariates are considered: $X_{ij} \overset{i.i.d}{\sim}$ Bernoulli(0.5), $i = 1, \ldots, n$, $j = 1, \ldots, 10$. The random error is set at two levels, $\epsilon_i \overset{i.i.d}{\sim} N(0, 1)$ and $N(0, 4)$. We assume $\{\epsilon_i\}'s$, $\{X_{ij}\}'s$, and $\{trt_i\}'s$ are mutually independent, for $i = 1, \ldots n$, $j = 1, \ldots, J$. The indicator function $I(E)$ takes a value of 1 if the event E is true and 0 otherwise. Here we set $n = 100$ for all cases. We illustrate scenarios (c), (e), and (f) with tree diagrams in Fig. 16.1, where X_i denotes the i-th covariate.

Under scenarios (a) and (b), there is actually no subgroup with heterogeneous treatment effects. Under scenarios (c) and (d), there are heterogeneous treatment effects, between group $\{i : I(X_{i1} = 0), 1 \leq i \leq n\}$ and the rest of the population. Since the subgroup is defined by a single covariate, these two approaches are expected to detect X_1 with a high probability. In scenario (d), there are three

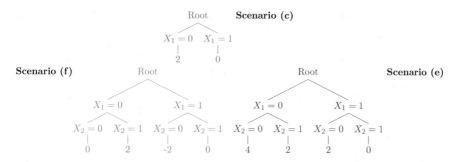

Fig. 16.1 Tree diagrams of scenarios (c), (e), and (f) along with the treatment effect size for each terminal node

prognostic variables X_1, X_2, and X_3, and it is desirable to test how these two methods perform when there are prognostic effects. For scenario (e), there are indeed three subgroups with heterogeneous treatment effects: $g_1 = \{i : I(X_{i1} = 0 \cap X_{i2} = 0), 1 \leq i \leq n\}$, $g_2 = \{i : I[(X_{i1} = 0 \cap X_{i2} = 1) \cup (X_{i1} = 1 \cap X_{i2} = 0)], 1 \leq i \leq n\}$, $g_3 = \{i : I(X_{i1} = 1 \cap X_{i2} = 1), 1 \leq i \leq n\}$, among which only g_3 is the subgroup with zero treatment effect. Under scenario (f), there is also qualitative treatment-subgroup interaction (i.e., there exists both subgroups with positive treatment effects and negative treatment effects), and there are two subgroups, $\{i : I(X_{i1} = 0 \cap X_{i2} = 1), 1 \leq i \leq n\}$ and $\{i : I(X_{i1} = 1 \cap X_{i2} = 0), 1 \leq i \leq n\}$, having non-zero treatment effects. We repeat each simulation independently for 200 times.

For the Bayesian tree method (Berger et al. 2014), the posterior probabilities P_i of having a non-zero treatment effect for individual $i = 1, \ldots, n$ are extracted as the major outcome for analysis. The simulations results are reported in Table 16.2. $\bar{P}^0$ represents the mean of P_i's of the patients whose treatment effects are 0. On the contrary, $\bar{P}^1$ represents the mean of P_i's of the patients whose treatment effects are non-zero. The medians of $\bar{P}^0$ across different scenarios are shown as the point estimate and the 95% confidence intervals for $\bar{P}^0$ are displayed below the point estimates correspondingly in the table. Similar results are shown for $\bar{P}^1$. Ideally, we should have $\bar{P}^0$ being close to 0, while $\bar{P}^1$ being close to 1 if a particular method performs well in distinguishing the patient group with non-zero treatment effects and the subgroup with the zero treatment effect.

For scenarios (c)–(f), since there are both subgroups with zero treatment effects and non-zero treatment effects, we can construct the receiver operating characteristic (ROC) curve and obtain the area under curve (AUC) for each scenario. To construct the ROC curve, we compare P_i with each of these threshold values $0, 0.01, 0.02, \ldots, 1.00$ to classify individuals to two groups: the group with non-zero treatment effect and the group with zero treatment effect. The AUC values are given in Table 16.2.

From Table 16.2, we see that for scenario (a), the mean of the patient's posterior probability of getting a non-zero treatment effect, $\bar{P}^0$, is 0.25, which is noticeably

Table 16.2 Simulation results under scenarios (a)–(f) obtained by the method (Berger et al. 2014)

	Scenario					
	(a)		(b)		(c)	
σ	1	2	1	2	1	2
$\bar{P}^0$	0.25	0.25	–	–	0.17	0.35
	(0.18, 0.59)	(0.18, 0.59)	–	–	(0.11, 0.59)	(0.18, 0.92)
$\bar{P}^1$	–	–	1.00	0.997	1.00	0.93
	–	–	(1.00, 1.00)	(0.82, 1.00)	(1.00, 1.00)	(0.27, 1.00)
AUC	–	–	–	–	1.00	0.89

	Scenario					
	(d)		(e)		(f)	
σ	1	2	1	2	1	2
$\bar{P}^0$	0.97	0.71	0.98	0.52	0.27	0.29
	(0.35, 1.00)	(0.19, 1.00)	(0.46, 1.00)	(0.24, 0.99)	(0.12, 0.97)	(0.15, 0.84)
$\bar{P}^1$	1.00	0.96	0.97	0.57	0.77	0.74
	(0.70, 1.00)	(0.59,1.00)	(0.65, 1.00)	(0.24, 0.99)	(0.64,0.99)	(0.45,0.93)
AUC	0.77	0.73	0.54	0.50	0.83	0.82

Symbol "–" is deployed when the criterion is not applicable for the cell

smaller comparing to $\bar{P}^1$ in scenario (b). Results from these two extreme scenarios give us some ideas about the "benchmark value" of P_i, regarding to patients with zero or non-zero treatment effects. Under scenario (c), there is only one binary predictive variable and no prognostic variable, this method performs very well in terms of AUC, and AUC drops a little when the noise level increases from $N(0, 1)$ to $N(0, 4)$. When adding prognostic variables to the model, we see from the results of scenario (d) that the AUCs are much smaller comparing to those under scenario (c). The point estimates of $\bar{P}^0$ are much closer to 1, which indicates that the method is not able to distinguish the group with zero treatment effect from the others. Since the method only considers up to one prognostic variable, when the prognostic effect structure is more complicated, it will affect the estimates of P_i's. For scenarios (e)–(f), there is no prognostic variable, while there are subgroups with non-zero treatment effects defined by more than one variable. Since we use the algorithm that allows at most one factor for split in Berger et al. (2014), the performance is not very good as expected. Results from scenario (f) are better comparing to those from (e), since there are more patients with zero treatment effect and it is easier to distinguish this "null group" from others.

Under the model space setting in Sivaganesan et al. (2011), the true models for scenarios (a) and (b) are M_{00} and M_{01}, namely, the overall null and the overall effect model. For scenarios (c) and (d), the true model is M_{11} indicating two subgroups defined by X_1, and the treatment effects in these two subgroups are zero and non-zero. For scenarios (e) and (f), based on the decision making algorithm, the models expected to be reported are M_{13} and M_{23}, representing there are heterogenous non-zero treatment effects defined by both X_1 and X_2. For the comparing threshold in the decision making algorithm, we set $c_0 = c_1 = c$ for simplicity, c varying from 0 to $\exp(25)$. Figures 16.2 and 16.3 show the probabilities of models reported under scenarios (a)–(f) for different values of c when $\sigma = 1$, 2. Note that c is chosen when type I error (TIE) is controlled and power is reached as big as possible. Under scenario (a) where TIE $= 1 - \Pr(M_{00}$ is reported$|M_{00})$, we observe from Figs. 16.2 and 16.3 that TIE is controlled at 0.1 for $\log(c) > 1.5$ and TIE is controlled at 0.05 for $\log(c) > 2$ approximately. From Fig. 16.3 when $\sigma = 2$, we notice that under scenarios (c)–(f), the probabilities of reporting true models are obviously lower than the probabilities when $\sigma = 1$. This indicates that selection accuracy of the method (Sivaganesan et al. 2011) is easily affected by data noise. In general, $\log(c) = 2$ controls TIE and achieves relatively high rates of reporting true models,

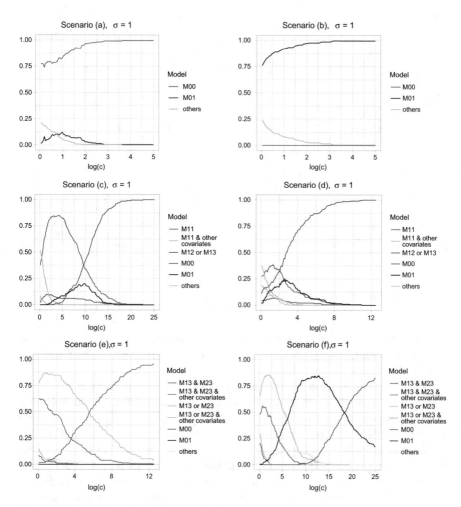

Fig. 16.2 Probabilities of models reported when $\sigma = 1$

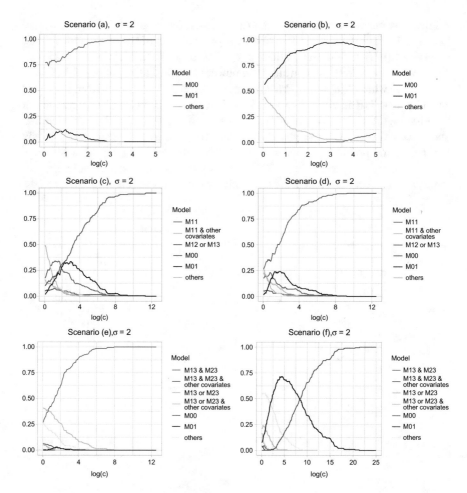

Fig. 16.3 Probabilities of models reported when $\sigma = 2$

therefore we choose $c_0 = c_1 = \exp(2)$ as the comparing threshold in the decision making algorithm, and the simulation results under this threshold value are shown in Table 16.3.

From Table 16.3 it can be seen that the method in Sivaganesan et al. (2011) performs quite well under scenarios (a) and (b) when there are no subgroup treatment effects for $\sigma = 1, 2$. However, by comparing results of scenarios (c) and (d), we find that prognostic variables cause great interference to selection results. This can also be seen from Fig. 16.2 and 16.3, where the reporting probability curve of M_{11} under scenario (d) is always lower than the curve of M_{11} in scenario (c). Under scenarios (e) and (f) when multiple covariates have subgroup effects, though reporting all true subgroup models is difficult, the probabilities of discovering at least one true subgroup model are notably higher.

Table 16.3 Simulation results under scenarios (a)–(f) obtained by the method (Sivaganesan et al. 2011) when $c_0 = c_1 = \exp(2)$

Scenario	(a)		(b)		(c)	
Ture model	M_{00}		M_{01}		M_{11}	
σ	1	2	1	2	1	2
P(TrueModel)	0.965	0.965	0.970	0.910	0.805	0.260

Scenario	(d)		(e)		(f)	
Ture model	M_{11}		M_{13}, M_{23}		M_{13}, M_{23}	
σ	1	2	1	2	1	2
P(TrueModel)	0.340	0.100	0.475	0.020	0.465	0.005
P^*	–	–	0.850	0.260	0.855	0.090

P^*=Pr(At least one true model is reported | multiple subgroup models) under scenarios (e)–(f)

16.4   A Real Data Example

We apply the method of Berger et al. (2014), the QUINT (Dusseldorp and Van Mechelen 2014) method (a frequentist approach), and the approach of Sivaganesan et al. (2011) to analyze the Breast Cancer Recovery Project (BCRP) dataset. BCRP dataset is publicly available in the R package *quint*. The test subjects were women with early-stage breast cancer. There were three treatment arms in the randomized trial: a nutrition intervention arm, an education intervention arm, and a standard care arm. We only study the patients from the education intervention (assign $trt = 1$) arm and the standard care arm ($trt = 0$). After removing missing values, we had 146 test subjects left, among which 70 patients were in $trt = 1$ group and 76 patients were in $trt = 0$ group.

The response variable was the improvement in depression score at a 9-month follow-up. There were nine covariates: age, nationality, marital status, weight change, treatment extensiveness index, comorbidities, dispositional optimism, unmitigated communion and negative social interaction, and we dichotomized each continuous or categorical variable by its median value so we can apply the Bayesian methods.

We use the default options to implement QUINT in R, and the final tree is just the trivial tree (i.e., no split is made), which indicates no notable qualitative treatment-subgroup interaction has been found. The posterior probabilities P_i's of having a non-zero treatment effect for all subjects are between 0.20 and 0.24, which also suggests no findings of subgroup effect. The method in Sivaganesan et al. (2011), where the decision-making is carried out based on $c_0 = c_1 = 2$ from our simulation results, also reports the overall null effect model, which means no subgroup is found. However, QUINT produces a non-trivial tree as reported in Liu et al. (2019) if the variables are not dichotomized. The information loss after dichotomization is also a main disadvantage for methods that are only applicable to binary variables.

16.5 Discussion

Overall, Bayesian subgroup analysis methods add in a lot of varieties and new aspects of thinking to the personalized medicine development. Bayesian methods such as Berger et al. (2014) and Sivaganesan et al. (2011) can provide inference over a model space rather than just one specific model, though it may not be easy to extend and apply these methods to accommodate categorical variables or continuous variables without information loss. In the aforementioned Bayesian tree methods, only the method developed by Zhao et al. (2018) can be applied directly to continuous variables, and considers the issue of splitting point selection implicitly when building the tree. Comparing to frequentist methods, Bayesian methods allow for the incorporation of prior information and expert's inputs as well as account for model uncertainty. Many of the Bayesian methods consider simple prognostic effect structures. When the dimension of the parameter space is high and there are various types of covariates, current Bayesian methods need to be improved to tackle these challenges.

Acknowledgements Mr. Liu's research was partially supported by NIH grant #P01CA142538. Ms. Geng's research was partially supported by NIH grant #GM70335. Dr. Chen's research was partially supported by NIH grants #GM70335 and #P01CA142538.

References

Berger JO, Wang X, Shen L (2014) A Bayesian approach to subgroup identification. J Biopharm Stat 24(1):110–129

Chen S, Tian L, Cai T, Yu M (2017) A general statistical framework for subgroup identification and comparative treatment scoring. Biometrics 73(4):1199–1209.

Chipman HA, George EI, McCulloch RE et al (2010) Bart: Bayesian additive regression trees. Ann Appl Stat 4(1):266–298

Dixon DO, Simon R (1991) Bayesian subset analysis. Biometrics 47(3):871–881

Dusseldorp E, Van Mechelen I (2014) Qualitative interaction trees: a tool to identify qualitative treatment–subgroup interactions. Stat Med 33(2):219–237

Foster JC, Taylor JM, Ruberg SJ (2011) Subgroup identification from randomized clinical trial data. Stat Med 30(24):2867–2880

Gu X, Yin G, Lee JJ (2013) Bayesian two-step lasso strategy for biomarker selection in personalized medicine development for time-to-event endpoints. Contemp Clin Trials 36(2):642–650

Hastie T, Tibshirani R et al. (2000) Bayesian backfitting (with comments and a rejoinder by the authors. Stat Sci 15(3):196–223

Jones HE, Ohlssen DI, Neuenschwander B, Racine A, Branson M (2011) Bayesian models for subgroup analysis in clinical trials. Clin Trials 8(2):129–143

Lipkovich I, Dmitrienko A, Denne J, Enas G (2011) Subgroup identification based on differential effect search–a recursive partitioning method for establishing response to treatment in patient subpopulations. Stat Med 30(21):2601–2621

Liu J, Sivaganesan S, Laud PW, Müller P (2017) A Bayesian subgroup analysis using collections of anova models. Biom J 59(4):746–766

Liu Y, Ma X, Zhang D, Geng L, Wang X, Zheng W, Chen M-H (2019) Look before you leap: Systematic evaluation of tree-based statistical methods in subgroup identification. J Biopharm Stat. https://doi.org/10.1080/10543406.2019.1584204

Loh W-Y, He X, Man M (2015) A regression tree approach to identifying subgroups with differential treatment effects. Stat Med 34(11):1818–1833

Schnell PM, Tang Q, Offen WW, Carlin BP (2016) A Bayesian credible subgroups approach to identifying patient subgroups with positive treatment effects. Biometrics 72(4):1026–1036

Sivaganesan S, Laud, PW, Müller P (2011) A Bayesian subgroup analysis with a zero-enriched polya urn scheme. Stat Med 30(4):312–323

Sivaganesan S, Müller P, Huang B (2017) Subgroup finding via Bayesian additive regression trees. Stat Med 36(15):2391–2403

Su X, Tsai C-L, Wang H, Nickerson DM, Li B (2009) Subgroup analysis via recursive partitioning. J Mach Learn Res 10:141–158

Tian L, Alizadeh AA, Gentles AJ, Tibshirani R (2014) A simple method for estimating interactions between a treatment and a large number of covariates. J Am Stat Assoc 109(508):1517–1532

Wang X (2012) Bayesian modeling using latent structures. Ph.D. Thesis, Duke University, Durham

Zhao Y, Zeng D, Rush AJ, Kosorok MR (2012) Estimating individualized treatment rules using outcome weighted learning. J Am Stat Assoc 107(499):1106–1118

Zhao Y-Q, Zeng D, Laber EB, Song R, Yuan M, Kosorok MR (2015) Doubly robust learning for estimating individualized treatment with censored data. Biometrika 102(1):151

Zhao Y, Zheng W, Zhuo DY, Lu Y, Ma X, Liu H, Zeng Z, Laird G (2018) Bayesian additive decision trees of biomarker by treatment interactions for predictive biomarker detection and subgroup identification. J Biopharm Stat 28(3):534–549

Chapter 17
Power of Statistical Tests for Subgroup Analysis in Meta-Analysis

Terri D. Pigott

Abstract Meta-analysis is used extensively in health and medicine to examine the average effect of a treatment and the variation among studies estimating this effect. However, the estimate of the average treatment effect is not the only concern in a systematic review. Researchers also need to understand how the treatment effect may vary. The potential challenges for subgroup analyses in meta-analysis parallel those in randomized trials where many have urged caution in the conduct and interpretation of subgroup analyses. This chapter will discuss the power of statistical tests for subgroup analysis in order to help in both the planning and interpretation of subgroup tests in a meta-analysis. The chapter will begin with an overview of subgroup analyses in a meta-analysis, reviewing recent research on the interpretation of these results. The chapter will then discuss the *a priori* power of subgroup analyses under the fixed and the random effects model and provide examples of power computations. The chapter will also present a general overview of power in meta-regression with directions for future research on power analysis for examining subgroup effects.

17.1 Overview of Subgroup Analysis in Meta-Analysis

Systematic review and meta-analysis examine the average treatment effect and its variance across a set of clinical trials. While meta-analysis can provide an estimate of the average treatment effect and its variance, researchers are often interested in whether the treatment effect differs for subgroups of patients or studies. Researchers conducting randomized trials often pose similar questions, asking whether the overall treatment effect is consistent across patients that differ on one or more

T. D. Pigott (✉)
Loyola University Chicago, Chicago, IL, USA

Georgia State University, School of Public Health and College of Education and Human Development, Atlanta, GA, USA
e-mail: tpigott@gsu.edu

© Springer Nature Switzerland AG 2020 347
N. Ting et al. (eds.), *Design and Analysis of Subgroups with Biopharmaceutical Applications*, Emerging Topics in Statistics and Biostatistics,
https://doi.org/10.1007/978-3-030-40105-4_17

baseline variables (Pocock et al. 2002). In a systematic review and meta-analysis, these questions focus on the variation in the estimated treatment effect across studies and whether that variation is associated with differences in study methods or subgroup differences.

When a researcher is conducting a meta-analysis, the first stage involves estimating the average treatment effect and its heterogeneity among the trials in the sample. Often the treatment effect differs across trials, and meta-analysis can help inform clinicians about potential reasons why the treatment effect varies. Guidelines for conducting subgroup analyses in randomized trials highlight the importance of pre-specification of subgroup analyses and the use of clinical knowledge to evaluate the credibility of these findings (European Medicines Agency 2014; Wang et al. 2007). These same cautions apply to subgroup analyses in meta-analysis. When the individual trials in a meta-analysis report on subgroup differences in the treatment and when researchers have an *a priori* hypothesis about the potential treatment differences, meta-analytic techniques can provide an estimate of the treatment effect for each subgroup, test whether the difference in that treatment effect is statistically significant, and examine the heterogeneity of those subgroup effects across studies.

Subgroup analyses in meta-analysis differ from the standard use of interaction effects in a randomized trial, the preferred method for testing the significance of subgroup effects (Brookes et al. 2004; Wang et al. 2007). When we are testing whether the mean effect size for a small number of subgroups differ, the meta-analysis techniques used are analogous to the use of one-way ANOVA models. This technique would be similar to the use of a single-predictor meta-regression. For example, let us say we are interested in testing whether there is a difference in treatment effect between men and women. Each study in the meta-analysis provides an effect size for the treatment effect for men and an independent effect size for the treatment effect for women. The current method for examining the subgroup effect is to assume that the effect sizes across studies are independent and to examine the difference in the mean effect size for men versus the mean effect size for women using an omnibus test for the equality of the two mean effects. When the group means differ and the effect sizes within groups are homogeneous, then there is evidence of a treatment effect difference. However, when the group means differ and the effect sizes within one or both groups are heterogeneous, the evidence for a treatment effect difference is unclear.

As in subgroup analyses within a clinical trial, subgroup analyses in meta-analysis are potentially both informative and potentially misleading as described by Oxman and Guyatt (1992). Subgroup analyses in meta-analysis should be planned in advance to avoid discovering a difference suggested by the data that may later prove spurious. Reviewers should clearly delineate among planned subgroup analyses and exploratory ones suggested by the data to avoid this difficulty. Planning for a small number of subgroup analyses is also important given the relative small numbers of trials in many systematic reviews in health.

17.1.1 Potential Role of Power in Subgroup Analysis in Meta-Analysis

Many researchers using meta-analysis have difficulty in the interpretation of the results of subgroup analysis in meta-analysis. Richardson et al. (2019) surveyed Cochrane Collaboration review authors who had included subgroup analyses in their completed reviews. Many of these authors were confused about the interpretation of the statistical significance of a subgroup difference particularly when one or more of the subgroups were heterogeneous. For example, Richardson et al. (2019) found that 28% of review authors incorrectly interpreted whether a statistically significant subgroup analysis was present in an example analysis. Richardson et al. (2019) provide criteria to assist review authors in interpreting subgroup analyses. When a subgroup analysis results in a statistically significant difference among subgroups, Richardson et al. (2019) urge researchers to discuss the plausibility and importance of the subgroup difference. In addition, they recommend that researchers also carefully examine the balance of studies contributing to each subgroup and whether there could be a confounding variable that could also explain the subgroup difference. This recommendation in consistent with guidelines for subgroup analysis in randomized trials that urge the use of clinical knowledge to evaluate the credibility of a subgroup finding as well as evidence from other sources of relevant trial data (European Medicines Agency 2014).

Richardson et al. (2019) provide a clear set of criteria for interpreting subgroup analyses in meta-analysis. Another strategy to assist reviewers in the interpretation of subgroup differences is *a priori* power computations. Knowing at the outset of a review the potential power for a range of subgroup differences can aid reviewers in their interpretation of the results. If there is adequate power (typically power of 0.8) to detect a potentially important treatment difference, then the reviewer can be more confident about the insights the meta-analysis may provide to clinical practice. Knowing the power of potential subgroup analyses may also researchers to resist the urge to conduct a number of exploratory subgroup analyses that are suggested by the data. As discussed by Brookes et al. (2004), power calculations for randomized trials focus on the overall treatment effect and rarely compute power for subgroup differences, a situation that also occurs in subgroup analysis in meta-analysis. This chapter will build on the suggestions of Richardson et al. by focusing in the power to detect a clinically important difference among subgroups in meta-analysis.

As in all statistical analyses, power for subgroup analysis depends on the size of the effect a researcher wishes to detect, the statistical test used, the sample size including the number of trials and the number of participants in the trial, and the balance of trials across the subgroups. The following sections provide an overview of computing power for subgroup analysis.

17.2   Power Computations in Meta-Analysis

In meta-analysis as in other statistical analysis, power analysis can provide the researcher with information about the potential power for conducting the analysis. The same is true in meta-analysis—power computations prior to collecting and coding studies for the systematic review allows the researcher to plan for the analysis and to avoid questionable research practices (Ioannidis 2005). This chapter will assume that the researcher has *a priori* hypotheses about variation in treatment effects across subgroups of patients and across studies using different methods.

Power computations in meta-analysis require the researcher to have educated guesses about the studies included in the review. Researchers conduct power analysis in meta-analysis under assumptions about the number of studies included in the review, the sample size of studies within the review, and other quantities related to the statistical test and form of the model (fixed or random). Hedges and Pigott (2001), Valentine et al. (2010) and Pigott (2012) provide a discussion of power analysis under the fixed and random effects model for the test of the average treatment effect and for the test of overall homogeneity. This chapter will focus on understanding the power of subgroup tests.

17.2.1   Power for Subgroup Analyses in Meta-Analysis: Test of Between-Group Homogeneity in Fixed Effects Models

In a fixed effects analysis, we assume that all variation among effect sizes in our sample of studies is due to random sampling error. Suppose that a researcher is interested in planning a meta-analysis that will examine how the treatment varies across studies. A common subgroup difference of interest relates to gender—how does a treatment vary for men and women? In meta-analysis, a researcher will typically examine the difference in the mean treatment effect for women and men, using either studies that report the treatment effect for women and men within a trial, or studies that use patients of only one gender. When there are a small number of subgroups, such as defined by gender or by a limited number of racial/ethnic categories, we use a test analogous to one-way ANOVA. For this test, we assume that the set of effect sizes are from independent samples within studies.

To set up our subgroup analysis, let us assume that there are k total studies in the meta-analysis. Within each study, we compute a value for the overall effect size within each study j, designated as T_j with variance v_j^2. This effect size may be a standardized mean difference, comparing the treatment and control group's performance on a continuous measure, or an odds ratio, comparing the number of events in the treatment versus the control groups. In meta-analysis, we compute the overall mean effect size as a weighted mean where the weights are given by $w_j = 1/v_j^2$, or the inverse of the variance of the effect size.

For our subgroup analysis, say we are interested in comparing the treatment effect sizes for a set of p subgroups. Since not all studies will provide an estimate of the effect size for the target subgroup, we will use m_i, where $i = 1, \ldots, p$, to indicate the number of studies within each of the i subgroups. We will designate the individual effect sizes by T_{ij}, $i = 1, \ldots, p$, and $j = 1, \ldots, m_i$. These effects sizes are for the treatment effect for each subgroup, computed within each of the j studies. Each effect size has variance, v_{ij}^2, $i = 1, \ldots, p$ and $j = 1, \ldots, m_i$. When we are interested in whether the treatment effect differs for men and women, p equals 2. If instead we are interested in subgroups defined by race or ethnicity, p may be greater than 2.

Our null hypothesis in a subgroup analysis in meta-analysis takes the form of

$$H_0 : \theta_1 = \theta_2 = \cdots = \theta_p \tag{17.1}$$

where θ_i, $i = 1, \ldots, p$ are the population treatment means for each subgroup. To test this hypothesis, we compute the between-groups omnibus test of homogeneity, Q_B, given by

$$Q_B = \sum_{i=1}^{p} w_i.(T_i. - T..)^2 \tag{17.2}$$

where $w_i.$ are the sums of the weights for the ith group or $w_i. = \sum_{j=1}^{m_i} 1/v_{ij}^2$. The subgroup mean $T_i.$ and the overall mean $T..$ are given by

$$T_i. = \frac{\sum_{j=1}^{m_i} w_{ij} T_{ij}}{\sum_{j=1}^{m_i} w_{ij}} \tag{17.3}$$

$$T.. = \frac{\sum_{i=1}^{p} \sum_{j=1}^{m_i} w_{ij} T_{ij}}{\sum_{i=1}^{p} \sum_{j=1}^{m_i} w_j} \tag{17.4}$$

The omnibus test of between-group homogeneity, or the test that all p means are equal, given by Q_B is compared to a chi-square distribution with $(p - 1)$ degrees of freedom. Note that in this overview, we have not specified the form of the variance of the effect sizes, v_{ij}^2, as these differ for each type of effect size.

When the null hypothesis in Eq. 17.1 is true, then Q_B given in Eq. 17.2 has the central chi-square distribution with $(p - 1)$ degrees of freedom. When $Q_B > c_\alpha$ where c_α is the $100(1 - \alpha)$ percentile point of the chi-square distribution with $(p - 1)$ degrees of freedom, we reject the null hypothesis.

When the null hypothesis is false, then at least one of the p subgroup means differs. Rejecting the null hypothesis also means that Q_B no longer has a central chi-square distribution. In this case, under the fixed effects model, Q_B has a non-central chi-square distribution with $(p - 1)$ degrees of freedom and non-centrality parameter λ_B given by

$$\lambda_B = \sum_{i=1}^{p} w_i.(\theta_i. - \theta_{..})^2 \tag{17.5}$$

where $\theta_i.$ are the p population subgroup means and $\theta_{..}$ is the overall population treatment mean. The non-centrality parameter expresses the extent to which the null hypothesis is false and is defined by a specific alternative hypothesis, in this case, about the difference among the group means. The power of the test Q_B is

$$1 - F(c_\alpha \mid p - 1; \lambda_B) \tag{17.6}$$

where $1 - F(c_\alpha \mid p - 1; \lambda_B)$ is the cumulative distribution of the non-central chi-square with $(p - 1)$ degrees of freedom and non-centrality parameter λ_B. The values of the non-centrality parameter λ_B will depend on the model used (fixed or random effects) and the values posed for the subgroup means. Recall that the power of a statistical test is the probability that we will reject the null hypothesis in favor of some other alternative hypothesis. In the case of subgroup analyses, our alternative hypothesis is typically that at least one subgroup mean differs from the others. In the simple case of gender differences, our alternative hypothesis will typically be that one of the means is larger than the other by some clinically significant quantity, such as $H_a : \theta_{\text{Female}} - \theta_{\text{Male}} = 0.5$. The value of the non-centrality parameter λ_B reflects our alternative hypothesis in the values chosen for the subgroup means.

17.2.2   Choosing Parameters for the Power of Q_B in Fixed Effects Models

In order to compute power *a priori* in planning a meta-analysis with a subgroup analysis, we need to pose values for parameters in the power computations. In a fixed effects model, we assume that the effect sizes from each study estimate a common effect (the fixed effect) and vary only as a result of random sampling variance.

In a fixed effects subgroup analysis, we are interested in the difference in treatment effect for a limited number of subgroups. If we are interested in two subgroups, such as the difference in the treatment effect for men and for women, we need to decide on the magnitude of a clinically important difference between men and women. In the context of standardized mean differences, we may say that a difference of 0.5 standard deviations between men and women is clinically important, and we will examine the power of our subgroup test for a difference of 0.5 standard deviations. Note that our choice of a clinically important difference will be informed by prior research and a deep understanding of the particular treatment of interest. Given that we will typically have a clinically important subgroup difference in mind *a priori*, we will use a one-sided test of statistical significance for the subgroup difference.

The non-centrality parameter λ_B also includes values for the sum of our subgroup weights, w_i. These weights are the inverse of the variance of our effect sizes. For both standardized mean differences and odds ratios, the variance of the effect size depends on the sample size of the individual studies. To conduct power analysis, we will need to provide a guess of the within-study sample size. We also need the number of studies within each group, m_i. Prior to conducting a systematic review and meta-analysis, these quantities may be difficult to predict. Many researchers with knowledge of the trials in the meta-analysis will have guesses about the typical sample size used in the trials and the number of trials that will be available for the review and can use this information to make a range of informed guesses about typical values for the parameters needed in power computations.

17.2.3 Example: Power for a Fixed Effects Analysis with the Standardized Mean Difference

Suppose we are interested in the power of detecting a difference of 0.5 standard deviation units between men and women in their response to a treatment using a fixed effects model. Let us assume that we have $k = 10$ studies. We will assume that all studies have $n = 40$ patients, equally divided into 20 men and 20 women. We will also assume that all $k = 10$ studies provide information about the treatment effect separately for men and women so that $m_1 = m_2 = 10$. Note that we are assuming that the treatment effect estimate within studies is independent—we have two estimates of the treatment effect within a study that are estimated from independent samples. We are also making the assumption that all studies have the same sample size, a simplifying assumption in order for us to compute power. In essence, we are assuming that the sample sizes in our included studies average to the value we are using for the power computations.

Suppose we are interested in the power to detect a difference of 0.5 standard deviation units between men and women in their treatment response. Let us assume for simplicity that the mean effect size for men is $\theta_{\text{Male}} = 0.0$ while the mean effect size for women is $\theta_{\text{Female}} = 0.5$ resulting in an overall mean effect size, $\theta_{..} = 0.25$. Note that we will assume a one-tailed test given that we will pose that one of the groups will have an effect size larger than the other group. To compute the value of the non-centrality parameter, λ_B, given in Eq. 17.5, we will also need a value for the $w_{i.}$, the sum of the weights within each group. We have assumed that each study has a total sample size of $n = 40$ with $n_{\text{Male}} = n_{\text{Female}} = 20$. The estimate of the variance, v^2, for the standardized mean difference in the fixed effects model within a study is

$$v^2 = \frac{n_T + n_C}{n_T n_C} + \frac{d^2}{2(n_T + n_C)} \tag{17.7}$$

where n_T is the sample size in the Treatment group, n_C the sample size in the Control group, and d is the effect size. We will assume a common value of v^2 across the groups. Let us use an effect size of $d = 0.25$, the overall mean effect size, as the value to compute the common variance across our studies with equal numbers in the treatment and control group ($n_T = n_C = 10$). The variance for the within-study effect size for both men and women is 0.20. With $m_i = 10$ studies for males and females, $w_{i.} = 10(1/0.20) = 50$. We can compute the value of the non-centrality parameter as

$$\lambda_B = \sum_{i=1}^{2} w_{i.}(\theta_{i.} - \theta_{..})^2 = 50(0.0 - 0.25)^2 + 50(0.5 - 0.25)^2 = 6.25 \tag{17.8}$$

The central chi-square distribution with $p - 1 = 2{-}1 = 1$ degree of freedom has a critical value equal to 3.84 for $\alpha = 0.05$. The power of the omnibus test that the two means are equal is given by Eq. 17.6, or $1 - F(c_\alpha \mid p - 1; \lambda_B) = 1 - F(3.84 \mid 1; 6.25)$. The value of the non-central chi-square at the critical value 3.84 with 1 degree of freedom and a non-centrality parameter value of 6.25 is 0.29 giving us power of $1 - 0.29 = 0.71$ to detect a difference in the treatment effect between men and women in this example.

Many standard statistical packages provide a function for obtaining values for the non-central chi-square distribution. SPSS, STATA and SAS provide the non-central chi-square distribution. In R, the function *pchisq* provides the values needed given values for the non-centrality parameter. Appendix 1 provides information about how to use the R function *pchisq*.

Figure 17.1 provides a comparison of power for the standardized mean difference under three different values for $m_1 = m_2$, namely 15, 10 and 5. Appendix 2 provides the R code to produce Fig. 17.1. The lower line in Fig. 17.1 shows that power is low when we have only five studies per group, and a within-study sample size of 20 for the subgroup effect sizes.

17.2.4 Example: Power for a Fixed Effects Analysis for the Log-Odds Ratio

Now suppose we are interested in detecting a difference of 1.0 in the log-odds ratio between men and women in a treatment. As in the prior example, the choice of the clinically important difference will should be guided by a deep understanding of the potential effects of the treatment in subgroups. In this case, let us assume that

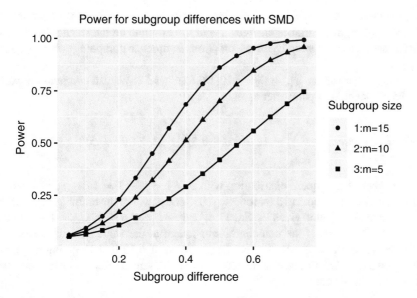

Fig. 17.1 Power for subgroup differences for the standardized mean differences

Table 17.1 Example of 2 × 2 table for an odds ratio

	Treatment	Control
Improvement	6	5
No improvement	4	5

we think the treatment difference from men will not be different from the control, but we think that women will have a treatment difference close to 1.0 in the log-odds metric. In log-odds ratios, we are assuming that the mean effect size for men is $\theta_{\text{Male}} = 0.0$ while the mean effect size for women is $\theta_{\text{Female}} = 1.0$ resulting in an overall mean effect size, $\theta_{..} = 0.5$ (given equal numbers of studies and equal sample sizes within each group). We will have the same assumptions as we have in the previous example, $k = 10$ studies, and within each study, we have $n = 40$ patients, equally divided into 20 men and 20 women.

As in the prior example, we will need to compute the common within-group variance for the effect size for men and for women. Estimating a common within-group variance is more difficult with the odds ratio than with the standardized mean difference. The variance of the log-odds ratio depends on the observed counts within each cell. For example, the Table 17.1 would result in an odds-ratio of 1.5, and a log-odds ratio of 0.40.

For the log-odds ratio, the variance of the effect size is $v^2 = \frac{1}{a} + \frac{1}{b} + \frac{1}{c} + \frac{1}{d}$ where $a, b, c,$ and d are the frequencies in each cell of the 2 × 2 table. The within-study variance for the log-odds ratio for the table above would be equal to $\frac{1}{6} + \frac{1}{4} + \frac{1}{5} + \frac{1}{5} = 0.82$. We will use the value 0.82 as the common value for the log-odds ratio for both groups as it corresponds closely to an log-odds ratio of 0.5, the hypothesized overall mean log-odds ratio. As the log-odds ratio deviates from 0 (equal events in each group), the variance changes. As in the example on the standardized mean difference, we make simplifying assumptions in order to compute statistical power *a priori.*

Using these values, $w_1. = w_2. = 10(1/0.82) = 12.2$. We can compute the value of the non-centrality parameter as

$$\lambda_B = \sum_{i=1}^{2} w_i.(\theta_i. - \theta..)^2 = 12.2(0.0 - 0.5)^2 + 12.2(1.0 - 0.5)^2 = 61 \qquad (17.9)$$

The central chi-square distribution with $p - 1 = 2 - 1 = 1$ degree of freedom has a critical value equal to 3.84 for $\alpha = 0.05$. The power of the omnibus test that the two means are equal is given by Eq. 17.6, or $1 - F(c_\alpha \mid p - 1; \lambda_B) = 1 - F(3.84 \mid 1; 6.10)$. The value of the non-central chi-square at the value 3.84 with 1 degree of freedom and a non-centrality parameter value of 6.10 is 0.31 giving us power of $1 - 0.31 = 0.69$ to detect a difference in the treatment effect between men and women in this example.

17.2.5   Power for Subgroup Analyses in Meta-Analysis: Test of Between-Group Homogeneity in Random Effects Models

In a random effects model, we assume that the variation among effect sizes is due to both random sampling variance and the underlying variance of the population, τ^2. Effect sizes in a random effects model are assumed drawn from a population of effect sizes. Thus, study-level effect sizes differ due to both sampling variance and the underlying population variance, τ^2. Power computations for subgroup analyses using a random effects model will need to include an estimate of τ^2.

As discussed by Borenstein et al. (2009), subgroup analyses using a random effects model must make assumptions about the variance component, τ^2. In this chapter, we will make the assumption of a common variance component within subgroups. Subgroup analyses in biopharmaceutical meta-analyses will likely have a small number of studies, making it more difficult to estimate a variance component without bias. Assuming that all groups share a variance component is the recommended strategy with small samples of studies (Borenstein et al. 2009). The power of the test of between-group heterogeneity follows the same steps as in the fixed effects case but with different values for the variance of effect sizes and the non-centrality parameter.

As in the fixed effects case, our null hypothesis under the random effects model is

$$H_0 : \theta_1^* = \theta_2^* = \cdots = \theta_p^* \qquad (17.10)$$

where the θ_i^* are the random effects means for the $i = 1, ..., p$ groups. We test this hypothesis using the random effects between-group heterogeneity test given by

$$Q_B^* = \sum_{i=1}^{p} w_{i\cdot}^* \left(T_{i\cdot}^* - T_{\cdot\cdot}^*\right)^2 \qquad (17.11)$$

where $w_{i\cdot}^*$ are the sums of the weights in the ith group, $T_{i\cdot}^*$ is the random effects mean effect size in the ith group, and $T_{\cdot\cdot}^*$ is the overall random effects mean effect size. These quantities are given as

$$w_{i\cdot}^* = \sum_{j=1}^{m_i} w_{ij}^* = \sum_{j=1}^{m_i} \frac{1}{\left(v_{ij}^2 + \tau^2\right)} \qquad (17.12)$$

$$T_{i\cdot}^* = \frac{\sum_{j=1}^{m_i} w_{ij}^* T_{ij}^*}{\sum_{j=1}^{km_i} w_{ij}^*} \qquad (17.13)$$

$$T_{\cdot\cdot}^* = \frac{\sum_{i=1}^{p} \sum_{j=1}^{m_i} w_{ij}^* T_{ij}^*}{\sum_{i=1}^{p} \sum_{j=1}^{m_i} w_{ij}^*} \qquad (17.14)$$

When the null hypothesis in Eq. 17.10 is true, then Q_B^* given in Eq. 17.11 has the central chi-square distribution with $(p - 1)$ degrees of freedom. When $Q_B^* > c_\alpha$ where c_α is the $100(1 - \alpha)$ percentile point of the chi-square distribution with $(p - 1)$ degrees of freedom, we reject the null hypothesis. When the null hypothesis is false, then at least one of the p subgroup means differs. In this case, Q_B^* has a non-central chi-square distribution with $(p - 1)$ degrees of freedom and non-centrality parameter λ_B^* given by

$$\lambda_B^* = \sum_{i=1}^{p} w_{i\cdot}^* \left(\theta_{i\cdot}^* - \theta_{\cdot\cdot}^*\right)^2 \qquad (17.15)$$

where $\theta_{i\cdot}^*$ are the p population subgroup means and $\theta_{\cdot\cdot}^*$ is the overall population treatment mean. The power of the test Q_B^* is

$$1 - F\left(c_\alpha \mid p - 1; \lambda_B^*\right) \qquad (17.16)$$

where $1 - F\left(c_\alpha \mid p - 1; \lambda_B^*\right)$ is the cumulative distribution of the non-central chi-square with $(p - 1)$ degrees of freedom and non-centrality parameter λ_B^*.

Jackson and Turner (2017) have recently examined the power of the random effects model in meta-analysis. Their analysis accounts for the fact that the variance component, τ^2, must be estimated, adding uncertainty to any random effects meta-analysis. They find in their simulation studies that ignoring the uncertainty introduced by estimating τ^2 is not a serious concern. However, the number of studies in the meta-analysis is an important factor in power for meta-analysis in random effects models as will be illustrated in the examples below.

17.2.6 Choosing Parameters for the Power of Q_B in Random Effects Models

As for power under the fixed effects model, we need to pose values for the typical within-study sample size, the number of studies eligible for the meta-analysis, and the clinically important difference we wish to test. We also need to estimate a value for the common variance component, τ^2, in order to estimate the variances and weights for our effect sizes. One way to arrive at a set of values for the variance component, τ^2, is to use conventions for I^2 as addressed in the Cochrane Handbook (Higgins and Green 2011). I^2 is a relative measure of variance in meta-analysis, the ratio of between-study variation as measured by τ^2 and total variation, or

$$I^2 = \frac{\tau^2}{\tau^2 + v^2} \tag{17.17}$$

where v^2 is the sampling variance for our common effect size. For example, an $I^2 = 0.75$ is in the range of considerable heterogeneity as indicated in the Cochrane Handbook. Given Eq. 17.17, a value of 0.75 for I^2 would result in $\tau^2 = 3v^2$, or a variance component that is three times the value of the sampling variance for the effect size. Thus, we can pose a number of values of τ^2 relative to the variance of the study level effect size that correspond to differing levels of heterogeneity. For this chapter, we will use $\tau^2 = 3v^2$ for a large degree of heterogeneity, $\tau^2 = v^2$ for a moderate degree of heterogeneity corresponding to an $I^2 = 0.5$ and, $\tau^2 = \frac{1}{3}v^2$ for a low degree of heterogeneity corresponding to an $I^2 = 0.25$.

17.2.7 Example: Power for a Random Effects Analysis with the Standardized Mean Difference

Returning to our example for the standardized mean difference, suppose we are interested in the power of detecting a difference of 0.5 standard deviation units between men and women. Recall that we have $k = 10$ studies, all studies have $n = 40$ patients, equally divided into 20 men and 20 women. We will also assume that all $k = 10$ studies provide information about the treatment effect separately for men and women so that $m_1 = m_2 = 10$. We will again assume that the mean effect size for men is $\theta_{\text{Male}} = 0.0$ while the mean effect size for women is $\theta_{\text{Female}} = 0.5$ resulting in an overall mean effect size, $\theta_{..} = 0.25$.

Within studies, we will assume a common variance for the within-study effect of 0.20 as described in the example for a fixed effects subgroup analysis for the standardized mean difference. To compute our weights $w_{i.}^*$ in Eq. 17.11, we will need to pose values for τ^2.

For a large degree of heterogeneity, we will use $3v$ as the value for τ^2 so that $\tau^2 = 3(0.20) = 0.60$. For both groups, the sum of the weights across the ten studies is $w_{i.}^* = 10 \left(\frac{1}{0.20+0.60} \right) = 12.5$. We can compute the value of the non-centrality parameter λ_B^* from Eq. 17.15 as

$$\lambda_B^* = 12.5(\, 0.0 - 0.25)^2 + 12.5(0.5 - 0.25)^2 = 1.56 \qquad (17.18)$$

The power of the subgroup test for a large degree of heterogeneity is given in Eq. 17.16 as $1 - F(c_\alpha \mid p - 1; \lambda_B^*)$ where $c_\alpha = 3.84$ for $\alpha = 0.05$ for a chi-square test with 1 degree of freedom. The power of the test is $1 - F(3.84 \mid 1; 1.56) = 1 - 0.76 = 0.24$. With a large degree of heterogeneity, we have little power to detect a difference of 0.5 standard deviations between men and women's response to treatment. In general, the power of tests under the random effects model is lower than under the fixed effects model.

17.2.8 Example: Power for a Random Effects Analysis for the Log-Odds Ratio

As in our prior log-odds ratio example, we are interested in detecting a difference of 1.0 in the log-odds ratio between men and women in a treatment. In log-odds ratios, we are assuming that the mean effect size for men is $\theta_{\text{Male}} = 0.0$ while the mean effect size for women is $\theta_{\text{Female}} = 1.0$ resulting in an overall mean effect size, $\theta_{..} = 0.5$ (given equal numbers of studies and equal sample sizes within each group). We will have the same assumptions as we have in the previous example, $k = 10$ studies, and within each study, we have $n = 40$ patients, equally divided into 20 men and 20 women and $m_1 = m_2 = 10$. We will use the value 0.82 as the common value for the log-odds ratio for both groups as it corresponds closely to a log-odds ratio of 0.5, the hypothesized overall mean log-odds ratio.

For this example, we will assume a moderate degree of heterogeneity so that $\tau^2 = v^2 = 0.82$. For both groups, the sum of the weights across the ten studies will be equal to is $w_{i.}^* = 10 \left(\frac{1}{0.82+0.82} \right) = 6.10$. We can compute the value of the non-centrality parameter λ_B^* from Eq. 17.15 as

$$\lambda_B^* = 6.10(\, 0.0 - 0.5)^2 + 6.10(1.0 - 0.5)^2 = 3.05 \qquad (17.19)$$

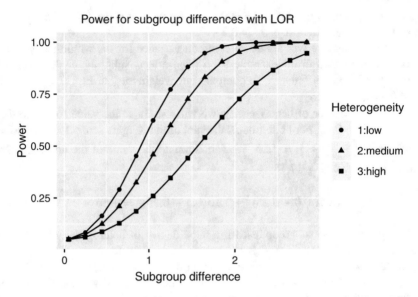

Fig. 17.2 Power for subgroup differences with log-odds ratio and varying degrees of heterogeneity

The power of the subgroup test for a large degree of heterogeneity is given in Eq. 17.16 as $1 - F(c_\alpha \mid p - 1; \lambda_B^*)$ where $c_\alpha = 3.84$ for $\alpha = 0.05$ for a chi-square test with 1 degree of freedom. The power of the test is $1 - F(3.84 \mid 1; 3.05) = 1 - 0.58 = 0.42$. With a moderate degree of heterogeneity, we have little power to detect a difference of 1.0 in the log-odds ratio between men and women's response to treatment.

Figure 17.2 provides a comparison of power in this example for a range of subgroup differences and for low, moderate and high degrees of heterogeneity. We have less power to detect a subgroup difference when there is a high degree of heterogeneity for both subgroups. Note that in this example, we have $m_1 = m_2 = 10$ studies within each subgroup. It is not uncommon for meta-analyses in health to have a smaller number of studies within each subgroup. In the case of small numbers of studies and a large degree of heterogeneity, Jackson and Turner (2017) caution that these meta-analyses, and by analogy these subgroup analyses, will have low power and should be interpreted with caution..

17.2.9 Example: Unbalanced Number of Studies Within Subgroups

Richardson et al. (2019), their tutorial on interpreting subgroup analysis in meta-analysis, urge researchers to examine the balance of the number of studies within each subgroup. The balance of the covariate across studies potentially influences

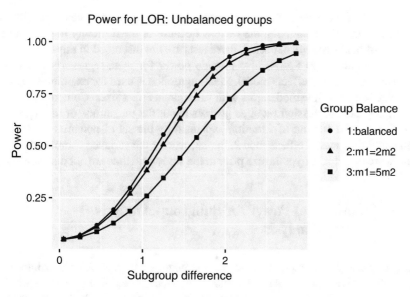

Fig. 17.3 Power for the log-odds ratio with varying numbers of studies within groups

power through the computation of the non-centrality parameter. Figure 17.3 provides power curves for three different scenarios in the prior example for power for random effects subgroup analysis with the log-odds ratio. The top curve is the power for the prior example with a log-odds ratio difference of 1.0 with a large degree of heterogeneity and equal numbers of studies within subgroups ($m_1 = m_2 = 10$). The middle curve shows power for $m_1 = 2m_2$, or $m_1 = 12$ and $m_2 = 6$. The curve with the lowest power has the most imbalance with $m_1 = 15$ and $m_2 = 3$. Jackson and Turner (2017) find that power for the random effects model in meta-analysis with less than five studies is low. Reviewers who anticipate that there will be an imbalance in the number of studies should use caution in interpreting subgroup analyses that are both imbalanced and have fewer than five studies within a subgroup.

17.2.10   Power for Other Tests of Moderators in Meta-Analysis

The discussion in this chapter has centered on power for subgroup differences, comparing the treatment effect in two or more groups. While the examples focused on two subgroups, note that power can be obtained for any number of subgroup differences. The researchers would need to hypothesize the size of clinically important differences among more than two means to compute power. Richardson et al. (2019) also highlight the need to examine the heterogeneity of groups when interpreting the subgroup difference in a meta-analysis. Hedges and Pigott (2004) and Pigott (2012) both discuss the power of the test of within-group homogeneity.

The power of these tests requires a guess about the amount of heterogeneity that may exist prior to conducting the review. Because of the difficulty in guessing the amount of heterogeneity expected, these tests are not discussed in this chapter.

Hedges and Pigott (2004) also provide power for meta-regression under both the fixed and random effects model. Meta-regression is used for examining the joint association of multiple moderators with variation in effect sizes. Computing *a priori* power for meta-regression requires guesses about the magnitude of the regression coefficients, the amount of variation explained by the set of potential moderators, and under the random effects model, the estimate of the variance component. These values are difficult to hypothesize prior to the study, and thus are not discussed here.

17.3 Summary of Power for Subgroup Analysis in Meta-Analysis

Subgroup analysis in meta-analysis has the potential to provide insight into treatment effect variation across studies and to contribute to clinical practice. However, Richardson et al. (2019) have demonstrated that reviewers find the interpretation of the results of subgroup analysis challenging. Computing *a priori* power for planned subgroup analysis can allow reviewers to understand the potential power of their analyses and help inform whether a subgroup analysis is warranted given the number of eligible studies.

Table 17.2 provides a summary of the steps for computing *a priori* power for a subgroup analysis. While power analysis does require a number of guesses about important parameters, reviewers could compute power for a range of values for the target subgroup difference, within-study sample size and number of eligible studies. Note that power will be low for small numbers of studies, particularly in the random effects model, as described recently by Jackson and Turner (2017).

Table 17.2 Steps for computing power in subgroup analysis in meta-analysis

1. Establish a critical value for statistical significance, c_α
2. Decide on the magnitude of the clinically important difference among subgroup means in the metric of the effect size
3. Assign values to the subgroup means, $\theta_1, \theta_2, \ldots, \theta_p$, corresponding to clinically important differences between the subgroup means
4. Estimate the number of studies within each group, $m_1, \ldots, m_p$
5. Compute v^2, the common value of the sampling variance for each effect size, given typical values for the within-study sample sizes
6. In the random effects model, provide values for τ^2 corresponding to different levels of heterogeneity expected in the meta-analysis
7. Compute values for the common weight for each study. In fixed effects, $w_i = 1/(v^2)$; in random effects, $w_i = 1/(v^2 + \tau^2)$
8. Compute the value of the non-centrality parameter for the appropriate model, λ_B
9. Compute power under the assumptions in steps 1–8

This chapter has focused exclusively on computing power *a priori*. The computation of observed power remains controversial as discussed by Hoenig and Heisey (2001). Subgroup analysis in meta-analysis should be planned in advance; any subgroup analyses suggested by the data gathered should be considered exploratory. Smith and Egger (2001) also note that individual patient data meta-analysis would provide more meaningful subgroup analyses rather than a reliance on aggregated data from clinical trials. As this volume suggests, subgroup analysis is potentially informative for understanding how treatment effects differ across subgroups but require careful planning and interpretation before applied to clinical practice.

Appendix 1   R Function *pchisq* to Compute Power

In R, the function *pchisq* gives the distribution function for the chi-square distribution, the area either above or below a particular value. To compute the power for the examples of the chapter, use the following code:

$$1\text{–pchisq}\,(x, df, ncp = y, \text{lower.tail} = \text{TRUE}, \log.p = \text{FALSE})$$

The value x is the critical value for the central chi-square for a test at the α level of significance with df degrees of freedom (the number of groups in the test – 1). The value y in $ncp = y$ is the value of the non-centrality parameter for the alternative hypothesis being tested. The argument lower.tail = TRUE gives the area in the non-central chi-square distribution that is less than the value of x, or the lower tail of the distribution. By default, the function returns the result in log units; setting the argument log.p = FALSE provides the value needed for computing power.

To obtain the critical value of the central chi-square distribution at the α level of significance, use the following code:

$$\text{qchisq}\,(p, df, ncp = 0, \text{lower.tail} = \text{TRUE}, \log.p = \text{FALSE})$$

The value p is $1 - \alpha$, and df are the degrees of freedom for the test (the number of groups in the test – 1). The expression $ncp = 0$ indicates that we are interested in the central chi-square distribution.

Appendix 2   R Code for Fig. 17.1, Power for Subgroup Differences with the Standardized Mean Difference

```
# Power analysis for fixed effects subgroup analysis of two groups - SMD
  library(ggplot2)
```

```
# Create a dataframe with N = 15 values for power analysis
# subdiff is a range of subgroup mean differences
# m is the common number of studies in both groups
# ssize is the within-study sample size
# vi is the common within-study variance for the effect size
N < -15
powerparms1 < - data.frame(subdiff = numeric(N), ssize = (N), vi = numeric
(N))
# add values for the power parameters
# seq is a function that creates a sequence of numbers for the subgroup
differences
# rep is a function that repeats a value for the common within-study sample size
powerparms1$subdiff <- c(seq(from = 0.05, to = 0.75, by = 0.05))
powerparms1$ssize <- c(rep(20, times = 15))
# use effsize and ssize to create vi for standardized mean difference
# I use 1/2 of the subgroup difference of interest for the overall mean effect size
powerparms1$vi < - ((powerparms1$ssize)/((0.5*powerparms1$ssize) ^ 2)) +
(((0.5*powerparms1$subdiff) ^ 2)/ (2*powerparms1$ssize))
# copy the dataframe 3 times to create parameters for 3 different
# scenarios for number of studies within groups m = 5, 10, 15
powerp1 < - rbind(powerparms1, powerparms1, powerparms1)
# create values for equal m within groups of 5, 10, 15
powerp1$m < - c(rep(5, times = 15), rep(10, times = 15), rep(15, times = 15))
powerp1$mlab <- c(rep("3:m = 5", times = 15), rep("2:m = 10", times = 15),
rep("1:m = 15", times = 15))
# get the sums of the weights for values of the vi and m
powerp1$w1sum < - powerp1$m*(1/powerp1$vi)
powerp1$w2sum < - powerp1$m*(1/powerp1$vi)
# get lambda the non-centrality parameter
powerp1$lambda <- (powerp1$w1sum*((0.5*powerp1$subdiff) ^ 2))+
(powerp1$w2sum*((0.5*powerp1$subdiff) ^ 2))
#get power
powerp1$power < - 1 - pchisq(3.84, 1, ncp = powerp1$lambda, lower.tail = T,
log.p = F)
# plot the power curves
ggplot(powerp1, aes(x = powerp1$subdiff, y = powerp1$power, group =
powerp1$mlab)) +
geom_line() +
geom_point(aes(shape = powerp1$mlab)) +
ggtitle("Power for subgroup differences with SMD") +
labs(y = "Power", x = "Subgroup difference") +
labs(shape = "Subgroup size")
```

Appendix 3   R Code for Fig. 17.2, Power for Subgroup Differences with Log-Odds Ratio and Varying Degrees of Heterogeneity

```
# Power analysis for random effects subgroup analysis of two groups - LOR
  library(ggplot2)
  # Create a dataframe with values for power analysis
  # subdiff is a range of subgroup mean differences
  # m is the common number of studies in both groups
  # ssize is the within-study sample size
  # vi is the common within-study variance for the effect size
  N < -15
  powerparms2 < − data.frame(subdiff = numeric(N), m = numeric(N),
ssize = numeric(N), vi = numeric(N))
  # add values for the power parameters
  # seq is a function that creates a sequence of numbers for the subgroup
differences
  # rep is a function that repeats a value
  powerparms2$subdiff <− c(seq(from = 0.05, to = 2.85, by = 0.20))
  powerparms2$ssize <− c(rep(20, times = 15))
  powerparms2$m < − c(rep(10, times = 15))
  # use an LOR = 1 for overall effect size to compute vi
  # ssize/4 for each cell to obtain common estimate of vi
  cellcounts <− powerparms2$ssize/4
  powerparms2$vi < − 4*(1/cellcounts)
  # copy the dataframe 3 times to create parameters for 3 different
  # scenarios for heterogeneity: low, medium, high
  powerp2 < − rbind(powerparms2, powerparms2, powerparms2)
  # create factor for each value of heterogeneity
  powerp2$levelh <− c(rep("1:low", times = 15), rep("2:medium", times = 15),
  rep("3:high", times = 15))
  # create values for tau for the three different levels of heterogeneity
  powerp2$tausq[1:15] < − powerp2$vi[1:15]/3
  powerp2$tausq[16:30] < − powerp2$vi[16:30]
  powerp2$tausq[31:45] < − powerp2$vi[31:45]*3
  # get the sums of the weights for values of the vi and m
  powerp2$w1sum < − powerp2$m*(1/(powerp2$vi + powerp2$tau))
  powerp2$w2sum < − powerp2$m*(1/(powerp2$vi + powerp2$tau))
  # get lambda the non-centrality parameter
  powerp2$lambda <− (powerp2$w1sum*((0.5*powerp2$subdiff) ^ 2))+
  (powerp2$w2sum*((0.5*powerp2$subdiff) ^ 2))
  # get power
  powerp2$power < − 1 - pchisq(3.84, 1, ncp = powerp2$lambda, lower.tail = T,
  log.p = F)
```

```
# plot the power curves
ggplot(powerp2, aes(x = powerp2$subdiff, y = powerp2$power, group =
powerp2$levelh)) +
geom_line() +
geom_point(aes(shape = powerp2$levelh)) +
ggtitle("Power for subgroup differences with LOR") +
labs(y = "Power", x = "Subgroup difference") +
labs(shape = "Heterogeneity")
```

Appendix 4   R Code for Fig. 17.3, Power for the Log-Odds Ratio with Varying Numbers of Studies Within Groups

```
# Power analysis for random effects subgroup analysis of two groups - LOR
  # Moderate heterogeneity and unbalanced groups
  library(ggplot2)
  # Create a dataframe with values for power analysis
  # subdiff is a range of subgroup mean differences
  # m is the common number of studies in both groups
  # ssize is the within-study sample size
  # vi is the common within-study variance for the effect size
  N < -15
  powerparms3 < − data.frame(subdiff = numeric(N), ssize = numeric(N),
vi = numeric(N))
  # add values for the power parameters
  # seq is a function that creates a sequence of numbers
  # rep is a function that repeats a value
  powerparms3$subdiff <− c(seq(from = 0.05, to = 2.85, by = 0.20))
  powerparms3$ssize <− c(rep(20, times = 15))
  # use an LOR = 1 for overall effect size to compute vi
  # ssize/4 for each cell to obtain common estimate of vi
  cellcounts <− powerparms3$ssize/4
  powerparms3$vi < − 4*(1/cellcounts)
  # copy the dataframe 3 times to create parameters for 3 different
  # scenarios for balance of subgroups - balanced, 1/2 and 1/5
  powerp3 < − rbind(powerparms3, powerparms3, powerparms3)
  # create factor for balance scenario
  powerp3$balance <− c(rep("1:balanced", times = 15), rep("2:m1 = 2 m2",
times = 15),
  rep("3:m1 = 5 m2", times = 15))
  # create values for m1 and m2
  powerp3$m2[1:15] < − 9
```

```
powerp3$m1[1:15] < − powerp3$m2[1:15]
powerp3$m2[16:30] < − 6
powerp3$m1[16:30] < − 2* powerp3$m2[16:30]
powerp3$m2[31:45] < − 3
powerp3$m1[31:45] < − 5* powerp3$m2[31:45]
# moderate level of tau: vi = tau
powerp3$tausq <− powerp3$vi
# get the sums of the weights for values of the vi and m
powerp3$w1sum < − powerp3$m1*(1/(powerp3$vi + powerp3$tau))
powerp3$w2sum < − powerp3$m2*(1/(powerp3$vi + powerp3$tau))
# get the value of the overall mean effect size given subdiff
# and the values of m1 and m2
powerp3$meaneff  <−  (powerp3$m2*powerp3$subdiff)/(powerp3$m1  +
powerp3$m2)
# get lambda assuming group 1 has mean 0 and group 2 has mean = subdiff
powerp3$lambda <− (powerp3$w1sum*((0 - powerp3$meaneff) ^ 2))+
(powerp3$w2sum*((powerp3$subdiff - powerp3$meaneff) ^ 2))
powerp3$power < − 1 - pchisq(3.84, 1, ncp = powerp3$lambda, lower.tail = T,
log.p = F)
# plot the power curves
ggplot(powerp3, aes(x = powerp3$subdiff, y = powerp3$power, group =
powerp3$balance)) +
geom_line() +
geom_point(aes(shape = powerp3$balance)) +
ggtitle("Power for LOR: Unbalanced groups") +
labs(y = "Power", x = "Subgroup difference") +
labs(shape = "Group Balance")
```

References

Borenstein M, Hedges LV, Higgins JPT, Rothstein HR (2009) Introduction to meta-analysis. John Wiley, Chichester

Brookes ST, Whitely E, Egger M, Smith GD, Mulheran PA, Peters TJ (2004) Subgroup analyses in randomized trials: risks of subgroup-specific analysis; power and sample size for the interaction test. J Clin Epidemiol 57:229–236

European Medicines Agency (2014) Draft guideline on the investigation of subgroups in confirmatory clinical trials. European Medicines Agency, London. https://www.ema.europa.eu/en/documents/scientific-guideline/draft-guideline-investigation-subgroups-confirmatory-clinical-trials_en.pdf

Hedges LV, Pigott TD (2001) The power of statistical tests in meta-analysis. Psychol Methods 6:203–217

Hedges LV, Pigott TD (2004) The power of statistical tests for moderators in meta-analysis. Psychol Methods 9:426–445

Higgins JTP, Green S (2011) Cochrane handbook for systematic reviews of interventions Version 5.1.0 [updated March 2011]. The Cochrane Collaboration, 2011. Available from www.handbook.cochrane.org

Hoenig J, Heisey D (2001) The abuse of power: the pervasive fallacy of power calculations for data analysis. Am Stat 55:19–24

Ioannidis JPA (2005) Why most published research findings are false. PLoS Med 2:e124

Jackson D, Turner R (2017) Power analysis for random-effects meta-analysis. Res Synth Methods 8:290–302

Oxman AD, Guyatt GH (1992) A consumer's guide to subgroup analyses. Ann Intern Med 116:78–84

Pigott TD (2012) Advances in meta-analysis. Springer, New York

Pocock SJ, Assmann SE, Enos LE, Kasten LE (2002) Subgroup analysis, covariate adjustment and baseline comparisons in clinical trial reporting: current practice and problems. Stat Med 21:2917–2930

Richardson M, Garner P, Donegan S (2019) Interpretation of subgroup analyses in systematic reviews: a tutorial. Clin Epidemiol Global Health 7(2):192–198

Smith GD, Egger M (2001) Going beyond the grand mean: subgroup analysis in meta-analysis of randomized trials. In: Egger M, Smith GD, Altman DG (eds) Systematic reviews in health care: meta-analysis in context. BMJ, London, pp 143–156

Valentine JC, Pigott TD, Rothstein HR (2010) How many studies do you need? A primer on statistical power for meta-analysis. J Educ Behav Stat 35:215–247

Wang R, Lagakos SW, Ware JH, Hunter DJ, Drazen JM (2007) Statistics in medicine—reporting of subgroup analyses in clinical trials. N Engl J Med 357:2189–2194

Chapter 18
Heterogeneity and Subgroup Analysis in Network Meta-Analysis

Jeroen P. Jansen

18.1 Background

Comprehensive healthcare decision-making requires a comparisons of the relevant competing treatment options for a particular disease state. Randomized controlled trials (RCTs) are considered the most credible evidence to obtain insight into the relative treatment effects of a medical intervention. However, an individual RCT rarely includes all competing interventions of interest. Typically, the evidence base consists of multiple RCTs where each of the available studies compares a subset of all the competing interventions of interest. If each of these trials has at least one intervention in common with another trial such that the evidence base can be represented with one connected network, a network meta-analysis (NMA) can provide relative treatment effects between all competing interventions of interest (see the network diagram in Fig. 18.1) (Ades 2003; Bucher et al. 1997; Dias et al. 2013a, 2018a; Hutton et al. 2015; Jansen et al. 2011, 2014; Lumley 2002; Lu and Ades 2004; Salanti et al. 2008). A NMA can be considered a generalization of conventional pairwise meta-analysis (Dias et al. 2018b, c). Rather than synthesizing the findings of multiple RCTs each comparing the same intervention with the same control, with a NMA we are simultaneously synthesizing the findings of multiple pair-wise comparisons across a range of interventions and obtaining estimates of relative treatment effects between all competing interventions based on direct and/or indirect evidence. Even if there was a conclusive RCT that included all competing interventions of interest, the available RCTs comparing a subset of the interventions

J. P. Jansen
Precision Health Economics & Outcomes Research, Oakland, CA, USA

Department of Health Research and Policy (Epidemiology), Stanford University School of Medicine, Stanford, CA, USA
e-mail: jpjansen@stanford.edu

© Springer Nature Switzerland AG 2020   369
N. Ting et al. (eds.), *Design and Analysis of Subgroups with Biopharmaceutical Applications*, Emerging Topics in Statistics and Biostatistics,
https://doi.org/10.1007/978-3-030-40105-4_18

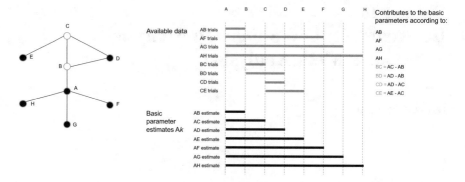

Fig. 18.1 Concept of network meta-analysis. An evidence network of connected randomized controlled trials for the competing interventions of interest (A, D, E, F, G, H) provide the data to estimate the relative treatment effects of each intervention relative to A (estimates for the basic parameters) given the assumption of consistency. These basic parameters are the basis for inferences regarding comparative effectiveness

provide relevant evidence as well. A NMA allows to estimate relative treatment effects based on the totality of the RCT evidence base.

RCTs of novel biopharmaceuticals are frequently performed in the context of regulatory approval. These trials are designed to demonstrate efficacy versus placebo or standard care, but not typically against each other. However, obtaining approval for drug licensing based on positive trials is not a guarantee for market access. Payers need to be convinced about the value of the new drug as well. As part of a health technology assessment (HTA), the value of the new intervention is assessed by examining its benefits, risks, and costs in comparison with existing standards of care for a given patient population of interest based on explicit and scientifically credible methods to inform healthcare and reimbursement decision-making. In many countries, the agencies tasked with HTA expect manufacturers of biopharmaceuticals to provide evidence regarding the comparative and cost-effectiveness of their drugs. With the evidence base characterized by multiple RCTs that only provide direct evidence regarding relative treatment effects for a subset of comparisons of interest, NMA is a core component of HTA submissions for new biopharmaceutical interventions.

It is well known that subgroups of patients within a population will derive value from a medical intervention that can differ systematically from the expected estimate of value for the overall population due to heterogeneous treatment effects (Espinoza et al. 2014; Sculpher 2008; Stevens and Normand 2004). As such, evaluation of treatment effects in subgroups are an integral part of the HTA review process, and treatment recommendations can be limited to specific subpopulations. In the context of relative treatment effects in RCTs, a subgroup effect can be understood as a categorical patient related covariate that modifies the treatment effect.

This chapter will discuss the estimation of relative treatment effects between competing interventions for specific subpopulations based on existing evidence by means of NMA methods. Methods that will be discussed include shrinkage estimation, network meta-regression, and a hierarchical approach to network meta-regression to combine study-level and patient level data.

18.2   Criteria for Valid Network Meta-Analysis

In order to appreciate the relevance of a NMA in the context of heterogenous and subgroup effects, it is important to highlight the criteria for a valid NMA first.

The purpose of a NMA is to estimate the relative treatment effects between competing interventions of interest for a specific target population based on available RCT evidence. In principle, this means that the study population in each of the RCTs that define the evidence base used for the NMA needs to be representative of the target population of interest. Individual RCTs are representative of the target population if there are no systematic differences in patient characteristics that influence the relative treatment effects, i.e. effect modifiers, between the study populations and the target population (Turner et al. 2009; Dias et al. 2018c). If this requirement for a relevant NMA is met, then there are no systematic differences in patient related effect-modifiers between the different RCTs in the network, and any of the relative treatment effects obtained with the NMA based on direct and/or indirect evidence is valid (Ades 2003; Dias et al. 2013a, 2018b, c; Jansen et al. 2012; Jansen and Naci 2013). Alternatively, if a subset of the trials in the network are not representative of the target population then there are differences in the distribution of effect-modifiers between the trials in the network and the estimated relative treatment effects based on indirect evidence are biased. If the study populations for all trials are different from the target population, but not different between trials, then the relative treatment effects of NMA are valid, but not representative of the target population of interest; we have *external bias* (Turner et al. 2009; Dias et al. 2018d). Obviously, effect-modifiers are not limited to patient characteristics. If there are study characteristics or contextual factors that act as effect modifiers and are different between the RCTs in the network, then the estimated relative treatment effects are biased as well.

In summary, for a credible NMA we need a connected network of RCTs where each trial has at least one intervention in common with another trial, without systematic differences in known and unknown effect modifiers between studies. For the findings to be relevant, there should not be systematic differences in effect-modifiers between the evidence base and the target population and setting of interest.

18.3 Standard Network Meta-Analysis Model

A NMA of RCTs relies on the same fundamental principle as a pairwise meta-analysis of RCTs. With a random effects pairwise meta-analysis, we assume that each study i aims to estimate study-specific relative treatment effects, $\delta_{i,AB}$, and are exchangeable, i.e. a priori the study-specific relative treatment effects are expected to be similar, yet non-identical (Dias et al. 2013a, 2018b). The study specific treatment effects come from a normal distribution with mean d_{AB} and variance σ^2_{AB} reflecting the between-study heterogeneity: $\delta_{i,AB} \sim N\left(d_{AB}, \sigma^2_{AB}\right)$. With a NMA we have multiple RCTs, each comparing a subset of all the interventions of interest, e.g. intervention A, B, C, and D (see Fig. 18.1). Now, we must assume the exchangeability of the study-specific relative treatment effects between any intervention k and b across the entire set of trials in the network: $\delta_{i,bk} \sim N\left(d_{kb}, \sigma^2_{kb}\right)$ (Dias et al. 2018b, c). We assume that the relative treatment effect $\delta_{i,AB}$ in trial i comparing B with A is a sample from the same random effects distribution as the other AB trials are estimating effects from, as well as the AC, AD, BC, and CD trials if these would have included intervention A and B as well. This notion extends to all the interventions in the network (Dias et al. 2018c). Accordingly, transitivity of the within-trial relative treatment effects of any intervention k relative to b in trial i can be described as $\delta_{i,bk} = \delta_{i,Ak} - \delta_{i,Ab}$. Consequently, the average treatment effects are related according to: $d_{bk} = d_{Ak} - d_{Ab}$. Typically, it is assumed that $\sigma^2_{kb} = \sigma^2_{Ak} = \sigma^2_{Ab} = \sigma^2$ (Dias et al. 2018b, c).

The general random-effects NMA model can be expressed as:

$$g\left(\gamma_{ik}\right) = \theta_{ik} = \begin{cases} \mu_i & k = b, \ b \in \{A, B, C, ..\} \\ \mu_i + \delta_{i,bk} & k \succ b \end{cases}$$

$$\delta_{i,bk} \sim Normal\left(d_{Ak} - d_{Ab}, \sigma^2\right) \qquad (18.1)$$

where g is an appropriate link function (e.g. the logit link for binary outcomes) and θ_{ik} is the linear predictor of the expected outcome with intervention k in trial i (e.g. the log odds). μ_i is the study i specific outcome with comparator treatment b. $\delta_{i,bk}$ reflects the study specific relative treatment effects with intervention k relative to comparator b and are drawn from a normal distribution with the pooled relative treatment effect estimates expressed relative to the overall reference treatment A: $d_{bk} = d_{Ak} - d_{Ab}$ (with $d_{AA} = 0$). Variance parameter σ^2 reflects the heterogeneity across studies. With a fixed effects NMA, $\delta_{ibk} \sim Normal(d_{Ak} - d_{Ab}, \sigma^2)$ is replaced with $\delta_{ibk} = d_{Ak} - d_{Ab}$ because σ^2 is assumed to be 0.

The primary parameters of interest are d_{Ak} and σ^2 and are estimated based on the available RCTs included in the evidence network. Estimates of d_{Ak} reflect the relative treatment effect of each intervention k relative to overall treatment of reference A based on direct and/or indirect evidence and facilitates decision-making regarding how interventions rank regarding their effects on the outcome of interest.

It is important to highlight that the model is applicable to many types of data, by just specifying an appropriate likelihood describing the data generating process and corresponding link function (Dias et al. 2018b). For example, when we have study level data and the measure of interest is response expressed as a proportion we use a binomial likelihood. When we have patient-level response (yes/no) data, we can use a Bernoulli distribution.

When the NMA is performed in a Bayesian framework, the parameters to be estimated, μ_i, d_{Ak}, and σ^2 need be given prior distributions. In principle, we like the model parameter estimates to reflect the observed data from the RCTs and will therefore consider non-informative, or minimally informative prior distributions, wherever possible (Dias et al. 2018b). For example, $\mu_i \sim Normal(0, 100^2)$, $d_{Ak} \sim Normal(0, 100^2)$, and $\mu \sim uniform(0, x)$ with x a reasonable upper bound dependent on the expected range of observed relative treatment effects.

18.4 Specific Challenges with Subpopulations

With a NMA we estimate relative treatment effects between competing interventions based on existing RCTs. The available trials for biopharmaceutical interventions have frequently been designed for regulatory approval and powered to detect a relative treatment effect for the overall study population. However, in the context of HTA, the target population of interest may be a subgroup of the overall study populations of the RCTs for the interventions of interest. This will pose challenges for the NMA when subgroup effects have not been reported or subgroup data are not available for the relevant trials. Even if the available RCTs do provide information on relative treatment effects for the subpopulation of interest, the studies may not have been powered to detect these subgroup effects and relative treatment effect estimates may be characterized by substantial uncertainty due to small sample sizes. In the following sections some of the methods that may be relevant for a NMA of treatment effects in subpopulations will be highlighted.

18.5 Shrinkage Estimation

Let us assume we have a connected network of RCTs that include all the competing interventions of interest, and all trials report results or provide data for mutually exclusive subgroups defined by observable patient characteristics. If the evidence base is rather weak due to a limited number of studies and/or small sample sizes in each of the subgroups, the NMA by subpopulation may not provide informative answers due to the uncertain estimates. As a potential solution we can consider using so called class effect models where the multiple interventions in the NMA are categorized into a smaller set of classes (Henderson et al. 2016; Kew et al. 2014; Lipsky et al. 2010; Mayo-Wilson et al. 2014; Warren et al. 2014). We assume that the treatment-specific relative effects within a class are exchangeable. For example, treatments with a similar mechanism of action fall into the same class and, *a priori*, their relative effects are more alike than effects of treatments from different classes (Dias et al. 2018d). For a class effects model with exchangeable treatment effects within a class, model 18.1 can be modified by defining that the basic parameters d_{Ak} are assumed to come from a distribution with a common mean and variance, if they belong to the same class:

$$d_{Ak} \sim Normal\left(m_{D_k}, \sigma^2_{D_k}\right) \qquad (18.2)$$

where D_k is defined as the class to which treatment k belongs. m_{D_k} is the mean class effect in class D_k, and $\sigma^2_{D_k}$ are the within-class variances. These models allow borrowing of strength across treatments in the same class: Unstable estimates for d_{Ak} due to limited subgroup data will be shrunken towards the class mean effect and become more precise than obtained with a model where d_{Ak} are assumed to be independent (Eq. 18.1). Depending on the sparseness of the available data, informative distributions may be needed for $\sigma^2_{D_k}$. It is recommended to perform multiple sensitivity analyses with different choices of values for the prior distributions of $\sigma^2_{D_k}$ (Dias et al. 2018d).

As an alternative to a NMA by subgroup, we can also define a model where all mutually exclusive subgroups are incorporated simultaneously and subgroup effects are exchangeable by treatment:

$$g\left(\gamma_{is,k}\right) = \theta_{is,k} = \begin{cases} \mu_{is} & k = b, \quad b \in \{A, B, C, ..\} \\ \mu_{is} + \delta_{is,bk} & k \succ b \end{cases}$$

$$\delta_{is,bk} \sim Normal\left(d_{s,Ak} - d_{s,Ab}, \sigma^2_s\right)$$

$$d_{s,Ak} \sim Normal\left(D_{Ak}, \sigma^2_k\right) \qquad (18.3)$$

where $\theta_{is,k}$ is the linear predictor for the expected outcome with intervention k in subgroup s of trial i. μ_{is} is the expected outcome with comparator treatment b in subgroup s of study i. $\delta_{is,bk}$ reflects the relative treatment effect with intervention k relative to comparator b in subgroup s of trial i and are drawn from a normal distribution with the pooled estimates expressed in terms of the overall relative treatment effects versus treatment A in that subgroup $d_{s,Ak}$. This model adds the assumption that the underlying subgroup specific treatment effects $d_{s,Ak}$ are drawn from a common normal distribution with mean D_{Ak} and treatment specific variance σ^2_k. With this model, highly uncertain relative treatment effects for each subgroup are stabilized by borrowing information from the data from other subgroups for that treatment (rather than from other treatments for the same subgroup) (Henderson et al. 2016).

The two "shrinkage models" presented here are a compromise between a model where treatment effects by subgroup are completely independent and a completely pooled analysis which ignores subgroup effects (Henderson et al. 2016).

18.6   Network Meta-Regression

Depending on the available data, a network meta-regression may be a relevant approach to estimate relative treatment effects between competing interventions for particular subpopulations. A meta-regression analysis can be used to explain between-study heterogeneity due to observed differences in the distribution of effect-modifiers between studies (Dias et al. 2018d). When there are differences between the target population of interest and the study populations of the individual studies included in the evidence network regarding effect-modifiers, a meta-regression can be used to adjust for this external bias (Turner et al. 2009; Dias et al. 2013b, 2018d). A network meta-regression analysis can be performed based on aggregate or study-level data, individual patient level data (IPD), or evidence networks where for a subset of the studies IPD is available and for other studies only aggregate level data (Efthimiou et al. 2016).

When the available evidence base only consists of aggregate level data, the model presented with Eq. 18.1 can be extended with covariates according to: (Cooper et al. 2009; Dias et al. 2013b, 2018d; Donegan et al. 2017, 2019; Nixon et al. 2007)

$$\theta_{ik} = \begin{cases} \mu_i & k = b, \quad b \in \{A, B, C, ..\} \\ \mu_i + \delta_{i,bk} + (\beta_{Ak} - \beta_{Ab})\left(m_i - x_{target}\right) & k > b \end{cases}$$

$$\delta_{i,bk} \sim Normal\left(d_{Ak} - d_{Ab}, \sigma^2\right) \tag{18.4}$$

m_i is the study-level covariate value for trial i, which can represent a subpopulation of interest. β_{Ak} represent the covariate effects with treatment k relative to the overall reference treatment A. x_{target} is the centered covariate value representing the target (sub) population of interest. d_{Ak} represent the relative effect of the treatment k compared to treatment A at the value x_{target}. As before, $d_{AA} = \beta_{Ak} = 0$. With this model we do not only assume consistency regarding relative treatment effects, but also regarding the parameters reflecting the impact of the covariates. Figure 18.2a, b illustrate the concept for an evidence base consisting of three AB and three AC trials for which only study-level data is available. If it is believed the Eq. 18.1 covariate of interest is not an effect modifier than the indirect BC estimate is relevant for the target population of interest (Fig. 18.2a). However, if the covariate is believed to be an effect modifier, a network meta-regression may be of interest to estimate the BC estimate for the target population (Fig. 18.2b). In model Eq. (18.4) the impact of the covariate on the relative treatment effects is assumed to be independent for each intervention k relative to A. However, we can also simplify the model by assuming the impact of the covariate is the same for every treatment k relative to A, $\beta_{Ak} = B$, or assume these to be exchangeable, $\beta_{Ak} \sim Normal\left(B, \sigma_B^2\right)$ (Cooper et al. 2009; Dias et al. 2013b, 2018d). This is useful when the number of studies for a certain intervention is limited or whether all the studies have the same covariate value.

Without access to IPD we only have information on trial-level covariates. The information on patient characteristics is aggregated at the trial level, such as proportion of patients with prior treatment or severe disease, or mean age of the study population (Dias et al. 2013b, 2018d). If the study population of a particular trial is homogeneous regarding a certain dichotomous characteristic (e.g. only treatment naïve or treatment experienced), we have a dichotomous between-trial covariate. If the trial population is heterogeneous regarding a dichotomous characteristic (e.g. a mixed population of treatment naïve and experienced) the between-trial covariate is continuous representing the proportion of individuals with the characteristics in the trial. For an aggregated continuously distributed patient characteristic, the between-trial covariate is continuous as well. If the precision of each trial is large and the number of studies is small, we may find a spurious relationship based on the between-trial comparisons to be statistically significant if the contrast in the between-trial level covariate between these studies is sufficiently large (Dias et al. 2018d). On the other hand, with continuously distributed patient characteristics, the within-trial variation is typically much larger than the variation in aggregated means used for the between-trial meta-regression, thereby not having the power to detect a true relationship (Dias et al. 2013b, 2018d). Using aggregated information regarding patient characteristics in a network meta-regression is vulnerable to ecologic bias: the parameter estimate indicating the impact of a patient characteristic on a relative treatment effect based on between-trial comparisons may be very different from the within-trial relationship, as illustrated by the different regression lines in Fig. 18.2b, c (Berlin et al. 2002; Higgins and Thompson 2004; Jansen 2012; Lambert et al. 2002; Riley et al. 2010; Schmid et al. 2004; Riley and Steyerberg 2010).

Even in the absence of study level confounding, ecological bias can exist in non-linear models (Schmid et al. 2004; Greenland 2002; Jackson et al. 2006, 2008; Jansen 2012; Riley and Steyerberg 2010). In Fig. 18.3, the relationship between a relative treatment effect on the log odds ratio scale is presented against a dichotomous patient level effect-modifier X ($x = 0$ and $x = 1$) that is aggregated across the individuals in a study and represented as the proportion with $x = 1$. Let us assume that the probability of response is 40% with treatment A in both AB and AC trials for $x = 0$ and $x = 1$. The probability of response with B in the AB trials is 60% when $x = 0$ and 80% when $x = 1$. The probability of response with C in the AC trials is 70% when $x = 0$ and 95% when $x = 1$. The solid non-linear lines in Fig. 18.3a reflect the true log odds ratio of AB and AC trials for different distributions of the dichotomous covariate X in a particular study given the probability of the outcome with intervention A, B and C. For the AB comparison there are 5 studies in which the proportions of subjects with $x = 1$ are 0.1, 0.15, 0.20, 0.25, and 0.30 (the blue dots). For the AC comparison the five studies have proportions of $x = 1$ of 0.20, 0.30, 0.40, 0.50, 0.60 (the red dots). The dashed line in Fig. 18.3b reflects a network meta-regression model where the study specific log odds ratios are modeled as a

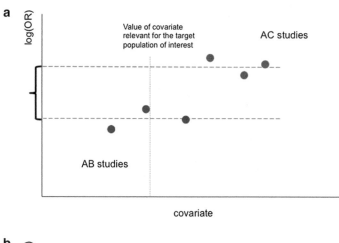

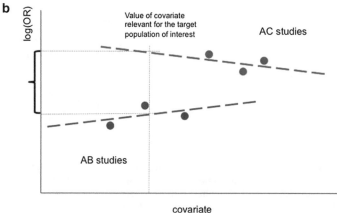

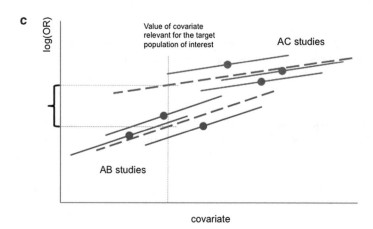

Fig. 18.2 Network meta-analysis of AB and AC studies with a continuous covariate. (**a**) Indirect estimate for BC comparison without adjustment for the imbalance in the effect modifier; (**b**) with adjustment based on aggregate level data; and (**c**) after adjustment with individual patient level data (modified from Jansen 2012)

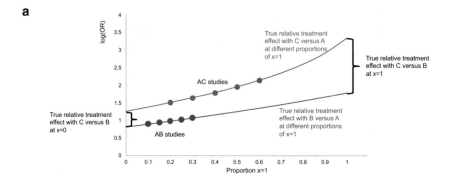

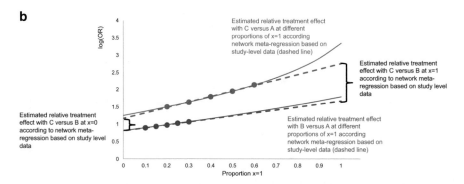

Fig. 18.3 Network meta-regression of AB and AC studies with a non-linear model with a dichotomous patient related covariate X to estimate subgroup effects $x = 0$ and $x = 1$ in the absence of study level confounding. (**a**) True indirect estimates for $x = 0$ and $x = 1$, can be obtained with patient level data for all studies; (**b**) Biased indirect estimates for $x = 0$ and $x = 1$ (b) based on aggregated level data network meta-regression (modified from Jansen 2012)

function the proportion of subjects with $x = 1$ in each study using study-level data. When we are interested in the AB, AC, and indirect BC estimate for the subgroups $x = 0$ and $x = 1$, the projected treatment effects with the network meta-regression model are biased: the estimated log odd ratios for $x = 0$ and $x = 1$ (the dashed lines) differ from the true log odds ratios (the solid).

The limitations of network meta-regression based on aggregate level data can be overcome with the use of IPD. In the context of adjusting for external bias in relation to patient characteristics, a network meta-regression analysis based on IPD can be considered the "gold-standard" (Cope et al. 2012; Dias et al. 2018d; Debray et al. 2018; Leahy et al. 2018). The solid curves in Fig. 18.3a show the estimates that an analysis with IPD would provide (Jansen et al. 2012). With IPD available for all the trials in the evidence network, a model can be defined according to:

$$\theta_{ijk} = \begin{cases} \mu_i + \beta_{0i}x_{ij} & k = b, \quad b \in \{A, B, C, ..\} \\ \mu_i + \delta_{i,bk} + \beta_{0i}x_{ij} + (\beta_{Ak} - \beta_{Ab})x_{ij} & k > b \end{cases}$$

$$\delta_{i,bk} \sim Normal\left(d_{Ak} - d_{Ab}, \sigma^2\right) \tag{18.5}$$

j reflects the individual in study i. β_{0i} is the main effect of covariate x on the outcome of interest in study i. x_{ij} is the value of the covariate for individual j in study i. Here we assume the interaction effect β_{Ak} is fixed across studies. We can also separate the within and between-trial interaction and define the model with a covariate for the mean value of the patient-related effect-modifier of each study and a covariate for the individual patient value of the effect-modifier minus the mean value in that study to describe the within-study variation (Riley and Steyerberg 2010; Donegan et al. 2013):

$$\theta_{ijk} = \begin{cases} \mu_i + \beta_{0i}x_{ij} & k = b, \quad b \in \{A, B, C, ..\} \\ \mu_i + \delta_{i,bk} + \beta_{0i}x_{ij} + \left(\beta_{Ak}^a - \beta_{Ab}^a\right)m_i & k > b \\ \quad + \left(\beta_{Ak}^w - \beta_{Ab}^w\right)(x_{ij} - m_i) \end{cases}$$

$$\delta_{i,bk} \sim Normal\left(d_{Ak} - d_{Ab}, \sigma^2\right) \tag{18.6}$$

β_{Ak}^a represent the between-study coefficient for the covariate effects with treatment k relative to the overall reference treatment A. β_{Ak}^w represent the within-study coefficient for the covariate effects with treatment k relative to the overall reference treatment A. If the within-trial and between-trial interactions are different, then ecological or confounding bias may be present and the relative treatment effects may be biased. In such cases, inferences regarding treatment effects for specific subpopulations of interest should be based on the within-trial interactions estimated by models that separate the two types of interaction (Dias et al. 2018d). Again, models (18.5) and (18.6) can be simplified by assuming that the impact of the effect modifier is the same for every intervention k relative to A.

Unfortunately, a meta-analyst may not have access to IPD for all trials, but only for a subset. Rather than excluding relevant aggregate level studies from the network meta-regression, a combination of IPD studies and aggregate level studies is of interest (Sutton et al. 2008; Riley et al. 2007, 2008). We need a model that can combine both sources of data, such as (Donegan et al. 2013):

IPD trials:

$$\theta_{ijk} = \begin{cases} \mu_i + \beta_{0i} x_{ij} & k = b, \quad b \in \{A, B, C, ..\} \\ \mu_i + \delta_{i,bk} + \beta_{0i} x_{ij} + \left(\beta_{Ak}^a - \beta_{Ab}^a\right) m_i & k \succ b \\ \quad + \left(\beta_{Ak}^w - \beta_{Ab}^w\right) (x_{ij} - m_i) \end{cases}$$

Aggregate level data trials:

$$\theta_{ik} = \begin{cases} \mu_i & k = b, \quad b \in \{A, B, C, ..\} \\ \mu_i + \delta_{i,bk} + \left(\beta_{Ak}^a - \beta_{Ab}^a\right) m_i & k \succ b \end{cases}$$

$$\delta_{i,bk} \sim Normal\left(d_{Ak} - d_{Ab}, \sigma^2\right) \tag{18.7}$$

With this model, the aggregate-level data studies as well as the IPD studies contribute to estimation of the between-study interactions. Now, let us assume we have a scenario where for a certain set of the direct comparisons in the network we have IPD and for the remaining set of direct comparisons we only have aggregate-level data. If treatment-by-covariate interactions are believed to be the same for each intervention k relative to A then depending on how IPD and study level data is divided over the available direct comparisons in the network, we may be able to "transfer" the within-trial interaction estimate for Ak comparison for which IPD is available to the Ak comparisons for which we only have aggregate level data. Unfortunately, this only applies to specific evidence structures. We can potentially improve the precision of the interaction effects for studies with only aggregate-level data for any structure based on the available IPD, if we define the model with the same interaction parameter for the within- and between-trial comparisons (Donegan et al. 2013; Jansen et al. 2012; Saramago et al. 2012). However, as mentioned, this will bias the estimates when there is study-level confounding, i.e. when not all effect-modifiers are accounted for, or when we have non-linear models (e.g. when adopting a logit link function).

18.7 Hierarchical Approach to Network Meta-Regression

As an alternative to using network meta-regression models that use the same interaction-effect parameters for the IPD studies and aggregate-level studies, we can use a type of model based on the *hierarchical related regression* approach to avoid aggregation or ecological bias (Jackson et al. 2006, 2008). A model for a hierarchical approach to a NMA (Jansen 2012; Phillippo et al. 2018) with a dichotomous covariate can be defined according to:

IPD trials:

$$g\left(\gamma_{ijk}\right) = \theta_{ijk} = \begin{cases} \mu_i + \beta_0 x_{ij} & k = b, \quad b \in \{A, B, C, ..\} \\ \mu_i + \delta_{i,bk} + \beta_0 x_{ij} & k \succ b \\ \quad + (\beta_{Ak} - \beta_{Ab}) x_{ij} \end{cases}$$

Aggregate level data trials:

$$\gamma_{ik} = \gamma_{ik}^o (1 - m_i) + \gamma_{ik}^1 m_i$$

$$g\left(\gamma_{ik}^o\right) = \theta_{ik}^o = \begin{cases} \mu_i & k = b, \quad b \in \{A, B, C, ..\} \\ \mu_i + \delta_{i,bk} & k \succ b \end{cases}$$

$$g\left(\gamma_{ik}^1\right) = \theta_{ik}^1 = \begin{cases} \mu_i + \beta_0 & k = b, \quad b \in \{A, B, C, ..\} \\ \mu_i + \delta_{i,bk} + \beta_0 + (\beta_{Ak} - \beta_{Ab}) & k \succ b \end{cases}$$

$$\delta_{i,bk} \sim Normal\left(d_{Ak} - d_{Ab}, \sigma^2\right) \tag{18.8}$$

The IPD part of this model is the same as the previous model (see Eq. 18.5), with the exception that the coefficient related to the covariate effect, β_0, is fixed across studies. For the aggregate-level data part of the model, γ_{ik} is the expected outcome in study i with intervention k and is determined by integrating the individual-level model over the joint within-study distribution of the binary covariate. γ_{ik} equals the sum of the proportion of subjects with covariate $x = 1$ in each aggregate level data study (m_i) multiplied with γ_{ik}^1 and the proportion of subjects with covariate $x = 0$ ($1-m_i$) multiplied with γ_{ik}^o. γ_{ik}^1 represent the marginal expected outcome with treatment k for a subject with the covariate x $= 1$ in study i. Similarly, γ_{ik}^o is the equivalent for a subject with $x = 0$. The essence of the approach is that an individual model is averaged over the population in study i to obtain the aggregate-level model for that study. Initial simulation studies have shown that relative treatment effects for subgroups with this approach may be less affected by bias than estimates obtained with network meta-regression for non-linear models with large treatment-by-patient-level-covariate interactions (Jansen 2012). Furthermore, it allows for the "transfer" of the within-trial interaction estimates to comparisons for which we only have aggregate level data available.

18.8   Conclusion

NMA provides a consistent framework to estimate relative treatment effects between competing interventions for a certain disease state based on RCT evidence. A NMA can be performed with study-level data, IPD, or a combination of both. In the context of health technology assessment, evaluation of treatment effects in subgroups are an integral part of the process to evaluate the value of a healthcare technology. Standard NMA to estimate subgroup effects may be challenging depending on the evidence available, and modified methods may be needed. If the evidence base is rather weak due to a limited number of RCTs and/or small sample sizes in each of the subgroups, the estimates obtained with a NMA may be stabilized with "shrinkage estimation" where intervention specific relative treatment effect are assumed exchangeable for interventions of the same class. As an alternative to a NMA by subgroup, we can also define a model where all mutually exclusive subgroups are incorporated simultaneously and subgroup effects are exchangeable by treatment, and thereby improving the precision of estimates. A network meta-regression may be a relevant approach to adjust for external bias when there are differences between the target population of interest and the study populations of the individual studies regarding effect-modifiers. When there is only access to study-level data, network meta-regression is vulnerable to ecologic bias when the subgroups of interest relate to patient characteristics. Even in the absence of study level confounding, ecological bias can exist in non-linear models. A NMA with IPD is the gold-standard to estimate relative treatment effects for different patient characteristics. Unfortunately, a meta-analyst may not have access to IPD for all trials, but only for a subset. Network meta-regression models can be defined where both sources of evidence are integrated. In order to improve the power to estimate interaction effects for comparisons for which only aggregate level data is available, we can use the same parameter for the within- and between-trial interaction effects at the expense of ecological or aggregation bias when there is study-level confounding or when we have non-linear models. A potential solution are hierarchical related regression models where the model for aggregate-level data is obtained by integrating an underlying IPD model over the joint within-study distribution of covariates. The methods presented in this chapter or only a selection of available techniques to perform evidence synthesis studies to estimate subgroup specific treatment effects. In principle, one could add comparative observational studies to the networks in an attempt to use a larger evidence base to estimate subgroup effects and effect-modification. However, additional research is needed regarding its benefit versus the potential risk of bias before any recommendations can be made. In general, the established NMA framework can be modified without violating its underlying assumptions of exchangeability and consistency in order to improve estimation of treatment effects for subpopulations based on the available RCT evidence. These methods are useful as part of drug development decisions and to support commercialization activities for novel drugs when preparing HTA submissions.

References

Ades AE (2003) A chain of evidence with mixed comparisons: models for multi-parameter synthesis and consistency of evidence. Stat Med 22(19):2995–3016

Berlin JA, Santanna J, Schmid CH, Szczech LA, Feldman HI (2002) Individual patient-versus group-level data meta-regressions for the investigation of treatment effect modifiers: ecological bias rears its ugly head. Stat Med 21(3):371–387

Bucher HC, Guyatt GH, Griffith LE, Walter SD (1997) The results of direct and indirect treatment comparisons in meta-analysis of randomized controlled trials. J Clin Epidemiol 50(6):683–691

Cooper NJ, Sutton AJ, Morris D, Ades AE, Welton NJ (2009) Addressing between-study heterogeneity and inconsistency in mixed treatment comparisons: application to stroke prevention treatments in individuals with non-rheumatic atrial fibrillation. Stat Med 28(14):1861–1881

Cope S, Capkun-Niggli G, Gale R, Lassen C, Owen R, Ouwens MJ, Bergman G, Jansen JP (2012) Efficacy of once-daily indacaterol relative to alternative bronchodilators in COPD: a patient-level mixed treatment comparison. Value Health 15(3):524–533

Debray TP, Schuit E, Efthimiou O, Reitsma JB, Ioannidis JP, Salanti G, Moons KG, GetReal Workpackage (2018) An overview of methods for network meta-analysis using individual participant data: when do benefits arise? Stat Methods Med Res 27(5):1351–1364

Dias S, Sutton AJ, Ades AE, Welton NJ (2013a) Evidence synthesis for decision making 2: a generalized linear modeling framework for pairwise and network meta-analysis of randomized controlled trials. Med Decis Mak 33(5):607–617

Dias S, Sutton AJ, Welton NJ, Ades AE (2013b) Evidence synthesis for decision making 3: heterogeneity—subgroups, meta-regression, bias, and bias-adjustment. Med Decis Mak 33(5):618–640

Dias S, Ades AE, Welton NJ, Jansen JP, Sutton AJ (2018a) Introduction to evidence synthesis. In: Network meta-analysis for decision-making. John Wiley, New York, pp 1–17

Dias S, Ades AE, Welton NJ, Jansen JP, Sutton AJ (2018b) The core model. In: Network meta-analysis for decision-making. John Wiley, New York, pp 19–58

Dias S, Ades AE, Welton NJ, Jansen JP, Sutton AJ (2018c) Validity of network meta-analysis. In: Network meta-analysis for decision-making. John Wiley, New York, pp 351–374

Dias S, Ades AE, Welton NJ, Jansen JP, Sutton AJ (2018d) Meta-regression for relative treatment effects. In: Network meta-analysis for decision-making. John Wiley, New York, pp 227–271

Donegan S, Williamson P, D'Alessandro U, Garner P, Smith CT (2013) Combining individual patient data and aggregate data in mixed treatment comparison meta-analysis: individual patient data may be beneficial if only for a subset of trials. Stat Med 32(6):914–930

Donegan S, Welton NJ, Tudur Smith C, D'Alessandro U, Dias S (2017) Network meta-analysis including treatment by covariate interactions: consistency can vary across covariate values. Res Synth Methods 8(4):485–495

Donegan S, Dias S, Welton NJ (2019) Assessing the consistency assumptions underlying network meta-regression using aggregate data. Res Synth Methods 10(2):207–224. https://doi.org/10.1002/jrsm.1327

Efthimiou O, Debray TP, van Valkenhoef G, Trelle S, Panayidou K, Moons KG et al (2016) GetReal in network meta-analysis: a review of the methodology. Res Synth Methods 7(3):236–263

Espinoza MA, Manca A, Claxton K, Sculpher MJ (2014) The value of heterogeneity for cost-effectiveness subgroup analysis: conceptual framework and application. Med Decis Mak 34(8):951–964

Greenland S (2002) A review of multilevel theory for ecologic analyses. Stat Med 21(3):389–395

Henderson NC, Louis TA, Wang C, Varadhan R (2016) Bayesian analysis of heterogeneous treatment effects for patient-centered outcomes research. Health Serv Outcomes Res Methodol 16(4):213–233

Higgins JP, Thompson SG (2004) Controlling the risk of spurious findings from meta-regression. Stat Med 23(11):1663–1682

Hutton B, Salanti G, Caldwell DM, Chaimani A, Schmid CH, Cameron C et al (2015) The PRISMA extension statement for reporting of systematic reviews incorporating network meta-analyses of health care interventions: checklist and explanations. Ann Intern Med 162(11):777–784

Jackson C, Best N, Richardson S (2006) Improving ecological inference using individual-level data. Stat Med 25(12):2136–2159

Jackson C, Best AN, Richardson S (2008) Hierarchical related regression for combining aggregate and individual data in studies of socio-economic disease risk factors. J R Stat Soc A Stat Soc 171(1):159–178

Jansen JP (2012) Network meta-analysis of individual and aggregate level data. Res Synth Methods 3(2):177–190

Jansen JP, Naci H (2013) Is network meta-analysis as valid as standard pairwise meta-analysis? It all depends on the distribution of effect modifiers. BMC Med 11(1):159

Jansen JP, Fleurence R, Devine B, Itzler R, Barrett A, Hawkins N et al (2011) Interpreting indirect treatment comparisons and network meta-analysis for health-care decision making: report of the ISPOR Task Force on Indirect Treatment Comparisons Good Research Practices: part 1. Value Health 14(4):417–428

Jansen JP, Schmid CH, Salanti G (2012) Directed acyclic graphs can help understand bias in indirect and mixed treatment comparisons. J Clin Epidemiol 65(7):798–807

Jansen JP, Trikalinos T, Cappelleri JC, Daw J, Andes S, Eldessouki R, Salanti G (2014) Indirect treatment comparison/network meta-analysis study questionnaire to assess relevance and credibility to inform health care decision making: an ISPOR-AMCP-NPC Good Practice Task Force report. Value Health 17(2):157–173

Kew KM, Dias S, Cates CJ (2014) Long-acting inhaled therapy (beta-agonists, anticholinergics and steroids) for COPD: a network meta-analysis. Cochrane Database Syst Rev 3:CD010844

Lambert PC, Sutton AJ, Abrams KR, Jones DR (2002) A comparison of summary patient-level covariates in meta-regression with individual patient data meta-analysis. J Clin Epidemiol 55(1):86–94

Leahy J, O'Leary A, Afdhal N, Gray E, Milligan S, Wehmeyer MH, Walsh C (2018) The impact of individual patient data in a network meta-analysis: an investigation into parameter estimation and model selection. Res Synth Methods 9(3):441–469

Lipsky AM, Gausche-Hill M, Vienna M, Lewis RJ (2010) The importance of "shrinkage" in subgroup analyses. Ann Emerg Med 55(6):544–552

Lu G, Ades AE (2004) Combination of direct and indirect evidence in mixed treatment comparisons. Stat Med 23(20):3105–3124

Lumley T (2002) Network meta-analysis for indirect treatment comparisons. Stat Med 21(16):2313–2324

Mayo-Wilson E, Dias S, Mavranezouli I, Kew K, Clark DM, Ades AE, Pilling S (2014) Psychological and pharmacological interventions for social anxiety disorder in adults: a systematic review and network meta-analysis. Lancet Psychiatry 1(5):368–376

Nixon RM, Bansback N, Brennan A (2007) Using mixed treatment comparisons and meta-regression to perform indirect comparisons to estimate the efficacy of biologic treatments in rheumatoid arthritis. Stat Med 26(6):1237–1254

Phillippo DM, Ades AE, Dias S, Palmer S, Abrams KR, Welton NJ (2018) Methods for population-adjusted indirect comparisons in health technology appraisal. Med Decis Mak 38(2):200–211

Riley RD, Steyerberg EW (2010) Meta-analysis of a binary outcome using individual participant data and aggregate data. Res Synth Methods 1(1):2–19

Riley RD, Simmonds MC, Look MP (2007) Evidence synthesis combining individual patient data and aggregate data: a systematic review identified current practice and possible methods. J Clin Epidemiol 60(5):431–4e1

Riley RD, Lambert PC, Staessen JA, Wang J, Gueyffier F, Thijs L, Boutitie F (2008) Meta-analysis of continuous outcomes combining individual patient data and aggregate data. Stat Med 27(11):1870–1893

Riley RD, Lambert PC, Abo-Zaid G (2010) Meta-analysis of individual participant data: rationale, conduct, and reporting. BMJ 340:c221

Salanti G, Higgins JP, Ades AE, Ioannidis JP (2008) Evaluation of networks of randomized trials. Stat Methods Med Res 17(3):279–301

Saramago P, Sutton AJ, Cooper NJ, Manca A (2012) Mixed treatment comparisons using aggregate and individual participant level data. Stat Med 31(28):3516–3536

Schmid CH, Stark PC, Berlin JA, Landais P, Lau J (2004) Meta-regression detected associations between heterogeneous treatment effects and study-level, but not patient-level, factors. J Clin Epidemiol 57(7):683–697

Sculpher M (2008) Subgroups and heterogeneity in cost-effectiveness analysis. PharmacoEconomics 26(9):799–806

Stevens W, Normand C (2004) Optimisation versus certainty: understanding the issue of heterogeneity in economic evaluation. Soc Sci Med 58(2):315–320

Sutton AJ, Kendrick D, Coupland CA (2008) Meta-analysis of individual-and aggregate-level data. Stat Med 27(5):651–669

Turner RM, Spiegelhalter DJ, Smith GC, Thompson SG (2009) Bias modelling in evidence synthesis. J R Stat Soc A Stat Soc 172(1):21–47

Warren FC, Abrams KR, Sutton AJ (2014) Hierarchical network meta-analysis models to address sparsity of events and differing treatment classifications with regard to adverse outcomes. Stat Med 33(14):2449–2466

Index

© Springer Nature Switzerland AG 2020
N. Ting et al. (eds.), *Design and Analysis of Subgroups with Biopharmaceutical
Applications*, Emerging Topics in Statistics and Biostatistics,
https://doi.org/10.1007/978-3-030-40105-4

Printed in the United States
by Baker & Taylor Publisher Services

Printed in the United States
by Baker & Taylor Publisher Services